Geographic Information Systems

Geographic Information Systems

Edited by **Marina De Lima**

⬒SYRAWOOD
PUBLISHING HOUSE

New York

Published by Syrawood Publishing House,
750 Third Avenue, 9th Floor,
New York, NY 10017, USA
www.syrawoodpublishinghouse.com

Geographic Information Systems
Edited by Marina De Lima

International Standard Book Number: 978-1-68286-088-5 (Hardback)

Printed in the United States of America.

Contents

Permissions

List of Contributors

Preface

Every book is a source of knowledge and this one is no exception. The idea that led to the conceptualization of this book was the fact that the world is advancing rapidly; which makes it crucial to document the progress in every field. I am aware that a lot of data is already available, yet, there is a lot more to learn. Hence, I accepted the responsibility of editing this book and contributing my knowledge to the community.

Geographic information system is an important tool which stores, manipulates and analyzes geographic or spatial data. This book consists of contributions made by international experts in the field of geoinformatics. It elucidates the models and techniques of geographic information system, spatial analysis and its applications such as data mining, etc. Also discussed within this book are the various studies that are constantly contributing towards advancing technologies and evolution of this field. This book is best suited for students and research scholars pursuing geoinformatics and associated disciplines.

While editing this book, I had multiple visions for it. Then I finally narrowed down to make every chapter a sole standing text explaining a particular topic, so that they can be used independently. However, the umbrella subject sinews them into a common theme. This makes the book a unique platform of knowledge.

I would like to give the major credit of this book to the experts from every corner of the world, who took the time to share their expertise with us. Also, I owe the completion of this book to the never-ending support of my family, who supported me throughout the project.

Editor

AUTOMATIC TOPOLOGY DERIVATION FROM IFC BUILDING MODEL FOR IN-DOOR INTELLIGENT NAVIGATION

S. J. Tang [b], Q. Zhu [bc], W. W. WANG [a], Y.T. ZHANG [b]

[a] Shenzhen research center of digital city engineering, Shenzhen, P.R. China-measurer@163.com
[b] State Key Laboratory of Information Engineering in Surveying Mapping and
Remote Sensing, Wuhan University, Wuhan, P.R. China-shengjun.tang@whu.edu.cn
[c] Faculty of Geosciences and Environmental Engineering of
Southwest Jiaotong University, Chengdu, P.R. China- zhuq66@263.net

Commission/WG

KEY WORDS: IFC, Geometry-Semantics-based, Topology Derivation, Navigation

ABSTRACT:

With the goal to achieve an accuracy navigation within the building environment, it is critical to explore a feasible way for building the connectivity relationships among 3D geographical features called in-building topology network. Traditional topology construction approaches for indoor space always based on 2D maps or pure geometry model, which remained information insufficient problem. Especially, an intelligent navigation for different applications depends mainly on the precise geometry and semantics of the navigation network. The trouble caused by existed topology construction approaches can be smoothed by employing IFC building model which contains detailed semantic and geometric information. In this paper, we present a method which combined a straight media axis transformation algorithm (S-MAT) with IFC building model to reconstruct indoor geometric topology network. This derived topology aimed at facilitating the decision making for different in-building navigation. In this work, we describe a multi-step deviation process including semantic cleaning, walkable features extraction, Multi-Storey 2D Mapping and S-MAT implementation to automatically generate topography information from existing indoor building model data given in IFC.

1. INTRODUCTION

As the growing complexity of Indoor environment and an increasing number of disasters (compartment fires, terrorist attack, etc.), there has been a practical requirements on indoor navigation approach, which is still one of the most challenging topics for research. A successful navigation depends mainly on the accuracy geometry and semantics of network (Li and He, 2008). As these complex geometry information can be easily obtained in outside by different 3D data capturing technologies (Shi et al., 2004), most researchers focus on the outdoor navigation environment. On the contrary, the information related to indoor still remains limited (Isikdag et al., 2013). Many cases of the indoor navigation process based on 2D maps or pure geometry information, which are inadequate for 3D topology deviation (Lee and Kwan, 2005). A successful navigation depends on correct semantic and structural information about building elements.

Many researches imply that Building Information Model (BIMs) can serve as a valuable information source for facilitating indoor navigation (Chen and Huang; Isikdag et al., 2013; Li, 2012; Rueppel and Stuebbe, 2008). A BIM model is a digital representation of all the physical and functional characteristics of a building through its entire life cycle (Isikdag and Zlatanova, 2009; Isikdag and Zlatanova, 2009; van Nederveen et al., 2014). The trouble, lack of indoor geometry and semantic information can be smoothed by employing IFC building model which contains detailed semantic and geometric information. Clear definition of building elements like storeys, door, windows, spaces in IFC model can provide pivotal information for

topology derivation and detailed physical, functional characteristics contained in each element is enable intelligent routing and other indoor applications. Integrated spatial relationship between building elements in IFC such as a stair can link two floors together and wall can be a container of openings can also facilitate the derivation of the topology network. Semantically rich 3D models can provide critical information for navigation.

In this paper, we present a method which combined a straight media axis transformation algorithm (S-MAT) with IFC building model to reconstruct indoor geometric topology network. This derived topology network aimed at facilitating the decision making for different in-building navigation. The remainder of this paper is organized as follows. Next section depicts the research background about IFC Building Model and topology deviation methods. Section 3 elaborates on a multi-step deviation process including semantic cleaning, walkable features extraction, Multi-Storey 2D Mapping and S-MAT implementation for generating topography information from existing indoor building model data given in IFC. Section 4 illustrates the applicability of the approach involving a typical IFC model. Sections 6 concludes by outlining the advantages of using the approach for topology derivations and overviews the future developments.

2. RESEARCH BACKGROUND

2.1 IFC Building Model

IFC is a standardized open data model developed for BIM developed by the International Alliance for Interoperability (IAI)

(BuildingSMART, 2013), which is used for facilitating interoperability in the building industry and sharing information among participants (El-Mekawy et al., 2012). For the semantics information, IFC data models are semantics rich as they cover all physical and functional characteristics of the building. All of the spatial relationship between building elements are maintained in IFC with a hierarchical manner. Figure 1 represents the semantics relationship of IFC building element. For the geometry representation, IFC model can usually be created by multiple geometry including B-rep (Foley et al., 1994), Sweep volumes and CSG.

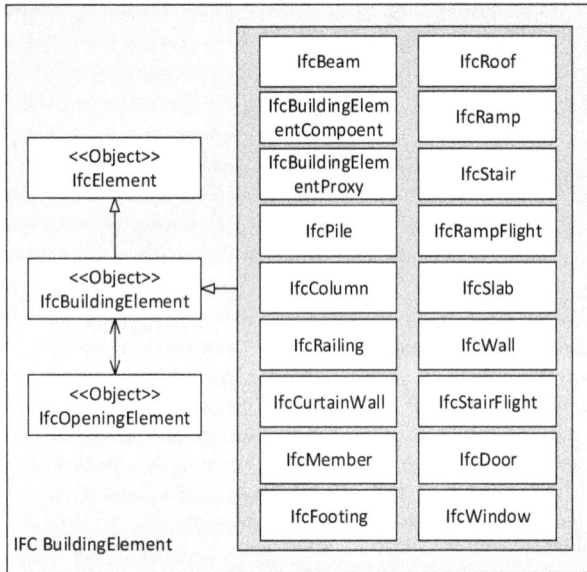

Figure 1. Semantics Relationship of IFC Building Element

2.2 Topology Deviation Method

Topology network explicitly represents the topological relationship among geographical features. These relationships includes connectivity, adjacency, inclusion and intersection (Egenhofer and Herring, 1990; Van Oosterom, 1993; Zlatanova et al., 2002). It is critical to explore a feasible way for building the connectivity relationships among 3D geographical features. Three kinds of method for topology network deviation (pure geometry- based, geometry semantics-based and pure semantics-based) are mainly used.

Pure Geometry-based: There is no semantics information involving in regeneration topology network in this method. (Lee, 2001) developed a topology data model which represents topological relationships, adjacency and connectivity relationships, between 3D spatial objects in built environments by using a straight MAT algorithm.

Geometry-Semantics-based: Most reaearches were based on both semantics and geometry information. Li and He (2008) and Li (2012) defined a 3D indoor navigation ontology for automatically conversion from a geometry model to a semantic-based topological network. Taneja and East (2011) used the straight MAT concept to transform IFC information into geometry networks, which need involve many manual operations. Chen and Huang (2014) present a method of constructing path networks directly from a revit building model. Two types of network construction algorithms, visibility graph and approximate MAT were presented and compared.which

combined network analysis with Building Information Model (BIM) to facilitate the decision making for rescue operations

Pure Semantics-based: Some researchers used a door-to-door method to derive a visibility which is a network whose vertices are the vertices of obstacles and whose arcs connected all vertex pairs that are visible to each other. Based on the constructed graph, the shortest paths between initial points and destination can be found by using the shortest path algorithms, such as Dijkstra's algorithm (Dijkstra 1959). Lorenz et al. (2006) propose a hybrid spatial model for indoor environments. The model consists of hierarchically structured graphs with edges and nodes.

3. TOPOLOGY DERIVATION FROM IFC MODEL

3.1 Semantics Cleaning

Semantics cleaning aimed at filtering these elements out that have meaningful usage in topology derivation. There are about 900 classed defined in IFC schema and these classes are used store different information, such as relationship, attribute, geometry representation and element type etc. (de Laat and van Berlo, 2011). Many researches have already investigated information models that can be used for indoor navigation (Brown et al., 2013; Diehl et al., 2006). Based on these investigations, very few of these classes are relevant for topology derivation. Therefore, a semantic filtering method was used to screening these meaningful objects out. Figure 2 presents the workflow diagram for the semantics cleaning.

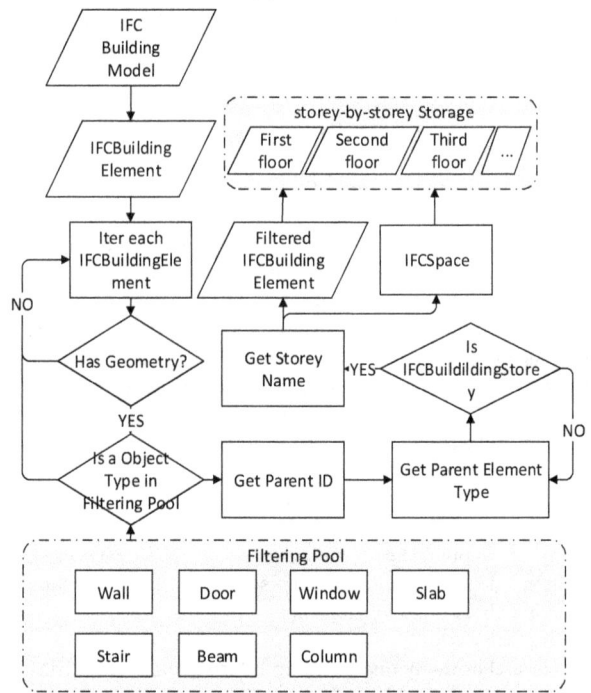

Figure 2. Workflow for the Semantics Cleaning

As shown in Figure 2, for each "IFCBuildingElement" existed in the IFC File, it should be checked whether it contains geometry. Those elements that do not have geometry would be ignored. Because very few of classes are relevant for topology derivation, a filtering pool consisted by seven types of the "IFCBuildingElement" is created in the filtering process. For topology derivation, each storey has different topology network, the element obtained from IFC file should be preserved storey-by-storey, so it is necessary to identified belonging information of each filtered elements. Most of the "IFCBuildingElement" like Wall, Stair, Slab, Beam and Column have a parent node with the type of "IFCBuildingStorey". However, door and window elements always beyond to an "IFCOpeningElement ", which's parent node type is wall type. So at least two times upward iteration should be applied to door and window element for finding which storey they are belonging. Meanwhile, the relevant class, "IFCSpace" can be filtering out with the rid of storey information.

3.2 Walkable Features Extraction

After semantics filtering, elements are classified storey-by-storey. The inner structure for each storey is very complicated, there are many rooms, corridor, office and other features. In order to facilitate the derivation of the inner topology network, walkable features extraction related to the corridor elements is the most important. Abundant semantic information in IFC model can provides an effective method to automatic extract walkable features. Based on the filtered "IFCSpace" of each storey, it is used to provided all information about the space as a functional area or volume within a spatial structure. There are several attributions like "Name", "Description", "Long Name" and "ObjectType" in "IFCSpace", "ObjectType" can be used for identified the functional category of the space.

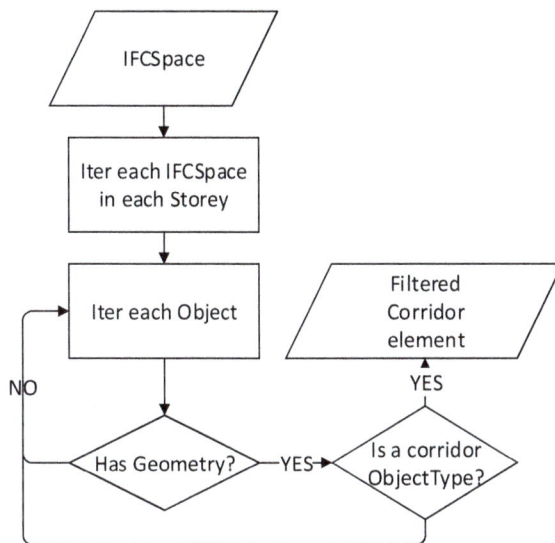

Figure 3. Workflow for walkable features extraction

Figure 3 presents the workflow diagram for walkable features extraction. For each "IFCSpace" in each storey, we should iterate all of the object and checked whether it contains geometry. Those elements that do not have geometry would be ignored. Then, when the obtained "ObjectType" is a "corridor" type, this object will be retained.

3.3 Multi-Storey 2D Mapping

Although 3D data models have been developed and implemented for geo-analyses and 3D visualization, they have some limitation with representing topological network because of the complex 3D geometric computational problems involved. In order to alleviate the problem, a combinatorial data model proposed by Lee and Kwan was employed to interpret the

topological term among 3D object by drawing a dual graph in this paper. In order to facilitate S-MAT implementation, the object represented in three-dimensions of each storey should be mapping to the 2D planar for picking up the xy coordinates of nodes and maintain topological consistencies. Meanwhile, the semantic information existed in the original element would be retained in their respective edge and node. These mapping result can represent the basic topological relationships and connectivity relationships and is the base of topology network construction.

3.4 Topology deviation from S-MAT

The straight MAT algorithm developed by Lee (2004) can produce the skeleton graph of the 3D indoor environment, which based on a series of spatial points derived from a closed polygon. The S-MAT is generated by the angular bisectors of each pair of neighbouring edges. A segment of the angular bisector of these two neighbouring edges is added in the set of line segments of the medial axis. Those bisectors generated from all convex vertices of polygon P is called rays. When two neighbouring rays intersect a point and environ a thiessen polygon, the end type vertex would be assigned to the corresponding rays. Make rays which starts from the intersection point of the two parent rays and is a segment of the bisector of edge e_i and e_{i+2} or e_{i-1} and e_{i+1}. After scanning all of the neighbouring bisector rays, a list of new rays is obtained. The process is ended when two rays of opposite direction remain.

After S-MAT implementation, the skeleton of the corridor can be obtained. An algorithm combining closest distance algorithm and door to door algorithm is employed to integrate door and stair to the prime network. For each door and stair, it should be checked whether it can connect to the skeleton of the corridor. Those elements that have no connection with the corridor should connect with its adjacent door or stair. At last, all of door and stair elements would be add into the derived topology for generating an integrated topology navigation network.

4. CASE STUDY

In this section, the proposed topology derivation method is tested with a two floor IFC building model named "091210Med_Dent_Clinic_Arch.ifc", which contains abundant geometry and semantics information. The elements in this model is sufficient for topology derivation.

4.1 First Step-semantics cleaning: the model before and after semantic cleaning were presented in Figure 4. This building is a two floor IFC building model, which contains abundant semantics information. The building model after semantic cleaning contains wall, door, window, stair, slab, beam and column objects.

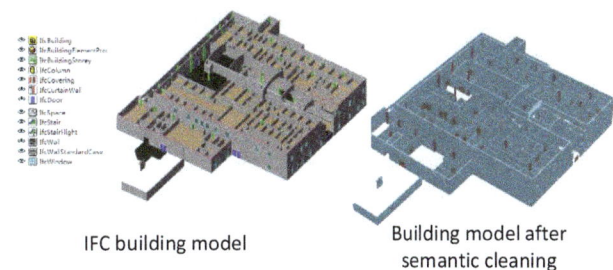

IFC building model   Building model after semantic cleaning

Figure 4. Original IFC model and model after semantic cleaning

4.2 Second Step-walkable feature extraction: Based on the extracted "IFCSpace", walkable features can be obtained. Figure 5 presents the results of walkable features extraction. The above pictures are "IFCSpace" elements of two floor. The pictures drawing with gray are the respective corridor features. The blue parts are those element with special functions. They can be separated easily according to the "objecttype" attribution.

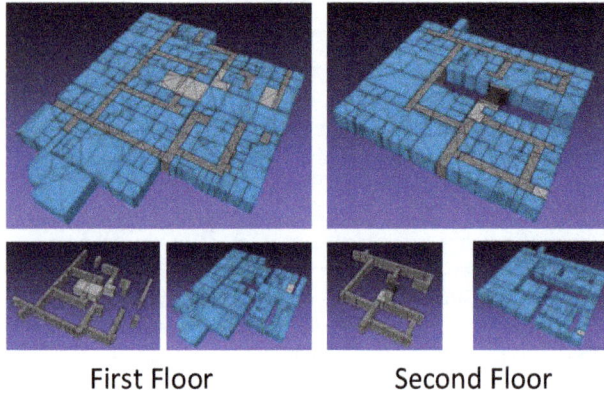

First Floor   Second Floor

Figure 5. Results of walkable features extraction

4.3 Third Step-Multi-storey 2D Mapping: After Multi-storey 2D mapping, we can obtained the x, y coordinates of the whole filtered object in each storey. The results of multi-storey 2D mapping were shown in Figure 6. In the figure, corridors are presented by green, doors are draw with blue, red lines represent stair elements and black lines are wall elements. Preparing for S-MAT implementation, 2D points of the walkable features polygon can be obtained. These points should organize by anticlockwise.

First Floor   Second Floor

Figure 6. Results of Multi-storey 2D Mapping

4.4 Fourth Step-S-MAT implementation: As shown in Figure 7, the MAT network of each floor can be obtained after S-MAT implement, which use the walkable features as input. In above pictures in Figure 7, we can see the polygon of the walkable features have a hole. In the process of the S-MAT, the inner polygon should be wipe off and the order of the points derived from the polygons should be anticlockwise. After adding door and stair elements, the whole topology network are presented in below pictures in Figure 7. The topology network of each storey can be connected by stairs. This derived topology network can support multi-storey indoor navigation.

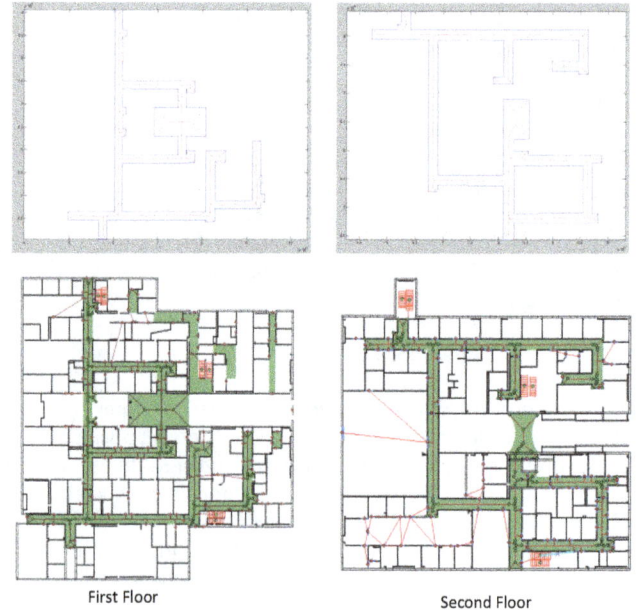

First Floor   Second Floor

Figure 7. Result of S-MAT implementation

5. CONCLUSION AND FUTURE WORK

In this paper, the authors presented a multi-step deviation process including semantic cleaning, walkable features extraction, multi-Storey 2D Mapping and S-MAT implementation to automatically generate topography information from existing indoor building model data given in IFC. The abundant information inside the BIM models can facilitate the topology derivation greatly. The cases study shows the performance of this method, which is enable to generate a multi-storey topology network aiming at indoor navigation. However, as described in walkable feature extraction, the corridors are obtained from "IFCSpace" which defined in a perfect IFC model. In the cases of lack of IFCSpace, a new method should be designed for obtaining the walkable features. The main goal of our future work is to creatively integrate current powerful IFC information to enable more intelligent application such as indoor emergency evacuation and indoor facilitate management.

ACKNOWLEDGEMENTS

The work is Supported by the Open Fund of Key Laboratory of Urban Land Resources Monitoring and Simulation，Ministry of Land and Resources (Project No. KF-2015-01-027).

REFERENCES

Brown, G., Nagel, C., Zlatanova, S. and Kolbe, T.H., 2013. Modelling 3D topographic space against indoor navigation requirementsProgress and New Trends in 3D Geoinformation Sciences. Springer, pp. 1-22.
BuildingSMART, 2013.
Chen, A.Y. and Huang, T., BIM-Enabled Decision Making for In-Building Rescue Missions, Computing in Civil and Building Engineering (2014). ASCE, pp. 121-128.
de Laat, R. and van Berlo, L., 2011. Integration of BIM and GIS: The development of the CityGML GeoBIM

extensionAdvances in 3D geo-information sciences. Springer, pp. 211-225.

Diehl, S., Neuvel, J., Zlatanova, S. and Scholten, H., 2006. Investigation of user requirements in the emergency response sector: the Dutch case, Second Symposium on Gi4DM, pp. 25-26.

Egenhofer, M.J. and Herring, J., 1990. A mathematical framework for the definition of topological relationships, Fourth international symposium on spatial data handling. Zurich, Switzerland, pp. 803-813.

El-Mekawy, M., Östman, A. and Hijazi, I., 2012. An Evaluation of IFC-CityGML Unidirectional Conversion. International Journal of Advanced Computer Science & Applications, 3(5).

Foley, J.D., Van Dam, A., Feiner, S.K., Hughes, J.F. and Phillips, R.L., 1994. Introduction to computer graphics, 55. Addison-Wesley Reading.

Isikdag, U. and Zlatanova, S., 2009. A SWOT analysis on the implementation of Building Information Models within the Geospatial Environment. Urban and Regional Data Management, CRC Press, The Netherlands: 15-30.

Isikdag, U. and Zlatanova, S., 2009. Towards defining a framework for automatic generation of buildings in CityGML using building Information Models3D geo-information sciences. Springer, pp. 79-96.

Isikdag, U., Zlatanova, S. and Underwood, J., 2013. A BIM-Oriented Model for supporting indoor navigation requirements. Computers, Environment and Urban Systems, 41: 112-123.

Lee, J., 2001. 3D data model for representing topological relations of urban features, Proceedings of the 21st Annual ESRI International User Conference, San Diego, CA, USA.

Lee, J. and Kwan, M.P., 2005. A combinatorial data model for representing topological relations among 3D geographical features in micro - spatial environments.

International Journal of Geographical Information Science, 19(10): 1039-1056.

Li, Y., 2012. Building Information Model for 3D Indoor Navigation in Emergency Response. Advanced Materials Research, 368: 3837-3840.

Li, Y. and He, Z., 2008. 3D indoor navigation: a framework of combining BIM with 3D GIS, 44th ISOCARP congress.

Lorenz, B., Ohlbach, H.J. and Stoffel, E., 2006. A hybrid spatial model for representing indoor environmentsWeb and Wireless Geographical Information Systems. Springer, pp. 102-112.

Rueppel, U. and Stuebbe, K.M., 2008. BIM-based indoor-emergency-navigation-system for complex buildings. Tsinghua Science & Technology, 13: 362-367.

Shi, Y., Nakagawa, M. and Shibasaki, R., 2004. Reconstruction of "Next-generation" 3D Digital Road Model from Three Linear Scanner Images. Proceedings of ACRS, Thailand.

Taneja, S. and East, E.W., 2011. Transforming an IFC-based Building Layout Information into a Geometric Topology Network for Indoor Navigation Assistance. ASCE Reston, VA.

van Nederveen, S., Wolfert, R. and van de Ruitenbeek, M., 2014. From BIM to life cycle information management in infrastructure. eWork and eBusiness in Architecture, Engineering and Construction: ECPPM 2014: 115.

Van Oosterom, P.J.M., 1993. Reactive data structures for geographic information systems. Oxford University Press New York.

Zlatanova, S., Rahman, A.A. and Pilouk, M., 2002. Trends in 3D GIS development. Journal of Geospatial Engineering, 4(2): 71-80.

RICE CROP MONITORING AND YIELD ESTIMATION THROUGH COSMO SKYMED AND TERRASAR-X: A SAR-BASED EXPERIENCE IN INDIA

S. Pazhanivelan[a],*, P. Kannan[a], P.Christy Nirmala Mary[a], E.Subramanian[a], S. Jeyaraman[a], Andrew Nelson[b], Tri setiyono[b],
Francesco Holecz[c],Massimo barbieri[c] and Manoj Yadav[d]
[a]Tamil Nadu Agricultural University, Coimbatore, Tamilnadu, India- pazhanivelans@gmail.com
pandian.kannan@gmail.com, chrismary@rediffmail.com, selvisubbug@yahoo.co.in, sjtnau@gmail.com
[b]International Rice Research Institute (IRRI), Los Banos 4031, Philippines - a.nelson@irri.org, t.setiyono@irri.org
[c]Sarmap, Purasca 6989, Switzerland - fholecz@sarmap.ch, mbarbieri@sarmap.ch
[d]Deutsche Gesellschaft für Internationale Zusammenarbeit (GIZ) GmbH, New Delhi 110029, India - manoj.yadav@giz.de

KEY WORDS: Rice, Food Security, SAR, Yield Estimation, ORYZA, COSMO Skymed, TerraSAR-X

ABSTRACT:

Rice is the most important cereal crop governing food security in Asia. Reliable and regular information on the area under rice production is the basis of policy decisions related to imports, exports and prices which directly affect food security. Recent and planned launches of SAR sensors coupled with automated processing can provide sustainable solutions to the challenges on mapping and monitoring rice systems. High resolution (3m) Synthetic Aperture Radar (SAR) imageries were used to map and monitor rice growing areas in selected three sites in TamilNadu, India to determine rice cropping extent, track rice growth and estimate yields. A simple, robust, rule-based classification for mapping rice area with multi-temporal, X-band, HH polarized SAR imagery from COSMO Skymed and TerraSAR X and site specific parameters were used. The robustness of the approach is demonstrated on a very large dataset involving 30 images across 3 footprints obtained during 2013-14. A total of 318 in-season site visits were conducted across 60 monitoring locations for rice classification and 432 field observations were made for accuracy assessment. Rice area and Start of Season (SoS) maps were generated with classification accuracies ranging from 87- 92 per cent. Using ORYZA2000, a weather driven process based crop growth simulation model; yield estimates were made with the inclusion of rice crop parameters derived from the remote sensing products viz., seasonal rice area, SoS and backscatter time series. Yield Simulation accuracy levels of 87 per cent at district level and 85- 96 per cent at block level demonstrated the suitability of remote sensing products for policy decisions ensuring food security and reducing vulnerability of farmers in India.

1.INTRODUCTION

1.1. Synthetic Aperture Radar for Mapping Rice Area

Rice is the most staple cereal food crop for ensuring food security in Asia (Maclean et al., 2013). Rice still accounts for 31% of the calorific intake being the largest single source of calories for more than 3.7 billion people in Asian countries even with rapid urbanization and diversification in consumption patterns, (FAOSTAT, 2014 and Timmer et al., 2010). Accurate and consistent information on the area under production is necessary for national planning in many countries, but conventional statistical methods cannot always meet the requirements of food security research and policy (Xiao et al., 2006 and Gumma et al., 2014). This information is vital to the policy decisions related to imports, exports and prices, which directly influence food security, especially amongst the poor (Balagtas et al., 2014, Mittal et al., 2009 and Dawe et al., 2012). Remote sensing has the scope for cost effective precise estimates of rice area to support, augment, improve or even replace survey and statistical methods (Gumma et al., 2014). But the technical challenges are many in the development of large scale dynamic remote sensing-based rice crop information systems. Rice cultivation during the monsoon season (Huke and Huke, 1997) which has wide cloud cover (NASA, 2014), wide range of conditions and environments, small land holdings and diverse and mixed cropping systems (Nguyen et al., 2012) are the most challenging factors in limiting the use of remote sensing as tool for rice crop monitoring.

Synthetic Aperture Radar (SAR) imagery is a promising option to overcome the issue of cloud cover and substantial research evidences are available on the suitability of SAR for rice crop mapping in the region. Optical images can complement SAR, but they cannot be relied upon as the main information source. The wide distribution of rice as a major food crop across India envisages large coverage to perfectly capture rice area and requires automated less supervised processing. Rice detection algorithms should be general and robust to suit wide range of practices and environments (Boschetti et al., 2014) ranging from irrigated to rainfed rice with different maturities (Maclean et al., 2013) and establishment practices, such as direct seeding or transplanting. The complex rice environments require high-resolution imageries and high-frequency acquisitions. Recent and planned launches of SAR sensors coupled with state-of-the-art automated processing can provide sustainable solutions to this challenge to map and monitor one of the world's most important crop. The objective of this study is to test a method of rice area mapping using a rule-based classification and parameter selection approach across multiple sites based on the agronomic knowledge on temporal development of rice crop under different conditions and its management in relation to backscatter. The far-reaching goal is to demonstrate that SAR-based operational mapping of rice crops across a diverse range of environments with multi-temporal SAR data and yield estimation by integrating these products into the ORYZA crop growth model.

---

* Corresponding author

1.2. Background of SAR Research and Applications for Rice Mapping

SAR data have a proven ability to detect lowland rice systems (both irrigated and rainfed) through the unique temporal signature of the backscatter coefficient (also termed sigma naught or $\sigma°$) exhibited by the crop. In the past years, a large number of publications have been dedicated to better understanding this relationship and applying it to rice detection and rice monitoring (Le Toan et al., 1997, Inoue et al., 2002, Suga and Konishi, 2008, and Bouvet et al., 2009). In summary, these studies have shown that lower frequencies (L- and C-band) penetrate deeper into the rice plant than higher frequencies, while only higher frequencies (X-band) interact with grain water content and grain weight sufficiently to show a dual-peak signal in $\sigma°$ during the rice season (Inoue et al., 2002, Suga and Konishi, 2008, Oh et al., 2009 and Kim et al., 2009). Further, short wavelengths (X-, Ka-, Ku-band), especially at large incident angles, are sensitive enough to detect even very small rice seedlings just after transplanting. The correlation between $\sigma°$ and rice biophysical parameters shows that lower frequencies are more closely related to total fresh weight, leaf area index (LAI) and plant height than other parameters (Inoue et al., 2009 and Kim et al., 2009).

Although $\sigma°$ from X-band is poorly correlated with LAI, it is best correlated with panicle biomass indicating the suitability for a direct assessment of rice grain yield (Inoue and Sakaiya, 2013 and Inoue et al., 2014). On the other hand, $\sigma°$ derived from C-band can provide information on par with the normalized difference vegetation index (NDVI) (Inoue et al., 2014). For X-band, the HH/VV polarization ratio continuously changes as a function of phenology during the vegetative and reproductive stages (Lopez-Sanchez et al., 2011). For X-band, the HH-VV phase difference is sensitive to early rice plant emergence. Moreover, the use of four polarimetric features derived from coherence coplanar dual-polarization X-band enables the estimation of five phenological stages from a single date scene (Inoue et al., 2002, Lopez-Sanchez et al., 2011 and Lopez-Sanchez et al., 2012).

It is clear from the literature that well-understood relationships exist between rice crop characteristics and backscatter coefficients from different wavelengths, and these relationships have been used to derive different types of algorithms for estimating rice crop characteristics from SAR data. Another approach for sparse time series is to extract temporal features from the data and relate those to the known temporal dynamics of the rice crop and use that knowledge to classify areas as rice or non-rice (Holecz et al., 2013). All of these approaches have been demonstrated successfully in the literature. Supervised classifiers rely on a substantial set of good-quality training data to ensure a good classification, and there is a risk of over-fitting the classification.

For this reason, a rule-based classification approach is tested for rice area mapping that is based on a small number of rules and parameters that can be quickly fine-tuned from site to site and season to season. Conceptually, the classification approach is based on rules that are agronomically meaningful and, thus, easily understood and easily fine-tuned based on the local knowledge of the rice-growing environment and the key rice-growing stages.

1.3. Rice Growing Stages and Key Characteristics for SAR Based Detection

Rice in subtropical India is mainly cultivated in irrigated or lowland semidry conditions. Rice varieties range in duration from 90 to more than 150 days and with three main crop stages: vegetative (from germination to panicle initiation, from 45 to 100 days), reproductive (from panicle initiation to flowering, around 35 days) and maturity (from flowering to mature grain, around 30 days) (Figure 1). The following aspects contribute to the change in space occupied by the rice plants within a three-dimensional space: (1) appearances and growth of leaves from the main stem (culm) and tillers; (2) stem development and elongation; (3) tillering, defined as the production of stems from rice plants; (4) leaf senescence; and (5) panicle and grain development. Prior to transplanting, the rice field is flooded with water at depths ranging from 2 to 15 cm (Le Toan et al., 1997). This deliberate agronomic flooding is a key element of most remote-sensing rice detection algorithms (Boschetti et al., 2014).

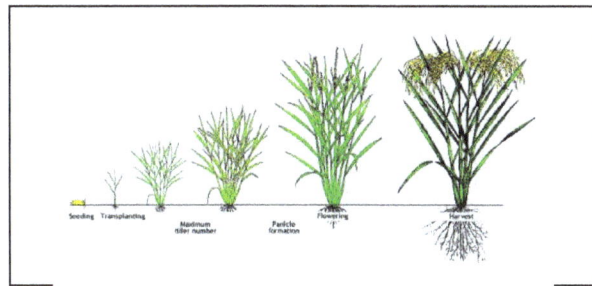

Figure 1. Rice crop stages. Image from the International Rice Research Institute (IRRI)-Rice Knowledge Bank.

2. SAR DATA, FIELD DATA AND STUDY SITES

The RIICE project—Remote sensing-based Information and Insurance for Crops in Emerging economies—tested SAR-based mapping of rice area across three sites in India (Cuddalore, Thanjavur ans Sivaganga) between late 2012 and early 2014 (RIICE, 2014). In this studyMulti-year and seasonal Synthetic Aperture Radar (SAR) data are acquired from all existing operational space borne systems which overcomes the spatial-temporal problem, hence assuring an appropriate temporal repetition at an adequate scale (i.e. spatial resolution) even over large areas and provides sensor independent operational monitoring with sufficient data redundancy to ensure information delivery. The crop growth simulation model ORYZA estimates yield and hence production using dedicated remote sensing products in addition to the usual meteorological, soil, and plant parameters. This remote sensing-crop model approach to yield estimation uses relevant remote sensing derived information on rice seasonal dynamic to initialize the model on the correct date and uses parameters derived from remote sensing as measurements of the crop's response to the environment and management thus reducing the reliance on other input data to the model that would be impossible to obtain over wide geographic areas. Further this approach considers the spatial distribution of rice fields and improves the yield estimation figures by forcing the model towards actual rather than attainable yields.

2.1. SAR Data

Multi-temporal X-band SAR Single Look Complex (SLC) data were obtained from the Italian Space Agency (ASI/e-GEOS) for COSMO-SkyMed (CSK) data and from InfoTerra GmbH for TerraSAR-X (TSX) data. In both the cases, data were obtained in HH polarization with consistent incidence angles in each multi-temporal stack, ranging from 41 to 44 degrees across sites. A large incidence angle is preferred, because (i) wind effects on water (in particular, during land preparation prior to transplanting) are significantly decreased, (ii) the dynamic of the radar backscatter is larger and (iii) the spatial resolution is higher. The image acquisition dates, locations, mode, pixel size, polarization and incidence angles are shown in Table 1. Image mode, extent, pixel size, polarization and incidence angles were constant for each footprint. CSK data are available from four X-band HH-SAR satellites with a 3.12-cm wavelength and a 16-day revisit period for the same satellite with the same observation angle. CSK data on Stripmap mode (3-m resolution) was used at two sites with a footprint of 40 × 40 km. Acquisition plans were made using one primary satellite from the constellation for each site with backup plans in place for the second, third and fourth satellites in the constellation in the event of a cancellation.TSX is provided by one X-band HH SAR satellite with a 3.11-cm wavelength and 11-day revisit period with the same observation angle. TSX data on Stripmap mode (3 m resolution) was used at one site with a footprint of 30 × 50 km.

Site No.	Study site	Start and end dates	# of images	Satellite	Scene center, area (sq km)	Mode, resolution (m)	Polarizat ion, angle (°)
1	Tamil Nadu, Cuddalore	16-08-13 07-01-14	10	CSK	11.74°N-79.56°E, 1,600	Stripmap, 3	HH, 44
2	Tamil Nadu, Thanjavur	16-08-13 26-12-13	9	CSK	10.87°N-79.25°E, 1,600	Stripmap, 3	HH, 41
3	Tamil Nadu, Sivaganga	18-08-13 19-01-14	13	TSX	9.86°N-78.50°E, 1,800	Stripmap, 3	HH, 44
Total No. of images and footprint area			32		5,000		

Table 1. SAR data acquisition summary: locations, dates and modes used for 2013 Samba season.

2.2. Field Observations for Calibration of the Rice Detection Algorithm and Map Validation

Field observations were performed throughout the season in up to 20 paddy fields within each footprint. These fields were selected, with the farmers' consent, prior to the start of the rice season and the image acquisition schedule. Observations were made on or as close to the image acquisition date as possible. Observations included latitude and longitude from handheld GPS receivers, descriptions and photos of the status of the field, plant height, water depth, weather conditions, crop stage and leaf area index (LAI). The same field data collection protocols were used at all sites. LAI measurements were taken only during visits between seedling and flowering

stages, and these were recorded non-destructively using AccuPAR LP-80 Ceptometer (Decagon Devices, Inc., Pullman, WA, USA). At the end of the season, the farmer was interviewed to collect information on the rice variety, water source, crop management and establishment practices, as well as inputs, such as pesticide and fertilizer.

In total, 58 locations were regularly monitored across the three footprints, with 432 separate visits made to these locations to collect in-season information on the status of the rice crop. A validation exercise was conducted for each footprint to assess the accuracy of the rice classification. A rapid land cover appraisal method was adopted to collect land cover information at approximately 100 locations throughout each footprint with these points split 50/50 between non-rice points and rice points. This conforms to the minimum number of samples per land cover class accounting. These map validation assessments were generally conducted in-season, in the reproductive or ripening stage before harvesting, but in some cases, the assessment was conducted post-season. Locations were chosen such that the land cover was homogeneous in a 15-m radius around each GPS point for sites using 3-m resolution imagery and a 50-m radius for sites using 10-m or 15-m resolution imagery.

2.3. Study Site Characteristics

Rice is the dominant crop among the three RIICE sites in India. In Cuddalore District, the samba season from mid-July 2013 to the first week of January 2014 was monitored. Rice fields in this district are predominantly under a well irrigation system; hence, most of the chosen locations were irrigated. The popular rice varieties grown were CR1009, BPT5204 and White Ponni, with maturity duration ranging from 135 to 160 days. Both transplanting and direct seeding of rice are common in this district, with the former establishment method being more dominant. In Sivaganga, the samba season lasted from September 2013 to January 2014, with rice cultivated on 86% of the total cropped area. Rice cultivation was broadly grouped into three types: transplanted, semi-dry and direct seeded. The transplanted system was practiced in the blocks of Thirupuvanam, Sivaganga, Manamadurai, Singampunari, Thirupattur, S. Pudur, Sakkottai, Kallal and Illayangudi. In the semi-dry rice system, seeds are pre-monsoon sown and are under rainfed conditions for 30-45 days. Later, the fields were converted into wet fields by irrigating from tanks and this type of cultivation was practiced in the blocks of Sakkottai, Kannankudi, Devakkottai, Kallal, Kalayarkovil, Sivagangai, Manamadurai, Illayangudi and Thirupathur. Direct-seeded rice cultivation mainly depends on rainfall and is mostly practiced in Illayangudi, Devakkotai, Kannankudi and Kalaiyarkovil blocks. Short-duration rice varieties such as ADT36, ADT45 and JGL were popularly grown in the monitoring locations. Thanjavur is popularly known as the "Rice Bowl" of Tamil Nadu and "Granary of South India" as it is the major district contributing to the food grain supply of the state. Most of the monitored locations were irrigated and farmers practiced transplanting and direct seeding as their crop establishment method. Medium- and long-duration varieties such as CR1009, BPT5204 and ADT (R) 50 were mainly grown, with duration from 135 to 150 days.

3. METHODS

The SAR time-series data underwent a series of basic processing steps to generate terrain-geocoded σ° values suitable for analysis. This multi-temporal stack was analyzed using a rule-based classifier to detect rice areas. The rules for the classifier were based on a small number of parameters that must be selected by the operator or user. Temporal feature descriptors were derived from temporal signatures in the monitored fields and used to guide the user in setting these parameters for each site. Finally, the accuracy of the rice area maps is assessed against field data.in the second stage the crop parameters derived from remote sensing were integrated into the ORYZA crop growth model and the rice yields were estimated and the yield maps were generated.

Site No.	Study site	Season	Period covered	Number of fields, visits	Crop establishment	Variety and maturity (days)
1	Tamil Nadu, Cuddalore	Samba	mid-Jul to Jan	20 fields, 160 visits	Transplanting Irrigated	CR1009 (160), BPT5204 (135), White Ponni (130), Co 50 (160)
2	Tamil Nadu, Thanjavur	Samba	Aug to Dec	20 fields, 162 visits	Transplanting/ direct seeding Irrigated	CR1009 (160), BPT5204 (135), ADT (R) 50 (160)
3	Tamil Nadu, Sivaganga	Samba	Sep to Jan	18 fields, 110 visits	Transplanting and direct seeding/ Semi-dry rice	ADT45 (110), JGL (100-110), ADT36 (110)

Table 2.Summary of site visits and observed rice crop characteristics during the monitored seasons.

3.1. Basic Processing of SAR Data for Multi-Temporal Analysis

A fully automated processing chain was developed to convert the multi-temporal space-borne SAR SLC data into terrain-geocoded σ° values. The processing chain is a module within the MAPscape-RICE software (Holecz et al., 2013). The basic processing chain included strip mosaicking , co-registration of Images acquired with the same observation geometry and mode and, Time-series speckle filtering to balance differences in reflectivity between images at different times (De Grandi et al., 1997) and terrain geocoding, radiometric calibration and normalization. Further Anisotropic non-linear diffusion (ANLD) filtering was done to smoothen homogeneous targets, while enhancing the difference between neighbouring areas. The filter uses the diffusion equation, in which the diffusion coefficient, instead of being a constant scalar, is a function of image position and assumes a tensor value (Aspert et al., 2007).

3.2. Multi-Temporal σ° Rule-Based Rice Detection

The multi-temporal stack of terrain-geocoded σ° images was put into a rule-based rice detection algorithm in MAPscape-RICE. The temporal evolution of σ° was analyzed from an agronomic perspective, based on prior knowledge of rice maturity, calendar and duration and crop practices from field information and knowledge of the study location. The temporal signature was frequency and polarization dependent and also relied on the crop establishment method and, to some extent, on crop maturity. The general rules were applied to detect rice, but that the parameters for these rules were adapted according to the agro-ecological zone, crop practices and rice calendar.

The choice of parameters was guided by a simple statistical analysis of the temporal signature of σ° values in the monitored fields. The mean, minimum, maximum and range of σ° were computed for the temporal signature of each monitored field. Further, (i) minima and (ii) maxima of those mean σ° values across fields; the (iii) maxima of the minimum σ° values across fields; the (iv) minima of the maximum. σ° value across fields; and the (v) minimum and (vi) maximum of the range of σ° values across fields were calculated. These six statistics, called as temporal features, concisely characterized the key information in the rice signatures of the observed fields, and each one related directly to one parameter. Hence, the value of the six temporal features from the monitoring locations at each site were used to guide the choice of the six parameter values based on which the rice pixels were classified and the rice area maps were generated.

3.3. Rice Map Accuracy Assessment

A standard confusion matrix was applied to the rice/non-rice validation points collected at each site. The overall accuracy of the rice/non-rice classification and the kappa value were recorded. The accuracy assessment was a comparison of the classified rice map against ground-truth data. The spatial resolution of the rice maps ranged from 3 m to 15 m. However, the ANLD filtering processes reduced the effective resolution by performing locally adaptive smoothing and edge detection. To account for this lower resolution and the horizontal accuracy of the handheld GPS units relative to the pixel size, the validation data were collected in areas that had homogeneous land cover in a 15m radius the around each GPS point for sites considering 3m resolution of the imagery.

3.4. Rice yield estimation

The yield was estimated using ORYZA2000, a crop growth simulation model developed by IRRI (Boumman et al., 2001). The simulations account for water and nitrogen dynamics based on climatic, soil conditions and management practices. Irrigation and nitrogen fertilizer inputs are assumed as recommended for achieving attainable yield. LAI values at 50 days after emergence provided by the SoS product are inferred from radar backscatter using cloud vegetation model (Attema and Ulaby, 1978) with parameters calibrated with in situ LAI measurements. Inferred LAI are finally used to calibrate the relative leaf growth rates parameters in ORYZA2000. For processing efficiency, the spatial units for yield simulation are aggregated to 150 meter resolution.

4. RESULTS AND DISCUSSION

4.1. Rice Area Maps

Figure 2 shows rice area maps derived from multi-temporal X-band SAR imagery for Sivaganga,Cuddalore and Thanjavur. Late rice and early rice were combined into one class and distinguished them from rice in the maps. Map accuracy considers any of the three rice subclasses as rice.

In the rice area map generated for Cuddalore, variability in rice crop establishment date was due to the uncertainty in the date of water availability. The clearly demarcated patches in the rice crop in the northern part are water tanks that were successfully excluded from the classification.

4.2. Rice Map Accuracy Assessment

The accuracy assessment for the rice maps was conducted on a rice/non-rice basis, where all other land cover types were grouped into a single non-rice class. Table 3 shows a summary of the validation data, rice area and rice classification accuracy. The total classified rice area across the 3 sites is more than 1.5 lakh hectares but the proportion of the footprint area that was classified as rice varied from 16% to 52% across footprints. The overall classification accuracy was consistently high (87% to 92%), with Kappa scores from 0.73 to 0.85. There was no relationship between the classification accuracy and either the rice area or the proportion of the footprint classified as rice. Large, homogeneous and landscape-dominating rice areas and small, fragmented, heterogeneous rice areas were all classified equally well. This rich non-rice dataset can be further exploited in the future to assess the SAR signatures of other land cover types commonly found in rice-growing areas. The same signatures can also be used to generate new bounding limits (based on the temporal signatures for other crops and urban and water surfaces, for example) to further guide parameter selection in the rule-based classifier.

Site No.	Study site	Validation points and date(s) of validation	Rice area (ha) and as % of footprint	Accuracy and Kappa
1	Tamil Nadu, Cuddalore	111, 12-02-2014 and 03-03-2014	26,015, 16%	92%, 0.85
2	Tamil Nadu, Thanjavur	102, 31-01-2014 and 01-02-2014	83,871, 52%	91%, 0.82
3	Tamil Nadu, Sivaganga	110, 14-02-2014 and 21-02-2014	41,825, 24%	87% 0.73
	Points and area (ha)	323	1,51,711	

Table 3. Summary of site validation visits, rice area and accuracy assessments.

4.3. Rice yield estimation

The yield was estimated using ORYZA2000, a crop growth simulation model. The model estimated yield based on input data such as daily weather data, soil properties, rice variety, water availability and crop management practices. The model was a 'point based' model and was run once for each location where a yield estimate was required. This resulted in many thousands of runs for an area covered by a typical remote sensing image. The ability of model to accurately estimate yield was improved by the inclusion of rice crop parameters derived from the remote sensing products viz., seasonal rice area, and start of season and rice growth rate information - extracted from the time series of images. With this information, the model can generate a rice yield estimate for each hectare where rice was grown in the season. These yield estimates were aggregated to get a yield estimate per block. In turn these estimates were compared against the average CCE yield per block to determine

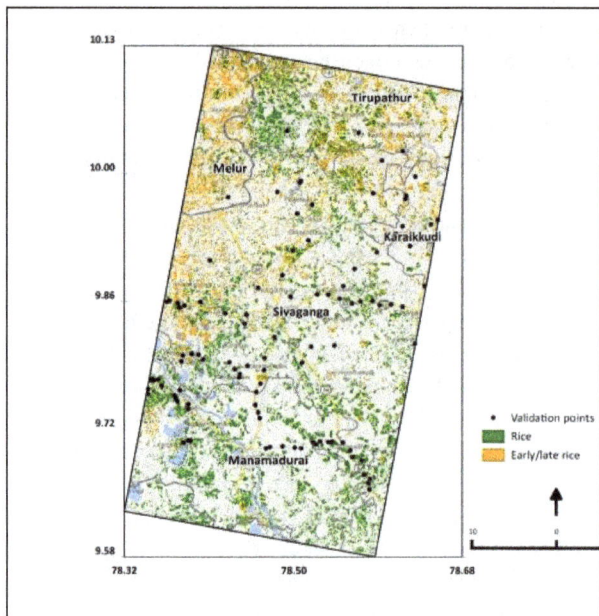

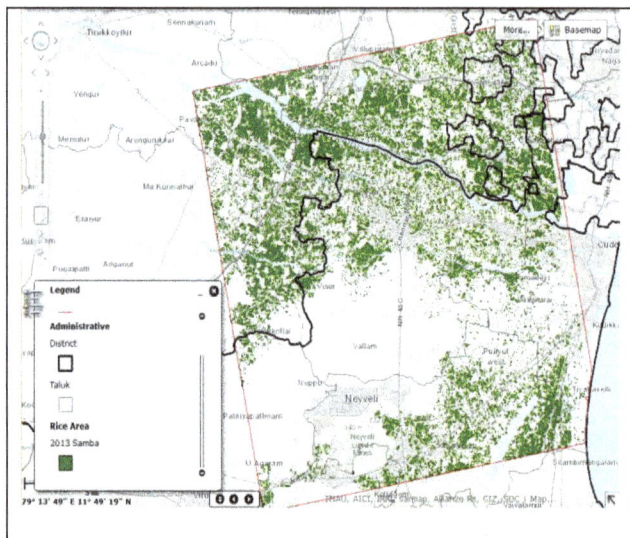

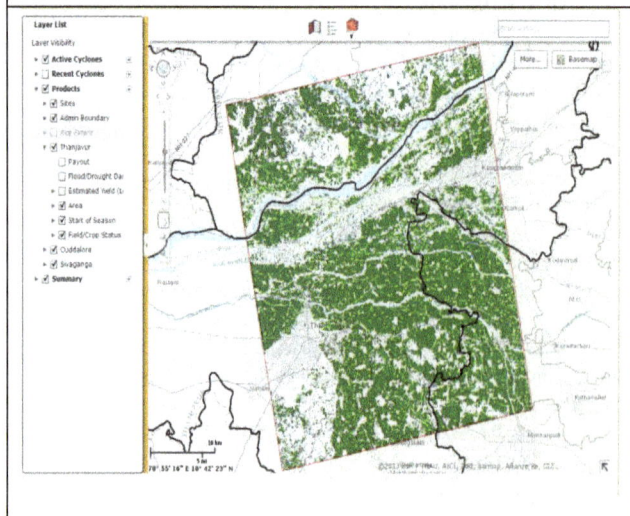

Figure 2. Rice Area map of Sivaganga, Cuddalore and Thanjavur 2013

the accuracy of the yield. In 2013 Samba season, District level, block level and field level rice yields were estimated and yield maps were generated across all the three sites.

District	RIICE estimate (kg/ha)
Cuddalore	3816
Viluppuram	3786
Thiruvarur	4866
Thanjavur	4918
Ariyalur	4932
Sivaganga	4079
Madurai	4010

Table 4. Summary of rice yield estimates at district level.

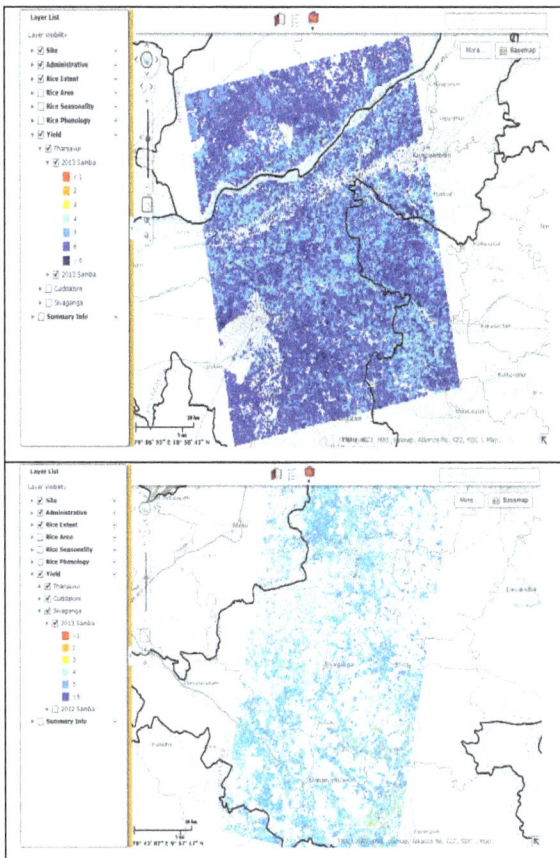

Figure.6.Rice yield map of Thanjavur and Sivaganga 2013

At district level an accuracy of 99% was achieved in Cuddalore followed Sivaganga and Thanjavur respectively with 88 and 86.7%. At block level it was interesting to come across an accuracy of 85-96% indicating the suitability of these products for policy decisions. At field level also, significantly higher accuracy was recorded for yield estimates derived using ORYZA model with Remote sensing based products indicating the scientific validation as compared to other methods of yield prediction.

The study demonstrated that rice area could be accurately classified with X-band HH polarization SAR images across multiple environments and management conditions and rice yields could be estimated by integrating ORYZA model with Remote sensing based products. With the Current and forthcoming SAR systems, such as CSK, TSX, RADARSAT-2,

RISAT-1, Sentinel-1 and ALOS-2, there exists a vast scope for rice crop monitoring and yield estimation at national level.

Block	District	RIICE estimate (kg/ha)	Block	District	RIICE estimate (kg/ha)
Cuddalore	Cuddalore	4064	Kumbakonan	Thanjavur	4892
Virudhachalam	Cuddalore	3542	Thanjavur	Thanjavur	5092
Panruti	Cuddalore	3763	Thiruvaiyaru	Thanjavur	4615
Kurinjipadi	Cuddalore	3667	Orattanadu	Thanjavur	5123
Viluppuram	Viluppuram	3911	Udaiyarpalaiyam	Ariyalur	5054
Tirukkoyilur	Viluppuram	3407	Ariyalur	Ariyalur	4881
Ulundurpettai	Viluppuram	3736	Karaikkudi	Sivaganga	4173
Pondicherry	Pondicherry	4063	Tiruppattur	Sivaganga	4132
Mannargudi	Thiruvarur	4943	Sivaganga	Sivaganga	4070
Valangaiman	Thiruvarur	4937	Manamadurai	Sivaganga	4066
Needamangalam	Thiruvarur	4630	Ilaiyankudi	Sivaganga	4002
Papanasam	Thanjavur	4844	Melur	Madurai	4010

Table 5. Rice yield estimates at block level.

Sl. No	District	RIICE estimate (Kg/ha)	CCE (Kg/ha)	RMSE (Kg/ha)	NRMSE (%)	Accuracy (%)
1	Sivaganga	4079	4635	555	12.0	88.0
2	Thanjavur	4918	5676	758	13.3	86.7
3	Cuddalore	3816	3854	38167	1.0	99.0
4	Thiruvarur	4866	5512	646	11.7	88.3

Table 6. RIICE CCE data vs. ORYZA2000 yield estimates at District level

Sl.No	Block	RIICE estimate (Kg/ha)	CCE (Kg/ha)	RMSE (Kg/ha)	NRMSE (%)	Accuracy (%)
1	Mannargudi	4943	5512	569	10.3	89.7
2	Papanasam	4844	5060	215	4.3	95.7
3	Kumbakonan	4892	5742	850	14.8	85.2
4	Cuddalore	4064	3854	209433	5.4	94.6
5	Tiruppattur	4132	4635	503	10.8	89.2

Table 7. RIICE CCE data vs. ORYZA2000 yield estimates at block level

5. CONCLUSIONS

The study demonstrates that regularly acquired X-band HH SAR imagery is suitable for rice crop monitoring across the major rice environments of South and Southeast Asia. The consistently high accuracy of the rice area classification across these sites demonstrates that the methodology is appropriate for rice detection across the most common rice agro-ecologies. The classification is based on a temporal analysis of the spectral signature, including a detection of agronomic flooding at the land preparation and/or seedling stage followed by a rapid increase in biomass relative to the duration of the vegetative stage of the varieties in the footprint. Yield Simulation accuracy of more than 87% at district level and 85-96% at block level from the study means that simulated yield matched observed yield perfectly indicating the suitability of these products for policy decisions ensuring food security besides reducing the vulnerability of smallholder rice farmers in India.

With the Current and forthcoming SAR systems, such as CSK, TSX, RADARSAT-2, RISAT-1, Sentinel-1 and ALOS-2, there exists a vast scope for rice crop monitoring and yield estimation at national level.

ACKNOWLEDGEMENTS

This work has been undertaken within the framework of the Remote Sensing-based Information and Insurance for Crops in Emerging economies (RIICE) program, financially supported by the Swiss Agency for Development and Cooperation (SDC). SAR data were provided by ASI/e-GEOS for COSMO-SkyMed and by InfoTerra GmbH for TerraSAR-X. The authors would like to thank International Rice Research Institute (IRRI), GIZ and Tamil Nadu Agricultural University for the institutional support and IRRI GIS Team and *sarmap* for the technical support.

REFERENCES

Aspert, F.; Bach-Cuadra, M.; Cantone, A.; Holecz, F.; Thiran, J.-P. Time-varying segmentation for mapping of land cover changes. In Proceedings of ENVISAT Symposium, Montreux, Switerland, 23–27 April 2007.

Attema E.P.W. and F.T. Ulaby. 1978. Vegetation modeled as a water cloud. Radio Science.13 (2): 357-364.

Balagtas, J.V.; Bhandari, H.; Cabrera, E.R.; Mohanty, S.; Hossain, M. Did the commodity price spike increase rural poverty? Evidence from a long-run panel in Bangladesh. Agric. Econ. 2014, 45,303–312.

Boschetti, M.; Nutini, F.; Manfron, G.; Brivio, P.A.; Nelson, A. Comparative analysis of normalised difference spectral indices derived from MODIS for detecting surface water in flooded rice cropping systems. PLoS One 2014, 9, e88741.

Bouman B.A.M. et al. 2001. ORYZA2000: modeling lowland rice, IRRI and Wageningen University.

Bouvet, A.; le Toan, T.; Lam-Dao, N. Monitoring of the rice cropping system in the Mekong delta using ENVISAT/ASAR dual polarization data. IEEE Trans. Geosci. Remote Sens. 2009, 47, 517–526.

Dawe, D.; Timmer, P.C. Why stable food prices are a good thing: Lessons from stabilizing rice prices in Asia. Glob. Food Sec. 2012, 1, 127–133.

De Grandi, G.F.; Leysen, M.; Lee, J.S.; Schuler, D. Radar reflectivity estimation using multiple SAR scenes of the same target: Technique and applications. In Proceedings of the 1997 IEEE International Geoscience and Remote Sensing "Remote Sensing—A Scientific Vision for Sustainable Development" (IGARSS '97), Singapore, 3–8 August 1997; Volume 2, pp. 1047–1050.

FAO FAOSTAT. Available online: http: //faostat3.fao.org/faostat-gateway/go/to/home/E (accessed on 1 June 2014).

Gumma, M.K.; Thenkabail, P.S.; Maunahan, A.; Islam, S.; Nelson, A. Mapping seasonal rice cropland extent and area in the high cropping intensity environment of Bangladesh using MODIS 500 m data for the year 2010. ISPRS J. Photogramm. Remote Sens. 2014, 91, 98–113.

Holecz, F.; Barbieri, M.; Collivignarelli, F.; Gatti, L.; Nelson, A.; Setiyono, T.D.; Boschetti, M.; Manfron, G.; Brivio, P.A.; Quilang, J.E.; et al. An operational remote sensing based service for rice production estimation at national scale. In Proceedings of the Living Planet Symposium 2013, Edinburgh, UK, 9–11 September 2013; ESA: Edinburgh, UK, 2013.

Huke, R.E.; Huke, E.H. Rice Area by Type of Culture: South, Southeast, and East Asia, A Revised and Updated Data Base; International Rice Research Institute: Manila, Philippines, 1997; p. 32 .

Inoue, Y.; Kurosu, T.; Maeno, H.; Uratsuka, S.; Kozu, T.; Dabrowska-Zielinska, K.; Qi, J. Season-long daily measurements of multifrequency (Ka, Ku, X, C, and L) and full-polarization backscatter signatures over paddy rice field and their relationship with biological variables. Remote Sens. Environ. 2002, 81, 194–204.

Inoue, Y.; Sakaiya, E. Relationship between X-band backscattering coefficients from high- resolution satellite SAR and biophysical variables in paddy rice. Remote Sens. Lett. 2013, 4, 288–295.

Inoue, Y.; Sakaiya, E.; Wang, C. Capability of C-band backscattering coefficients from high-resolution satellite SAR sensors to assess biophysical variables in paddy rice. Remote Sens. Environ. 2014, 140, 257–266.

Kim, Y.H.; Hong, S.Y.; Lee, Y.H. Estimation of paddy rice growth parameters using L-, C-, X-bands polarimetric scatterometer. Korean J. Remote Sens. 2009, 25, 31–44.

Le Toan, T.; Ribbes, F.; Wang, L.F.; Floury, N.; Ding, K.H.; Kong, J.A.; Fujita, M.; Kurosu, T. Rice crop mapping and monitoring using ERS-1 data based on experiment and modeling results. IEEE Trans. Geosci. Remote Sens. 1997, 35, 41–56.

Lopez-Sanchez, J.M.; Ballester-Berman, J.D.; Hajnsek, I. First results of rice monitoring practices in Spain by means of time series of TerraSAR-X dual-pol images. IEEE J. Sel. Top. Appl. Earth Obs. Remote Sens. 2011, 4, 412–422.

Lopez-Sanchez, J.M.; Cloude, S.R.; Ballester-Berman, J.D. Rice phenology monitoring by means of SAR polarimetry at X-band. IEEE Trans. Geosci. Remote Sens. 2012, 50, 2695–2709.

Maclean, J.L.; Hardy, B.; Hettel, G.P. Rice Almanac, 4th ed.; International Rice Research Institute: Los Baños, Philippines, 2013; p. 298.

Mittal, A. The 2008 Food Price Crisis: Rethinking Food Security Policies; United Nations: New York, NY, USA, 2009; pp. 1–40.

NASA Cloud Fraction. Available online: http: //neo.sci.gsfc.nasa.gov/view.php?datasetId= MODAL2_M_CLD_FR (accessed on 1 June 2014).

Nguyen, T.T.H.; de Bie, C.A.J.M.; Ali, A.; Smaling, E.M.A.; Chu, T.H. Mapping the irrigated rice cropping patterns of the Mekong delta, Vietnam, through hyper-temporal SPOT NDVI image analysis. Int. J. Remote Sens. 2012, 33, 415–434.

Oh, Y.; Hong, S.-Y.; Kim, Y.; Hong, J.-Y.; Kim, Y.-H. Polarimetric backscattering coefficients of flooded rice fields at L- and C-bands: Measurements, modeling, and data analysis.

IEEE Trans. Geosci. Remote Sens. 2009, 47, 2714–2721.

RIICE. Available online: http://www.riice.org (accessed on 1 December 2014).

Suga, Y.; Konishi, T. Rice crop monitoring using X-, C- and L-band SAR data. Proc. SPIE 2008, 7104, 710410.

Timmer, C.P. The Changing Role of Rice in Asia's Food Security; ADB Sustainable Development Working Paper Series, No 15; Asian Development Bank (ADB): Manila, Philippines, 2010; p. 18.

Xiao, X.; Boles, S.; Frolking, S.; Li, C.; Babu, J.Y.; Salas, W.; Moore, B. Mapping paddy rice agriculture in South and Southeast Asia using multi-temporal MODIS images. Remote Sens. Environ. 2006, 100, 95–113.

3

A FRAMEWORK OF COGNITIVE INDOOR NAVIGATION BASED ON CHARACTERISTICS OF INDOOR SPATIAL ENVIRONMENT

Ruochen SI * and Masatoshi ARIKAWA

Center for Spatial Information Science, The University of Tokyo, Kashiwa, Japan - (si, arikawa)@csis.u-tokyo.ac.jp

Commission/WG

KEY WORDS: Indoor Spatial Environment, Indoor Spatial Cognition, Interactive Positioning, Indoor Map Alignment, Active Navigation

ABSTRACT:

People are easy to get confused in indoor spatial environment. Thus, indoor navigation systems on mobile devices are expected in a wide variety of application domains. Limited by the accuracy of indoor positioning, indoor navigating systems are not common in our society. However, automatic positioning is not all about location-based services (LBS), other factors, such as good map design and user interfaces, are also important to satisfy users of LBS. Indoor spatial environment and people's indoor spatial cognition are different than those in outdoor environment, which asks for different design of LBS. This paper introduces our design methods of indoor navigation system based on the characteristics of indoor spatial environment and indoor spatial cognition.

1. INTRODUCTION

Indoor spatial environment is labyrinthine, and indoor navigation systems are needed to guide people in indoor environment. On one hand, indoor positioning is one of the key problems in indoor LBS. Different equipment and methods are used for indoor positioning, such as WiFi, Bluetooth, LED lighting, dead reckoning, and so on (Evennou, F. and Marx, F., 2006; Altini, M., Brunelli, D., Farella, E. and Benini, L., 2010; Yang, S.-H., Jeong, E.-M., Kim, D.-R., et al, 2013; Fischer, C., Muthukrishnan, K., Hazas, M. and Gellersen, H., 2008; Renaudin, V., Yalak, O., Tome, P. and Merminod, B., 2007). However, the indoor positioning accuracy is still not high and stable, compared with outdoor GPS positioning results. On the other hand, the quality of LBS is also affected by factors other than positioning accuracy, such as map design, user interface design and so on. That's to say, LBS quality can be improved by good design and use of maps and navigation systems. With this idea, we analyze the characteristics of indoor spatial environment and people's indoor spatial cognition features, to achieve good indoor LBS.

The rest parts of the paper are organized as follows: section 2 analyzes characteristics of indoor spatial environment; section 3 analyzes features of indoor spatial cognition; section 4 introduces a framework of indoor LBS system that adapts to the indoor spatial and cognitive environment; section 5 introduce related prototype and experiment results; section 6 draws a conclusion.

2. INDOOR SPATIAL ENVIRONMENT

Indoor spatial environment is different than outdoor spatial environment. Five main characteristics of indoor spatial environment are as follows:

Simple spatial pattern. The outdoor roads and paths form complex network (see Figure 1). There are usually many ways to connect two points of interest (POI) in the road network. The topology of paths in indoor environment is simpler. Figure 2 shows two basic indoor spatial patterns: linear pattern and tree

pattern. For the linear pattern, POIs are stringed by a main path (include the case of ring). Linear pattern is often seen in office buildings and shopping malls. For tree pattern, one path leads to multiple paths at diverse points. Tree pattern are usually seen in trains and underground stations. Some complex indoor environment did form network topology, but many of them can be treat as compositions of the above two types of spatial patterns.

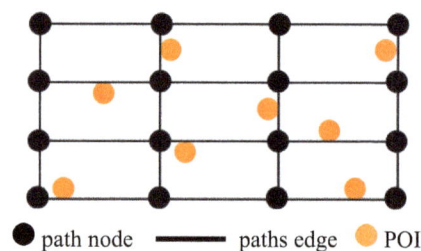

● path node ── paths edge ● POI

Figure 1. Outdoor spatial pattern

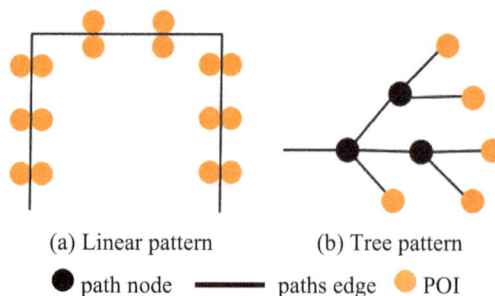

(a) Linear pattern   (b) Tree pattern

● path node ── paths edge ● POI

Figure 2. Indoor spatial patterns

Multi-floors. The outdoor spatial environment is dominant with one ground surface, and tiles of maps are connected parallely. While indoor spatial environment can contain more than one surfaces, and layers of maps are overlaid vertically. This characteristic of indoor environment adds a discrete vertical dimension for indoor objects and brings differences in topological relations on indoor maps. For outdoor maps, the entrances and exists are on the edge of the maps; objects can cross multiple tiles of maps; one must move across the edge of the map to enter and leave a map, and must cross neighbour tiles of maps to reach points on further maps (see Figure 3). While for indoor maps, the entrances and exists can be both on the edge of the map and in the map; objects cannot cross multiple layer of maps (except for layer connecters such as stairs and elevators); one does not necessarily enter and leave at the edge of maps, and can skip neighbour layers of maps to reach points on further layer of maps (see Figure 4).

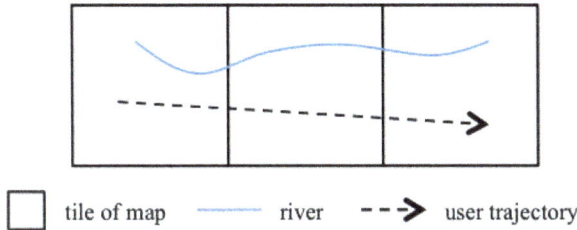

Figure 3. Topology relations on outdoor maps

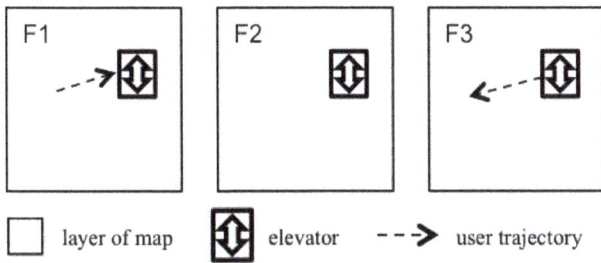

Figure 4. Topology relations on indoor maps

Narrow view. Blocked by walls and corners, the view field in indoor is narrow. People can hardly get a whole view of indoor environment without a floor map. The narrow sight view makes it difficult for people to recognize an overall structure of the indoor environment.

Unique objects. In outdoor environment, many objects are not unique by name, such as chain shops. On the contrary, indoor environment is rich of unique spatial references, such as numbered rooms and shops. The unique references can help to locate people's position in indoor environment.

3. INDOOR SPATIAL COGNITION

Affected by the characteristics indoor spatial environment, indoor spatial cognition also has different features from outdoor cognition.

Allocentric positioning. Allocentric floor maps are widely used for indicating users positions. The allocentric map alignment makes it easy for users to understand the overall structure of the indoor saptial environment and the relations of their positions to the indoor space.

Egocentric directing. In indoor environment, people rely more on egocentric relative relations (e.g. left to and right to) for directing. Absolute spatial directions (e.g. north to and south to) are difficult to recognize because of the narrow view of indoor environment.

Self-location description. Although people are easily get lost in indoor environment, it is convenient to describe self-location by unique spatial references, such as "in Room 101", "in front of ABC Shop", and so on. And by checking the near by spatial objects and self-moving trajectories, people can tell obvious

mistakes of positioning results of indoor navigation systems and make corrections.

Cognitive collages. People recognize indoor space in the units of cognitive collages. A cognitive collage is an area in the indoor environment in which spatial cognition is continuous and smooth. Cognitive collages are separated by indoor spatial references, usually the rooms. Two points within same collage can be treated as same place while two points that do not belong to a same collage are treated as different places. LBS results are sensitive between different collages but insensitive within a same collage. Cognitive collages can be partly or completely overlaid. As shown in Figure 5, collage A and collage B are partially overlaid, and the point P can be described as either "in front of ABC Shop" or "in front of DEF Store".

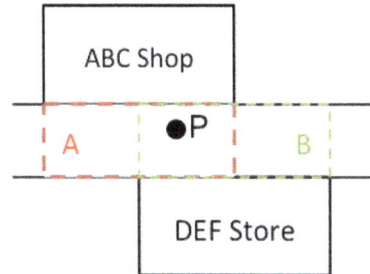

Figure 5. Indoor area is divided into cognitive collages by references, and cognitive collages can be overlaid.

4. FRAMEWORK OF COGNITIVE INDOOR NAVIGATION SYSTEM

We propose a framework of cognitive indoor navigation system that is designed based on the characteristics of indoor spatial environment and indoor spatial cognition. As shown in Figure 6, the framework contains three main parts: mapping, positioning, and navigation.

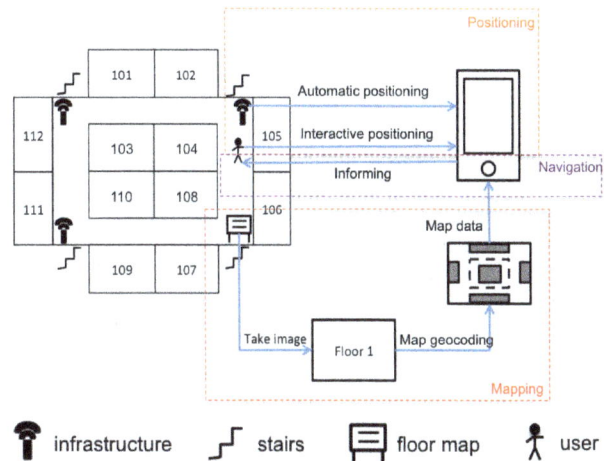

Figure 6. Framework of cognitive indoor navigation system

4.1 Mapping

We use the images of floor maps as resources to create map data for indoor navigation. The map images need to be geocoded to enable LBS. The following three kinds of information need to be geocoded onto map images.

Cognitive information. Virtual elements will be added for indoor navigation, including the center line of paths, positions of entrances of rooms, and so on. And we need to break the indoor space into subareas of cognitive collages by indoor spatial references.

Positioning information. Indicate the positions of indoor infrastructures that are used for positioning on the map images.

Illustrating information. Add multimedia content to enrich the indoor information, such as pictures of shops, texts of the discount information, audio guide of suggested POIs, short video of related events, and so on.

Table 1 shows the spatial types for geocoding the above information.

Information type	Information item	Spatial type
Cognitive information	Center line	Polyline
	Room entrance	Point
	Cognitive collage	Polygon
Positioning information	Infrastructure	Point
Illustrating information	Picture	Point
	Text	Point
	Audio	Point/Polyline/Polygon
	Video	Point/Polyline/Polygon

Table 1. Geocode different kinds of spatial information

4.2 Positioning

We use two positioning methods to get user's position: interactive positioning and automatic positioning.

Interactive positioning requires users to report their current positions to the system. Two ways are provided for users to report their positions. One is report via referencing objects. There are many unique spatial references in indoor spatial environment, and users can use nearby reference's name to report their positions, such as inputting "ABC Shop". The other way to report self-position is by indicating directly on map. When users find their positions obviously wrong on the navigation system, they can directly tap on the map image to correct their positions.

Automatic positioning uses infrastructures' signals to calculate user's position automatically. As mentioned in section 1, there are many indoor positioning methods, but one common problem of current indoor positioning methods is the stability of positioning accuracy. Usually, indoor positioning accuracy is reasonable at check points near the infrastructures, but the accuracy decreases when leaving the infrastructures. For our system, infrastructures should be set at key places, such as decision points and the places that are difficult for people to report their positions.

The interactive and automatic positioning will work together to ensure reasonable positioning results all over the indoor maps.

4.3 Navigation

To adapt the navigation system to indoor spatial environment, we enable automatic map alignment function and active leading function for the system.

Automatic map alignment function is to rotate and align map images for convenient map cognition. Allocentric and egocentric map alignment are mainly used for floor maps. We use a combination of allocentric alignment and geocentric alignment for map presentation. As shown in figure 7, we divide four directions according to an allocentric floor map. When user is facing up, the map is not rotated; when user is facing left, the map is rotated 90 degree clockwise; when user is facing down, the map is rotated 180 degree; when user is facing right, the map is rotated 90 degree anticlockwise. This map aligning method keeps the map allocentric, which is easy for

users to recognize their position and direction related to the whole indoor spatial environment. It also ensures that user is facing to the upside of the aligned map, which is maintain the cognitive features of ego-centric alignment and is good for navigation.

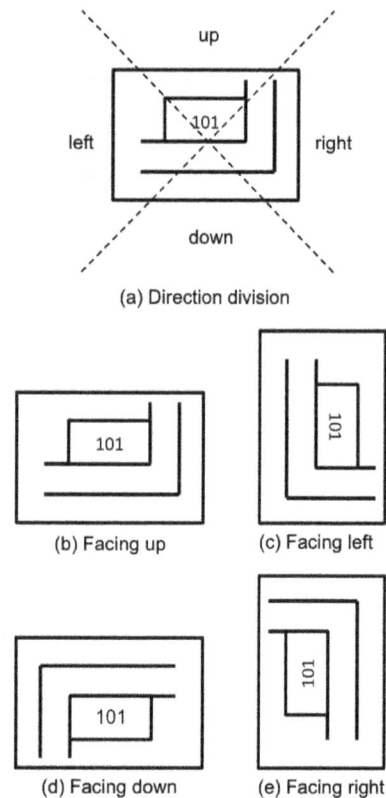

Figure 7. Map alignments for indoor navigation

The instability of positioning accuracy results in an unsmooth user trajectory during navigation, which leaves negative impressions to users. To solve the problem, we use a "You should be here point" instead of "You are here point" for indoor navigation. A leading point will be shown on the map image and automatically moving along the suggested path, and users will follow the leading point. Users can adjust the moving speed of the leading point to adapt to their walking speed. And the leading point can be enriched by illustrated information such as pictures and audios. The movement of the leading point is stable and smooth, which will leave users positive impressions on the system.

5. PROTOTYPE AND EXPERIMENT

So far, we do not have a prototype of the cognitive indoor navigation system. However, we have developed an outdoor navigation system that share parts of similar functions of the proposed indoor system, including taking and geocoding map images, positioning on map images and auto-alignment of map images. Experiment on the outdoor navigation prototype achieved good results in positioning (Figure 8) and map aligning (Figure 9) which optimize us for realizing the prototype of the indoor navigation system

In the future, we will develop a prototype of the proposed indoor navigation system and do experiments on it and improve it according to the results of the experiments.

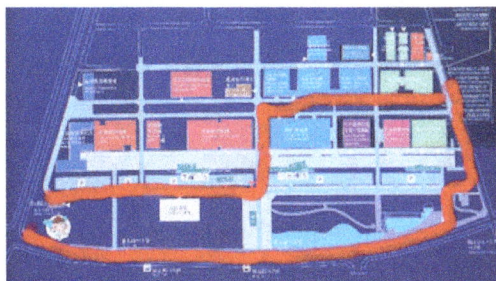

Figure 8. Positioning results of user's trajectory on related prototype of outdoor navigation system

Figure 9. Auto-alignment of map image on related prototype of outdoor navigation system

6.   CONCLUSIONS AND FUTURE WORK

In this paper, we analysed the characteristics of indoor spatial environment and indoor spatial cognition. We found indoor spatial pattern is basically made up by linear pattern and tree patter, which is simpler than outdoor network spatial pattern; indoor space has narrow view and rich of unique objects that can be used as location references. We also found that allocentric map alignment is good for indoor positioning while egocentric map alignment is good for indoor directing; people can easily report their positions by the nearby spatial references and recognize indoor space with pieces of cognitive collages.

Based on the analysis of indoor spatial environment and indoor spatial cognition, we proposed a framework of cognitive indoor navigating system. The system uses images of floor maps for LBS. It applies both interactive positioning and automatic positioning, and provides automatic map alignment and positive leading function for indoor navigation.

However, the usage situation is limited for some functions. For example, the interactive positioning function is fit for users who have free time, but not fit for users who are in a hurry.

In the future, we are going to develop a prototype of the proposed indoor navigation system, and do experiments to test its performance for improving the framework.

References

Evennou, F. and Marx, F., 2006. Advanced Integration of WiFi and Inertial Navigation Systems for Indoor Mobile Positioning. EURASIP Journal on Applied Signal Processing, 2006(01), pp 1-11.

Altini, M., Brunelli, D., Farella, E. and Benini, L., 2010. Bluetooth Indoor Localization with Multiple Neural Networks. 2010 5th International Symposium on Wireless Pervasive Computing (ISWPC), pp 295-300.

Yang, S.-H., Jeong, E.-M., Kim, D.-R., et al, 2013. Indoor three-dimensional location estimation based on LED visible light communication. ELECTRONICS LETTERS, 49(1), pp 54-56.

Fischer, C., Muthukrishnan, K., Hazas, M. and Gellersen, H., 2008. Ultrasound-Aided Pedestrian Dead Reckoning for Indoor Navigation. Proceedings of the first ACM international workshop on Mobile entity localization and tracking in GPS-less environments, pp 31-36.

Renaudin, V., Yalak, O., Tome, P. and Merminod, B., 2007. Indoor Navigation of Emergency Agents. European Journal of Navigation, 5(3), pp 36-45

4

DATA AND TECHNIQUES FOR STUDYING THE URBAN HEAT ISLAND EFFECT IN JOHANNESBURG

C. H. Hardy*, A. L. Nel

Department of Mechanical Engineering Science, University of Johannesburg, Johannesburg, South Africa - caroline.h.hardy@gmail.com, andren@uj.ac.za

KEY WORDS: Urban Heat Island, Remote Sensing, Thermal Infrared, Land Surface Temperature, Multi-temporal

ABSTRACT:

The city of Johannesburg contains over 10 million trees and is often referred to as an urban forest. The intra-urban spatial variability of the levels of vegetation across Johannesburg's residential regions has an influence on the urban heat island effect within the city. Residential areas with high levels of vegetation benefit from cooling due to evapo-transpirative processes and thus exhibit weaker heat island effects; while their impoverished counterparts are not so fortunate. The urban heat island effect describes a phenomenon where some urban areas exhibit temperatures that are warmer than that of surrounding areas. The factors influencing the urban heat island effect include the high density of people and buildings and low levels of vegetative cover within populated urban areas. This paper describes the remote sensing data sets and the processing techniques employed to study the heat island effect within Johannesburg. In particular we consider the use of multi-sensorial multi-temporal remote sensing data towards a predictive model, based on the analysis of influencing factors.

1. INTRODUCTION

Johannesburg is South Africa's most populous city and the provincial capital of the Gauteng province. According to (Statistics South Africa, 2012), in 2011 Johannesburg had a population of 4 434 827 people living in an area of only 1 645 km^2. Johannesburg's population is growing at a rapid pace, in the decade between Census 2001 and Census 2011 the population increased by more than one million inhabitants. The City of Johannesburg Metropolitan Municipality is both the financial hub and the core of the Gauteng City-Region (GCR) conurbation. The GCR conurbation has developed into a contiguous urban region comprised of five constituent municipalities; the City of Johannesburg Metropolitan, City of Tshwane Metropolitan, Ekurhuleni Metropolitan, West Rand District and Sedibeng District.

Johannesburg and the GCR have been subject to urban growth and environmental change, particularly in the region's informal settlements as people migrate to cities within the GCR. Urbanization has an influence on the natural environment as natural surfaces are replaced by man-made structures and increased anthropogenic activity leads to greater emissions of aerosols and trace gases into the atmosphere. The result is the development of urban heat islands (UHIs) within the city and informal settlements.

In this paper we describe the data and methods used to the study of the urban heat island effect (UHIE) in Johannesburg.

1.1 Remote Sensing and the urban heat island

The surface urban heat island (SUHI) refers to urban areas where the surface temperature of the area is higher than that of surrounding non-urban areas (Voogt and Oke, 2003). The SUHI may be present both during the day and at night, and is influenced by factors such as anthropogenic activities, urban function, urban landscape and form, temporal aspects, climate, weather, topography and location (Arnfield, 2003, Taha, 1997).

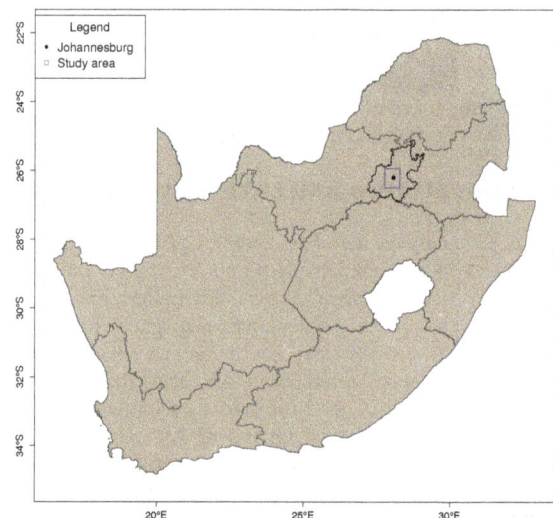

Figure 1: Map of South Africa showing the location of Johannesburg (denoted by a dot) and the study area as indicated by a blue rectangle.

The urban landscape is not a homogeneous surface as it is comprised of different materials and structures; the surface properties, thermal characteristics and placement of these materials and natural features such as rivers and hills describe the urban form. Temporal aspects refer to the time of day and seasonal variation. Urban function is characterized by the primary land use and activities undertaken in the urban area. The urban function of an area affects the amount of energy used and the level of aerosols and gas emissions. Climate and weather influence the SUHI as clouds block solar radiation from reaching the Earth's surface, while winds may cause the heat which is usually trapped in the city to dissipate.

In order to effectively study the SUHI, thermal sensors are required to detect heat radiating from the urban surface. Remotely sensed thermal observations of the Earth's surface can be used to

---
*Corresponding author.

study the thermal state of the surface and to analyze the UHI if the observations provide sufficient resolution. The spatial resolution of the data describes the level of spatial detail recorded while the temporal resolution refers to the time between consecutive images. If the temporal resolution of the data is too low, the gaps between image acquisitions may be too large to perform a reliable long-term study. Low spatial resolution data is unable to reflect changes in small objects in the scene. Spatial and temporal resolution are inversely related, thus it is not possible to obtain data with a high temporal resolution while maintaining a high spatial resolution.

1.2 Research Objective

This paper describes the preliminary work undertaken towards the development of a predictive UHI model for Johannesburg to determine how the UHIs within Johannesburg evolve over time.

2. MATERIALS AND METHODS

2.1 Study Area

The primary focus area of this study is the Greater Johannesburg Metropolitan area. The study area, depicted in Figure 2, covers approximately 60km by 70km which includes the primary focus area and surroundings. The area was selected to study the UHIE, since Johannesburg is South Africa's most populous city.

Figure 2: Landsat 7 image (15 March 2003) of the study area. The white boundary line indicates the extent of Greater Johannesburg Metropolitan area. The study area is centred over the Johannesburg CBD.

UHI studies have been conducted in many cities around the world (Peng et al., 2012, Roth and Chow, 2012, Arnfield, 2003) but few studies have been conducted in African cities (Tyson et al., 1972, Goldreich, 1985, Adebayo, 1987). The urban form of Johannesburg differs from many other cities, as Johannesburg is not a compact city and it was not founded near a river or the ocean. Unlike

many large cities, Johannesburg does not consist of a dense network of city blocks. Instead Johannesburg is spread out and has many residential suburbs and distributed business centres. Johannesburg also contains over 10 million trees and is therefore referred to as an urban forest. Studies have shown vegetation to have a mitigating effect on the UHI. However not all parts of Johannesburg contain high levels of vegetation. Areas which are most deficient in vegetation are the city's urban centres, such as Sandton and Randburg, and informal settlements.

2.2 Remotely sensed data

Cloud-free thermal infrared data is required for UHI studies. High spatial resolution data is preferred over lower resolution data for urban studies. Advanced Spaceborne Thermal Emission and Reflection Radiometer (ASTER) and Landsat offer high resolution thermal infrared data; ASTER thermal infrared bands have a resolution of 90m, while Landsat 4/5 Thematic Mapper (TM) thermal infrared data have a 120m resolution and the Landsat 7 Enhanced Thematic Mapper Plus (ETM+) thermal infrared data has a resolution of 60m. Landsat 8, which was launched in February 2013, acquires thermal infrared data at 100m pixel-resolution. All thermal infrared data from Landsat 4/5 TM, Landsat 7 ETM+ and Landsat 8 thermal infrared data are resampled to 30m.

Landsat 4/5 TM and 7 ETM+ data of Johannesburg is only available for day time scenes, acquired at approximately 10:00AM local time. ASTER data is not appropriate for this study, except for validation purposes, because the available ASTER data is both temporally sparse and does not cover the full study area. Satellites which offer high resolution data typically have a long repeat cycle; ASTER and all Landsat satellites return to the same path once every 16 days. With long revisit times, there is a low chance of acquiring high resolution cloud-free data over areas which are frequently cloudy. Johannesburg's rainy season occurs in summer, thus many Landsat scenes over Johannesburg in summer contain cloud cover. In order to increase the chances of acquiring cloud-free data, low resolution sources must be considered as these satellites typically have short revisit times of between one and three days.

Only historical data is available from the ENVISAT Advanced Along Track Scanning Radiometer (AATSR) program as ENVISAT is no longer operational, while current and historical low resolution thermal infrared data is available from Moderate Resolution Imaging Spectroradiometer (MODIS) and Advanced Very High Resolution Radiometer (AVHRR). Thermal infrared data from MODIS, AVHRR and AATSR all have a nominal pixel resolution of 1km.

ENVISAT AATSR data, which has a revisit time of 3 days and provides both day and night scenes (acquired at approximately 10:00AM and 10:00PM local time), been used in this study. A disadvantage low spatial resolution data is that not possible to resolve finer scale details, such as vegetation levels and accurate land cover. This problem is overcome by combining AATSR data with classified Landsat data from a date representative of the AATSR scene, in this way annual and seasonal land cover and land use change can be factored into the LST values derived from the AATSR data.

2.3 Case study

In this study data from Landsat 5 TM, Landsat 7 ETM+ and ENVISAT AATSR have been employed to study the UHIE in Johannesburg. The Geospatial Data Abstraction Library (GDAL) and the Basic ERS & Envisat (A)ATSR and Meris (BEAM) Toolbox were used to preprocess and reproject all AATSR Level 2 (ATS NR 2P) and Landsat data. The preprocessing procedure entails; reprojecting to UTM zone 35S using the WGS84 ellipsoid, image coregistration and spatial subsetting to clip the data to the extent of the study area. The extent of the study region is depicted in Figure 2.

After preprocessing, all processing of the AATSR and Landsat data was performed using R (R Core Team, 2015) and the R landsat package (Goslee, 2011). Using the landsat package, the Landsat thermal bands were converted to temperature maps. The ATS NR 2P data is a AATSR product which contains land surface temperature (LST) and normalized difference vegetation index (NDVI) data. Algorithm details for AATSR operational LST retrieval can be found in (Prata, 2002). An example of the AATSR LST data can be seen in Figure 4.

Unsupervised k-means classification was performed on the Landsat data to classify the data according to seven land cover classes. Example output of the classification process is shown in Figure 3. Landsat 7 ETM+ data which contain gaps due to the Landsat 7 scan line corrector (SLC) error require gap filling prior to classification, gap filling has been covered extensive in the literature (Chen et al., 2011, Weiss et al., 2014, Maxwell et al., 2007).

Thirty four measurements points were selected according to the Landsat land cover classification map and then manually reclassified into six land use classes according to the original land cover class and geographic location; outlying non-urbanized/rural areas, airports/impervious surface, urban centres (i.e. Sandton, Fourways and Randburg), urban green areas, Johannesburg CBD and high density residential settlements. The 34 points were used to to extract LST values from the combined Landsat and AATSR dataset.

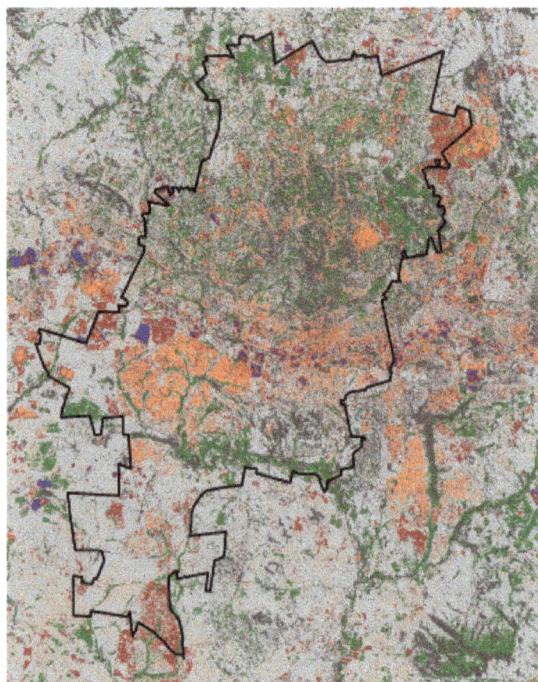

Figure 3: Result of k-means classification on Landsat 7 image (15 March 2003) of the study area.

It is not possible to acquire a full cloud-free dataset; thus to maximize the number of days for which valid data is available and simultaneously minimize cloud contamination of LST values in partially cloudy scenes we extract mean LST values from the data. The mean LST values is extracted from each LST image in the dataset from a predefined neighbourhood around the geographic coordinates of the centre pixel for each of the 34 measurement points.

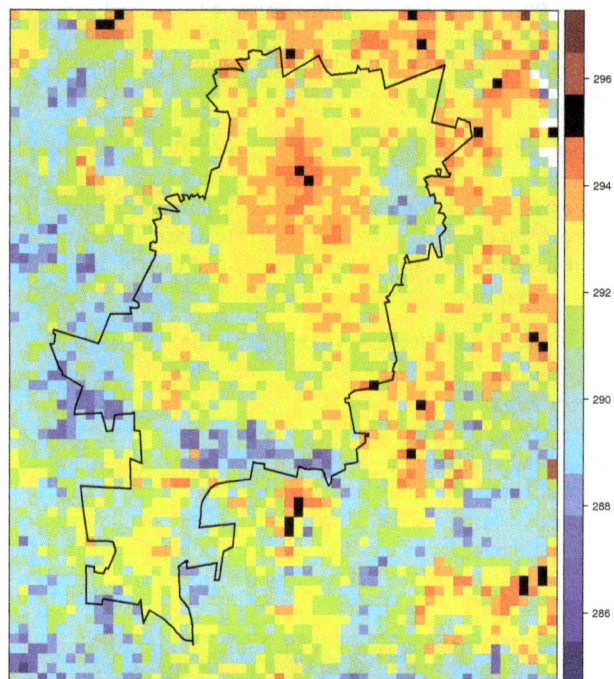

Figure 4: A land surface temperature map from AATSR data shows the night-time Johannesburg urban heat island on 14 March 2003 at 22:25 local time. The data has a nominal pixel size of 1km by 1km.

3. RESULTS

The Landsat classification map was used to identify measurements points representative of the seven land cover classes identified. The urban heat island intensity (UHII), which is described by the equation $\Delta T_{u-r} = T_u - T_r$, was calculated by subtracting the mean temperature of the non-urbanized class from the mean temperature of the each of the other land use classes for each day represented in the satellite data dataset. For the period of operation of ENVISAT, from 2002 and 2012, the mean night-time UHII over all land use classes ranged between $2°C$ and $3.5°C$, the highest night-time UHII values were associated the urban centres and high density residential settlement classes.

4. CONCLUSIONS

Comparison of the LST data with classified Landsat data revealed that areas with high levels of vegetation are cooler and thus have lower UHII values than less vegetated areas, while built-up areas and densely populated areas are hotter and have higher UHII values.

This study found that UHIs exist within the study area, at night the UHIE is strongest over Johannesburg's Northern Suburbs (as seen in Figure 4) and in areas where building density is high. The UHIE is weakest in areas that contain water bodies (such as dams and streams) or high levels of vegetation. Remotely sensed thermal infrared data is a valuable asset in studies of the urban environment, especially in areas where ground-based weather stations are spatially sparse. The AATSR dataset of Johannesburg is best used to study UHIs occurring at night since many day-time observations are not usable due to cloud cover, while the Landsat data for Johannesburg as only available for day-time scenes and thus the Landsat thermal data and Landsat derived land surface temperature can only be used to study the day-time heat island.

Classification maps (similar to that shown in Figure 3), derived from k-means classification applied over the optical and near infrared Landsat bands, can be used to supplement the AATSR night-time thermal data for a better understanding of the underlying urban environment and to determine the associated land use and land cover changes. Due to the combination of cloud cover, satellite overpass time and repeat cycle length, it is not possible to create a comprehensive time series of the complete study area. The use of satellite data from multiple sensors is required in order to study the dynamic nature of UHIs within the city.

Landsat data provides a high level of detail for spatial features, which is sufficient for the purposes of tracking and monitoring land use and land cover change associated with spatial variation in the UHIE. However, the largest shortcoming is data temporal resolution; the inclusion of thermal infrared and LST data from AVHRR and MODIS to the dataset may provide more cloud-free data to close the temporal gaps in the dataset, thus enabling better temporal analysis of the dataset and thus improving future UHII estimation.

REFERENCES

Adebayo, Y. R., 1987. Land-use approach to the spatial analysis of the urban 'heat island' in Ibadan, Nigeria. Weather 42(9), pp. 273–280.

Arnfield, A. J., 2003. Two decades of urban climate research: a review of turbulence, exchanges of energy and water, and the urban heat island. International Journal of Climatology 23(1), pp. 1–26.

Chen, J., Zhu, X., Vogelmann, J. E., Gao, F. and Jin, S., 2011. A simple and effective method for filling gaps in landsat ETM+ SLC-off images. Remote Sensing of Environment 115(4), pp. 1053–1064.

Goldreich, Y., 1985. The structure of the ground-level heat island in a central business district. Journal of Climate and Applied Meteorology 24(11), pp. 1237–1244.

Goslee, S., 2011. Analyzing remote sensing data in R: The landsat package. Journal of Statistical Software 43(4), pp. 1–25.

Maxwell, S. K., Schmidt, G. L. and Storey, J. C., 2007. A multiscale segmentation approach to filling gaps in landsat ETM+ SLC-off images. International Journal of Remote Sensing 28(23), pp. 5339–5356.

Peng, S., Piao, S., Ciais, P., Friedlingstein, P., Ottle, C., Bréon, F.-M., Nan, H., Zhou, L. and Myneni, R. B., 2012. Surface urban heat island across 419 global big cities. Environmental Science & Technology 46(2), pp. 696–703.

Prata, F., 2002. Land Surface Temperature Measurement from Space: AATSR Algorithm Theoretical Basis Document. Kartographische Nachrichten 55(1), pp. 3–11.

R Core Team, 2015. R: A Language and Environment for Statistical Computing. R Foundation for Statistical Computing, Vienna, Austria.

Roth, M. and Chow, W. T., 2012. A historical review and assessment of urban heat island research in Singapore. Singapore Journal of Tropical Geography 33(3), pp. 381–397.

Statistics South Africa, 2012. Population Census 2011.

Taha, H., 1997. Urban climates and heat islands: albedo, evapotranspiration, and anthropogenic heat. Energy and Buildings 25(2), pp. 99–103.

Tyson, P., du Toit, W. and Fuggle, R., 1972. Temperature structure above cities: Review and preliminary findings from the Johannesburg Urban Heat Island Project. Atmospheric Environment 6(8), pp. 533–542.

Voogt, J. and Oke, T., 2003. Thermal remote sensing of urban climates. Remote Sensing of Environment 86(3), pp. 370–384.

Weiss, D. J., Atkinson, P. M., Bhatt, S., Mappin, B., Hay, S. I. and Gething, P. W., 2014. An effective approach for gap-filling continental scale remotely sensed time-series. ISPRS Journal of Photogrammetry and Remote Sensing 98, pp. 106–118.

5

SATELLITE RADIOTHERMOVISION OF ATMOSPHERIC MESOSCALE PROCESSES: CASE STUDY OF TROPICAL CYCLONES

D.M. Ermakov[a, b, *], E.A. Sharkov[b], A.P. Chernushich[a]

[a] Institute of Radioengineering and Electronics of RAS, Fryazino department (FIRE RAS),
Vvedenskogo sq., 1, Fryazino, Moscow region, 141120, Russian Federation – pldime@gmail.com
[b] Space Research Institute of RAS (IKI RAS), 84/32 Profsoyuznaya str, Moscow, 117997, Russian Federation – e.sharkov@mail.ru

KEY WORDS: Radiothermovision, tropical cyclones, spatiotemporal interpolation, dynamics, latent heat, advection

ABSTRACT:

Satellite radiothermovision is a set of processing techniques applicable for multisource data of radiothermal monitoring of ocean-atmosphere system, which allows creating dynamic description of mesoscale and synoptic atmospheric processes and estimating physically meaningful integral characteristics of the observed processes (like avdective flow of the latent heat through a given border). The approach is based on spatiotemporal interpolation of the satellite measurements which allows reconstructing the radiothermal fields (as well as the fields of geophysical parameters) of the ocean-atmosphere system at global scale with spatial resolution of about 0.125° and temporal resolution of 1.5 hour. The accuracy of spatiotemporal interpolation was estimated by direct comparison of interpolated data with the data of independent asynchronous measurements and was shown to correspond to the best achievable as reported in literature (for total precipitable water fields the accuracy is about 0.8 mm).

The advantages of the implemented interpolation scheme are: closure under input radiothermal data, homogeneity in time scale (all data are interpolated through the same time intervals), automatic estimation of both the intermediate states of scalar field of the studied geophysical parameter and of vector field of effective velocity of advection (horizontal movements). Using this pair of fields one can calculate the flow of a given geophysical quantity though any given border. For example, in case of total precipitable water field, this flow (under proper calibration) has the meaning of latent heat advective flux.

This opportunity was used to evaluate the latent heat flux though a set of circular contours, enclosing a tropical cyclone and drifting with it during its evolution. A remarkable interrelation was observed between the calculated magnitude and sign of advective latent flux and the intensity of a tropical cyclone. This interrelation is demonstrated in several examples of hurricanes and tropical cyclones of August, 2000, and typhoons of November, 2013, including super typhoon Haiyan.

1. INTRODUCTION

Progress in algorithms of retrieval of geophysical parameters from satellite passive microwave measurements provides accuracy which is appropriate for a wide range of applications of remote sensing of the Earth, see, e.g., (Wentz, 1997). However, the developed principles of remote sensing data processing usually interpret these data as a set of independent measurements, ignoring their spatial and temporal coherence. The account of this coherence, i.e. interpretation of remote sensing data and derived products as dynamic fields enables obtaining more valuable information.

A possible approach to the interpretation of satellite radiothermal data as dynamic fields is considered in this paper. Conceptually close, but significantly different in the implementation is the approach (Nerushev and Kramchaninova, 2011) designed for the data of the Earth monitoring in the visible and infrared range from geostationary satellites. A known alternative to the authors' approach to interpret the data from polar-orbiting satellites as dynamic fields (Wimmers and Velden, 2011), advanced to the level of operational technology, has some fundamental differences that, in particular, may be source to some errors, as briefly discussed in the paper.

The authors' approach called "satellite radiothermovision" is built up of a set of algorithms (including creation of reference fields, spatiotemporal interpolation and estimation of physically

meaningful integral characteristics) which form a closed scheme of satellite data processing applicable to investigation of a wide range of mesoscale (and synoptic) atmospheric processes. This concept is demonstrated on a case study of evolution of tropical cyclones (TC) in August, 2000, and November, 2013. Other important opportunities of the approach are briefly discussed in the concluding remarks.

2. SATELLITE RADIOTHERMOVISION BASICS

In remote sensing of the Earth from polar-orbiting satellites meridional coverage is carried out by the rotation of the satellite in orbit, while the zonal coverage is accomplished due to the rotation of the Earth relative to the orbital plane. In case of sun-synchronous orbits the result is the natural periodicity of measurements of about 12 hours, considering both ascending and descending orbits. This coverage, generally, is partial at global scale: the divergence of satellite swaths on the equator gives rise to "lacunae" – extended areas not covered by the measurements.

A full set of daily measurements from a polar-orbiting sun-synchronous satellite (or estimated from them geophysical parameters) will be called a "source field» W_i^S. Here, the symbol "W" represents a geophysical parameter (or measurement), the index "S" allows to distinguish between data sources (different satellites as well as ascending/descending orbits), and the index

---

* Corresponding author

"i" allows to order the data by the local date and time of orbit nodes. Fields W_i^S characterize the state of the observed region at the moments of measurements and, taken separately, do not contain direct information on the dynamics of the observed processes. Such information may be obtained informally by some indirect estimates, with the use of additional data and the assistance of experts. The technique of spatiotemporal interpolation outlined in this paper allows formalizing and unifying the extraction of such information on the basis of a processing scheme closed in respect to the W_i^S remote data. The basic scheme consists of the three steps considered below in operator form.

2.1 Creating reference fields

The first step is the formation of the "reference fields" W_i, each of which combines the measurements in a small range of local date and time (within 1 hour) of the orbit node for the entire set of data to be processed (e.g., weekly, monthly, and annual measurements). The absence of a superscript notation differs the reference field W_i from source fields W_i^S. The calculation can be considered as the result of application of a special operator L_R to one or more W_i^S (for the case of a single instrument the W_i^S are well-known daily fields built up separately from ascending and descending swaths):

$$L_R\left(W_i^{S_1}, W_i^{S_2}, ...\right) = W_i \qquad (1)$$

The operation is to combine the measurements on a single regular rectangular grid and fill the remaining lacunae with a smooth extension of the W_i in the blank grid nodes. Mathematical model and specific algorithmic implementation of this step is described in detail in (Ermakov et al, 2013c). Physical meaning of the operation is search of the field isolines in vicinity of a lacuna and smooth extrapolation of the field values along them to fill the lacuna. It was established experimentally that in the context of considered tasks the measurements "close enough in time" are those separated by time intervals of about 1 hour or less.

The created reference fields W_i are arranged in chronological order by local time. The need for interpolation on a regular grid and for "lacunae stapling" is caused by the conditions of applicability of the next two operators: motion estimation L_M and compensation L_C.

2.2 Motion estimation

Motion estimation operator L_M is applied in the second step of the calculation scheme for all pairs (W_i, W_{i+1}) of the reference fields adjacent on the time in the formed chronological sequence and generates, for each pair, the displacement vector field $\mathbf{V}_i$, describing in linear approximation the transformation (transition) of W_i to W_{i+1} on the grid:

$$L_M\left(W_i, W_{i+1}\right) = \mathbf{V}_i \qquad (2)$$

The operation is performed for each pair (W_i, W_{i+1}) iteratively at hierarchically decreasing spatial scales down to a minimum, defined by the distance between grid nodes. As a result consistently restored are, at first, macroscale "background" motions, and then, as corrections, transition, rotation and deformation at smaller scales. Motion estimation is quite a known operation in the domain of video stream processing and

analysis. A detailed description of some realizations of motion estimation algorithm can be found in (Richardson, 2003).

2.3 Motion compensation

Motion compensation operator L_C is applied in the third step of calculation and generates, basing on the reference field W_i and the displacement vector field $\mathbf{V}_i$, the estimate of the intermediate state of the field $W_{i+1/2}$, equidistant in (local) time from W_i and W_{i+1}:

$$L_C\left(W_i, \mathbf{V}_i\right) = W_{i+1/2} \qquad (3)$$

The resulting intermediate fields $W_{i+1/2}$ together with the reference fields W_i form a new chronological sequence, which can be processed iteratively by operators (2) and (3). Some principles and instructions for software realization of these steps are described in (Ermakov et al, 2011). In the end, with reasonable accuracy (as shown below), the evolution of the W field can be estimated with time sampling of about 1.5 hour.

2.4 Further analysis

It should be noticed that along with spatiotemporal interpolation of the W fields a simultaneous estimation of their dynamics (in terms of $\mathbf{V}_i$ fields) is carried out. After appropriate geometric calibration and normalization the latter give the velocity vectors of the effective advection observed in the W field. This enables direct calculation of some integral physically meaningful characteristics, such as advective flux of W and/or derived products through any predetermined contour with horizontal scales of 50 – 100 km and larger over long periods of continuous observations with time sampling of about 1.5 hour and spatial sampling as fine as 0.125 degrees. This allows speaking of satellite radiothermovision (investigation in dynamics with high spatiotemporal resolution) of mesoscale and synoptic atmospheric processes. On a more qualitative level this was demonstrated in (Ermakov et al, 2013b, 2013d). However a more advanced quantitative analysis is possible, as shown further.

3. MULTISENSOR DATA ANALYSIS

The task of adapting mathematical models and computational schemes for the data of different satellite instruments is of inevitable relevance. In addition to replacement of defective or obsolete instruments with new, often with fundamentally improved characteristics (e.g. spatial resolution) in the context of the radiothermovision technique it is crucially important to have as much amount of available satellite data as possible at the step of creating reference fields (see previous section). Unfortunately, not all the available information can be used directly due to the discrepancy in the observation time. For example, while the data from SSMIS instruments on F16 and F17 satellites of DMSP series can be separated by acceptable time intervals of 1 hour or less, the discrepancy with AMSR-2 measurements (onboard GCOM-W1) can be as much as 2.5 – 3.5 hours making their direct merging for the further interpolation with time sampling of 1.5 hour impossible.

A solution proposed by the authors is to introduce an additional processing loop into the basic scheme described above. In the first approach only "satisfactory synchronized" data products (e.g. SSMIS F16 data products supplemented with SSMIS F17 and/or WindSat ones to partially fill the lacunae) are interpolated. Since the interpolated W fields are obtained with the time sampling of 1.5 hour it is possible to find in the resulting sequence those stages W_i which are as close to the appropriate AMSR-2 data product by local time of measurement as 45 minutes or less. Thus a new sequence of reference fields can be formed from AMSR-2 data product supplemented in the areas of lacunae and other data gaps by the interpolated data products from other satellite instruments. Application of the same processing scheme to this new data sequence results in obtaining the interpolated W field with partly better spatial resolution, since AMSR-2 provides better resolution than that of SSM/I, WindSat, and SSMIS sensors.

In addition, the assimilation of the data from different sensors enhances the opportunities to evaluate the accuracy of the interpolation scheme by direct matching of the interpolated W field against the actual data of independent measurements by other sensor at corresponding time. One method of such an assessment was described in (Wimmers and Velden, 2011) and is taken as a basis for assessing the accuracy of the radiothermovision technique.

4. ACCURACY ASSESSMENT

As a basis (source fields) for the assessment of the accuracy of the spatiotemporal interpolation we have used the data products provided by Remote Sensing Systems, USA. Interpolated products based on SSMIS F16 data have been matched against original products based on SSMIS F17 data. The choice was primarily due to the homogeneity of the data origin (identical devices) and data pre-processing technique, thus allowing minimizing the discrepancy introduced into the data products before interpolation. Also the selected products (total precipitable water, TPW) were identical to those used in (Wimmers and Velden, 2011). The selected time interval, about a month global observations by both devices (November, 2013) allowed accumulating a representative sample of order of 10^7 pairs of measured and interpolated values.

Following the procedure in (Wimmers and Velden, 2011) let us consider as a local measure of interpolation error the residual data δ in a given node, i.e. the absolute difference between the "actual" TPW value obtained from SSMIS F17 measurements and the closest in time interpolated TPW value originated from SSMIS F16 data. Let us assume then that the integral characteristic of the accuracy e_δ is the average δ throughout the volume of the whole sample. In order to investigate the effect of spatiotemporal interpolation let us analyze e_δ as a function of "proximity in time" Δt between "actual" and interpolated value, considering not only the minimal achievable Δt (about 30 minutes), but also other Δt values (with 1.5 hour step) within the daily interval.

The overall results of calculation of e_δ as a function of Δt (Δt negative values correspond to the earlier times of interpolation compared to the time of SSMIS F17 measurements over the same points) are shown in Figure 1. One can see that the e_δ values vary in a relatively small range of 0.8 – 3.8 mm (while the range of TPW values is 0 – 75 mm) decreasing approximately linearly as Δt approaches to 0. The value of e_δ at minimal achievable $|\Delta t|$, which determines the accuracy of interpolation, also reaches its minimum of 0.8 mm.

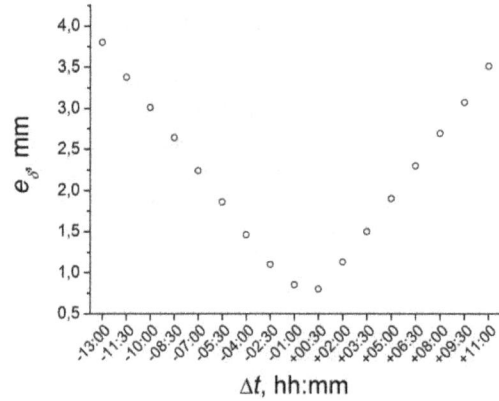

Figure 1. Interpolation accuracy (circles) of TPW fields as function of time discrepancy, see Section 4

The described features of e_δ are in good agreement with the results obtained in (Wimmers and Velden, 2011), and in particular, shown in Figure 8 of this work (not represented here). However, the final estimate of the accuracy in the cited paper is 0.5 - 2.0 mm. This is due to the fact that the algorithm implements a non-uniform temporal interpolation scheme: different portions of data must be "advected" through different time intervals in order to form a composite field coherent in time. Thus, the best accuracy of 0.5 mm is in fact not attainable for the most of the data because it reflects the residual errors in the source data from different sensors. When data requires some temporal interpolation this algorithm, according to Figure 8, gives an error ≥ 1.0 mm, and the error grows up to 2.0 mm as the algorithm attempts to compensate the time discrepancy of about 7 hours. Let us notice that a source to this error (in addition to those indicated by the authors) can partly reside in the approach to evaluate the displacement vector fields $\mathbf{V}_i$. While in presented approach $\mathbf{V}_i$ are calculated directly from the source data, in (Wimmers and Velden, 2011) they are constructed as a weighted sum of wind fields (from ancillary numerical model) at a set of horizons. Such an approach implies a fixed vertical water vapor profile while it is not quite exact for complex atmospheric conditions over wide areas. However, the achieved accuracy is considered very reasonable for most applications.

5. ESTIMATION OF LATENT HEAT ADVECTION

When choosing TPW product as W fields the calculated $\mathbf{V}_i$ fields being geometrically calibrated and normalized (Ermakov et al, 2013a) have the sense of effective (integral by the height of the atmosphere) velocity of the water vapour advection. Calculated with the use of these velocity vectors flux of W through a given contour multiplied by a constant factor $q = 2.26$ MJ/kg has a sense of latent heat flux Q into or out of the area enclosed by the border with vertical sidewalls which projections on the Earth surface repeat the given contour.

For brevity let us give only the final expressions. If a displacement vector with the beginning at a node (x, y) has the coordinates (m_x, m_y) on a regular grid with a step of s degrees, the corresponding velocity of advection is given by:

$$\vec{v} = \frac{Rs\left(\sin(s \cdot y) \cdot m_x, -m_y\right)}{\Delta t} \qquad (4)$$

where $R = 6371$ km, radius of the Earth
          Δt = time step of interpolation (1.5 or 3 hours).

Advective latent heat flux across an elementary border crossing the node (x, y) at some angle so that its normal has coordinates $(\cos\alpha, \sin\alpha)$ in grid projection and seen at angle $d\alpha$ from the observation point is given by:

$$dQ(x,y) = -\frac{R^2 s^2 r\left(\sin^2(sy)m_x \cos\alpha + m_y \sin\alpha\right)}{\Delta t\sqrt{\sin^2(sy)\cos^2\alpha + \sin^2\alpha}} \times \qquad (5)$$
$$\times q \cdot W(x,y)\sqrt{\sin^2(sy)\sin^2\alpha + \cos^2\alpha}\,d\alpha$$

where r = distance from border to observation point
          $W(x, y)$ = TPW value at the node (x, y).

A "polar" representation of the flux through a given elementary border is especially useful for the case of circular contours when integration is carried out under fixed value of r for α from 0 through 2π. Consider an example in which established is a system of contours in the form of concentric circles of different radii, drifting along with the tropical cyclone (TC). In a number of cases studied (Ermakov et al, 2014) in the different basins of the World Ocean, when a TC eye is sufficiently distant from a coastline it is possible to identify an interrelation between the variations of TC intensity (maximum sustainable wind in the wall of the eye) and the advective latent heat flux, calculated as described in (Ermakov et al, 2013a).

Below considered are the two case studies carried out for the tropical cyclones of August, 2000, including hurricane Alberto, and those in November, 2013, including super typhoon Haiyan.

5.1 Case study: Tropical cyclones in August, 2000

For the interval of August, 2000 the most appropriate satellite microwave data were those provided by SSM/I instruments on F13, F14, and F15 satellites of DMSP series. The data in the form of fields of brightness temperatures were extracted from IKI RAS database (Ermakov et al, 2007) and recalculated into TPW fields by a simple algorithm (Ruprecht, 1996). Predominantly the data from F14 and F15 satellites were used due to the proximity of measurements in time (about 0.5 hour in August, 2000). F13 data (measured about 2.5 hours earlier than F14 data over the same region) were used sporadically to diminish few vast data gaps in absence of fast atmospheric motions.

The source data were spatiotemporally interpolated as described in Section 2 on the regular grid with 0.2° spatial step and 3-hour time sampling. A system of concentric circular contours of radii ranging from 2° to 8° was established. With the use of the interpolated TWP fields and displacement vector fields recalculated into velocity vectors the advective fluxes Q of latent heat through these contours were calculated, while at every 3-hour time step the system of contours was centered at the eye of the TC under investigation (a total of 7 TC all over the World Ocean were studied). In order to suppress possible positioning errors the system of contours was each time shifted one node (0.2°) in every direction, the corresponding values of Q were calculated for each radius. The resulting value was the average Q per every radius.

The estimates of the advective latent heat fluxes considered as functions of time were matched against the TC intensity (maximum sustainable wind) as function of time. The information on initial approximation of TC center as well as on TC intensity was extracted from the "Global-TC" database of IKI RAS (Pokrovskaya and Sharkov, 2006). The results of the investigation are discussed in the next section.

5.2 Case study: Tropical cyclones in November, 2013

For the interval of November, 2013 the most appropriate satellite microwave data were those provided by AMSR-2 onboard GCOM-W1 satellite and of SSMIS instruments on F16, and F17 satellites of DMSP series. In this case the ready TPW products were used as source data, namely RSS daily data products (separated by ascending/descending orbits) based on SSMIS data, and JAXA data products based on single AMSR-2 swaths. Temporal proximity of F16 nad F17 measurements was about 1 hour in November 2013, while the AMSR-2 measurements were 3.5 – 4.5 hours ahead. Due to this reason the advanced multisensory data processing scheme (see Section 3) was applied.

In the first processing stage the TWP fields based on SSMIS F16 data were supplemented with TWP fields from SSMIS F17 data and spatiotemporally interpolated on the regular grid with 0.25° spatial step and 1.5-hour time sampling (TPW SSMIS fields). AMSR-2 daily data (separately for ascending/descending swaths) were interpolated and merged on a regular grid with 0.125° spatial step. Then the interpolated TPW SSMIS fields closest in time to the local time of AMSR-2 measurements (attainable time discrepancy was about 0.5 hour) were used to supplement the AMSR-2 TWP fields in lacunae and other data gaps. To merge the data on a common grid the appropriate TPW SSMIS fields were additionally bi-linearly interpolated from 0.25° to 0.125° grid. The resulting TPW fields were spatiotemporally interpolated on the 0.125° grid with the time sampling of 1.5 hour.

The approach of the further estimations of latent heat fluxes Q and their combined analysis with intensity variations of tropical cyclones and tropical storms was principally similar to that described in paragraph 5.1. A minor difference was that the analysis was carried out with better spatiotemporal resolution. It was also supplemented with consideration of sea surface temperature (SST) along the TC tracks. To this end the RSS daily SST composites were used. For every position of a TC eye center along its track the average SST within a 1.5°-diameter circle one day before TC, one day after TC, and 3 days after TC were calculated, as well as their differences, and were analyzed as functions of time together with other characteristics (advective latent heat flux and TC intensity).

The main focus of investigation was the super typhoon Haiyan. Also considered was a tropical storm Podul which was generated several days after Haiyan and developed over similar ocean conditions but did not approach the typhoon stage. The results of the investigation are discussed in the next section.

6. DISCUSSION OF RESULTS

6.1 Tropical cyclones in August, 2000

Among tropical cyclones of August, 2000 the hurricane Alberto is of greatest interest. It moved over the Central and the North

Atlantics on 08/03 – 08/23 by a complex trajectory distant from coastlines and experienced three stages of rapid intensification.

Figure 2 illustrates the TC Alberto track. Figure 3 demonstrates the evolution of its intensity V, m/s (thick line, left axis) together with the synchronized estimations of the advective latent heat flux Q, MW (thin lines, right axis) during the whole TC lifetime. Positive values of Q correspond to flux inside the contour (convergence). The enumerated arrows in Figures 2 and 3 indicate the same moments of time, namely the first intensity maximum (1), the beginning of the second intensification (2), the second intensity maximum (3) and the third intensity maximum (4). The fluxes Q shown in Figure 3 were calculated through the two circular contours of nearly the same radii of about 8° (shown in Figure 2) and plotted together in order to estimate the stability of calculations in respect to small changes of a contour.

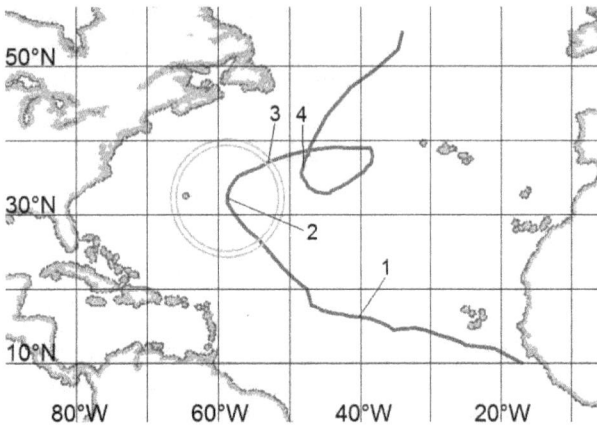

Figure 2. TC Alberto track

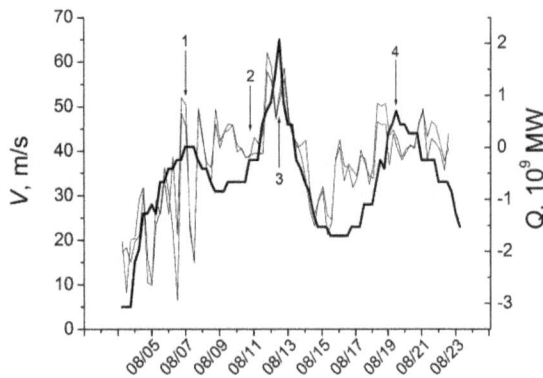

Figure 3. TC Alberto evolution: here and further on thick line – TC intensity V; thin lines – advective latent heat fluxes Q

It can bee seen from Figure 3 that the estimations of Q is quite stable in respect to small changes of the contour of integration. The overall pattern of Q plots reflects remarkably well the evolution of TC intensity: positive values and increase of Q (convergence of latent heat) generally correspond to intensification of TC while negative values and decrease of Q correspond to dissipation of TC. It is also worth noting that the variations of Q are of order of several petawatts (10^9 MW) which correlates well with some estimates of maximum TC power (Ermakov et al, 2014).

The proximity to the coastline could possibly affect the estimates in the beginning of the TC Alberto trajectory (08/03 – 08/05). In order to demonstrate the influence of the coastline proximity let us consider another example: TC Ewiniar (08/11 – 08/18) which developed in the North-West Pacific and experienced two stages of rapid intensification, while the second half of its track lied in vicinity of the coastline of Japan, see Figure 4.

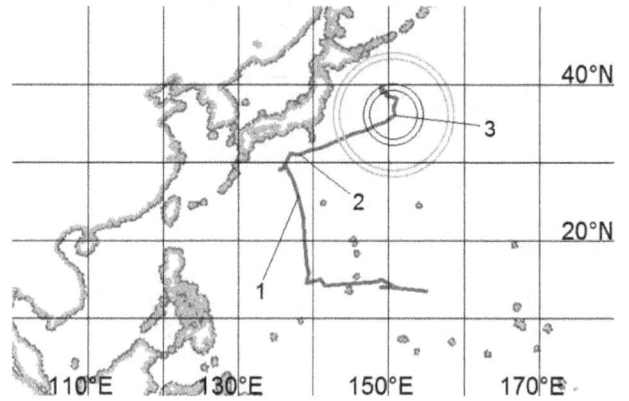

Figure 4. TC Ewiniar track

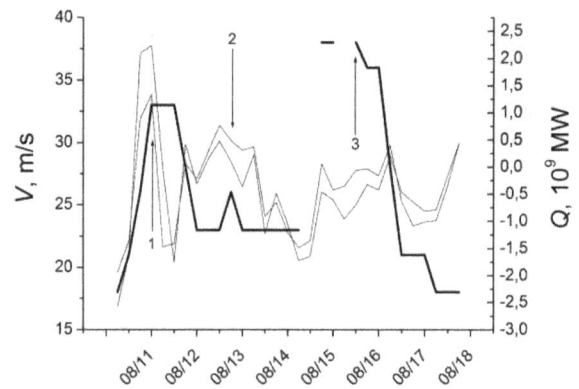

Figure 5. TC Ewiniar evolution, large contours for Q

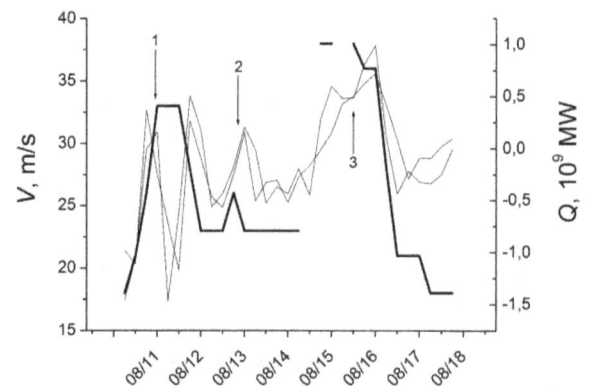

Figure 6. TC Ewiniar evolution, small contours for Q

Figures 5 and 6 demonstrate the evolution of TC Ewiniar intensity V, m/s (thick line, left axis) together with the synchronized estimations of the advective latent heat flux Q, MW (thin lines, right axis) during the whole TC lifetime. The enumerated arrows in Figures 4, 5, and 6 indicate the same moments of time, namely the first intensity maximum (1), the dissipation phase (2), and the second intensity maximum (3).

The values of Q plotted in Figure 5 were calculated through the two circular contours of nearly the same radii of about 8° (shown in Figure 4 with thick gray circles). It can be seen that Q plots in Figures 5 can satisfactorily reflect only the first intensity peak of the TC Ewiniar. The estimates for the second half of its track are significantly disturbed by the proximity of the coastline.

Due to the fact the TC Ewiniar was about twice less intense and smaller than the TC Alberto the reasonable estimates of Q can be obtained with the use of reduced radii of contours of integration. Those plotted in Figure 6 were calculated through the two circular contours of nearly the same radii of about 4° (shown in Figure 4 with thin black circles). It can be seen in Figure 6 that the pattern of Q plots has significantly changed in its right half and now reflects remarkably well both the first intensity peak and the maximum TC intensity on 08/15.

It should be noticed that the estimations are quite stable to small changes of the sizes of the contour of integration, and the variations of Q are again of the order of 10^9 MW though somewhat smaller than those calculated for the TC Alberto, especially in the case of reduced radii.

In sum all the hurricanes and typhoons of August, 2000 were analyzed in the same way, and principally the same behavior of Q plot patterns was observed (intensification of a TC corresponds to convergent latent heat flux, while dissipation corresponds to divergent one) in all cases wherever the coastlines did not disturb significantly the results of the estimations (Ermakov et al, 2014).

6.2 Tropical cyclones in November, 2013

Among tropical cyclones of November, 2013 the super typhoon Hiayan is of great interest (Mori et al, 2014). It was formed in the North West Pacific and evolved over the deeply warmed ocean (Lin et al, 2014) on 11/02 – 11/11 moving quickly ashore by almost straight westward trajectory. The fast drift of Haiyan is believed to determine (along with appropriate ocean conditions) extreme characteristics of Haiyan (Lin et al, 2014) due to smaller cooling effect. However this also means that the ocean conditions remained almost undisturbed by Haiyan which is also proved by the absence of a cold trail and significant sea surface temperature (SST) anomalies in the next days (see below). Nevertheless the next several tropical storms (TS) including TS Podul generated a week after Haiyan in the same region and travelling in almost the same direction but somewhat to the south off the Haiyan track on 11/09 – 11/15 were not even close to evolve into a super typhoon. Figure 7 illustrates the tracks of the TC Haiyan and the TS Podul.

Figure 8 demonstrates SST measured along the TC Haiyan track one day before (black squares), one day after (open circles), and 3 days after (open triangles) the TC passed corresponding areas (see paragraph 5.2) together with the TC intensity, V (thick gray line, right axis). The same quantities for TS Podul track are plotted in Figure 9. It can be seen that SST was not affected significantly by both Haiyan and Podul, varying in almost the same range with maximum of 29.7 - 30°C, and its temporal

variations all along the considered tracks were less than 1°C, both negative and positive.

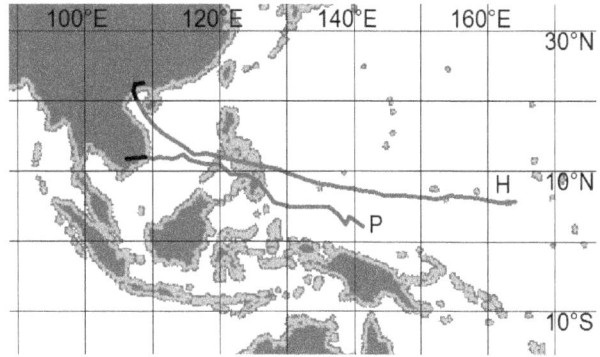

Figure 7. Tracks of the TC Haiyan (H) and the TS Podul (P)

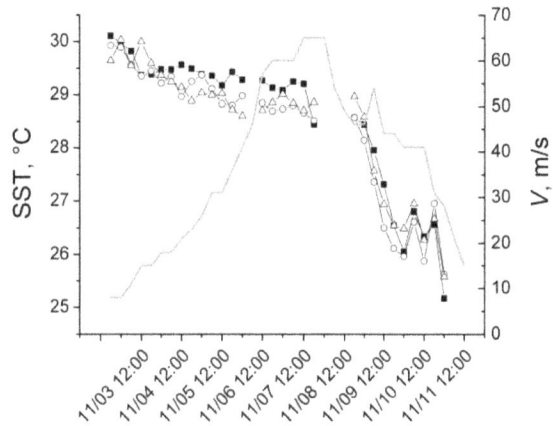

Figure 8. SST along TC Haiyan track (marked lines) and the TC intensity (thick gray line), see paragraph 6.2 for details

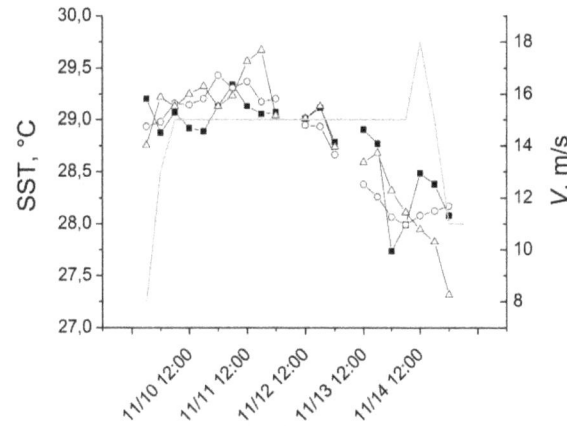

Figure 9. SST along TS Podul track (marked lines) and the TS intensity (thick gray line), see paragraph 6.2 for details

Figure 10 illustrates the evolution of TC Haiyan intensity V, m/s (thick line, left axis) together with the synchronized estimations of the advective latent heat flux Q, MW (thin lines, right axis) during the whole TC lifetime. The integration contours used to estimate Q were two concentric circles with radii of about 8° (see paragraph 6.1). Figure 11 demonstrates the plots of the same quantities for TS Podul. The smaller contours of 4° radii were used due to vicinity of the coastline and much smaller intensity of the TS.

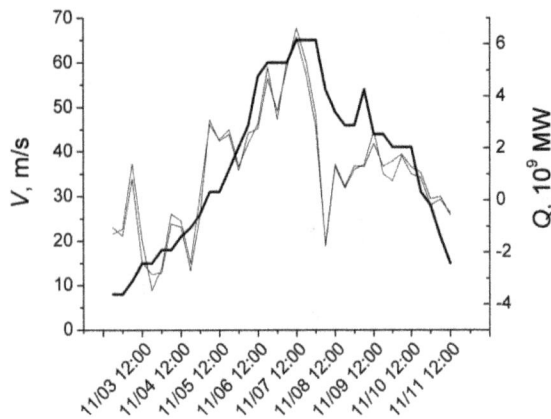

Figure 10. The TC Haiyan evolution

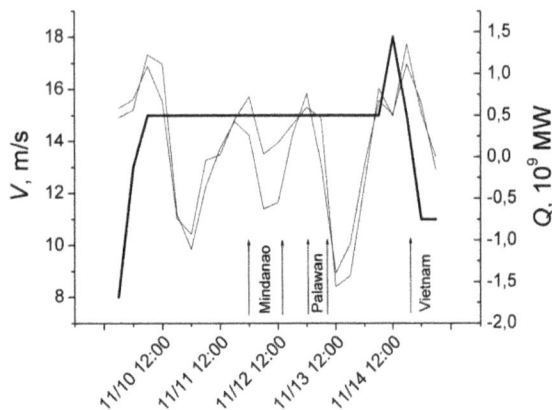

Figure 11. The TS Podul evolution

It can be seen from these figures that the Q plots in both cases reflect quite well the intensity variations of the TC Haiyan and the TS Podul. The rapid intensification of the former corresponds to significant increase of Q up to 6 PW. On the contrary in the case of the TS Podul the Q plot oscillates around 0 with the amplitude less than 1 PW, falling down significantly each time when the TS crosses coastlines (indicated with arrows and toponym in Figure 11). This corresponds well to almost unchanged intensity of the TS of about 15 m/s. Hence, both cases prove the same concept formulated in paragraph 6.1: a TC intensity is found to be interrelated with the advection of heat flux, calculated through the contour that surrounds the TC wall: increase of a TC intensity corresponds to positive (convergent) flux, while its decrease corresponds to negative (divergent) one. This flux can be estimated with the use of the described calculation scheme closed in respect to the satellite microwave data.

7. CONCLUSIONS

The approach of satellite radiothermovision is presented which consists of a set of processing techniques applicable for multisource data of passive microwave monitoring of ocean-atmosphere system, and allows creating dynamic description of

mesoscale and synoptic atmospheric processes and estimating physically meaningful integral characteristics of the observed processes (like avdective flow of the latent heat through a given border).

The approach is demonstrated on a case study of tropical cyclones of August, 2000, and November, 2013. The investigation reveals a remarkable interrelation between the intensity of a TC and the estimate of advective latent heat flux through a contour surrounding the TC: convergent flux corresponds to the TC intensification while divergent flux corresponds to its dissipation. This interrelation requires more thorough examination on a large number of tropical cyclones travelling distantly enough from coastlines.

However the presented approach opens wider opportunities for a researcher. One of those is a combined analysis of a set of geophysical parameters interpolated to a common grid and "synchronized" on a timeline with a good accuracy (one hour and less). Other interesting opportunity is to consider different types of contours and borders to investigate latent heat flux and other integral characteristics. For example introducing a system of borders lying along the parallels at different latitudes one can study the process of polar transport over the selected basin or the whole World Ocean.

One of the advantages of the presented approach is that the calculation scheme is closed in respect to the satellite microwave data, hence, conceptually simplifies the analysis by requiring no additional information to be gathered and processed.

ACKNOWLEDGEMENTS

The SSMIS data used for the preparation on this paper are produced by Remote Sensing Systems and sponsored by the NASA Earth Science MEaSUREs Program and are available at www.remss.com. AMSR2 data used for the preparation of this paper was supplied by the GCOM-W1 Data Providing service, Japan Aerospace Exploration Agency. The design of the specific data processing software used to process the satellite data products was partly supported by RFBR grant N15-07-04422.

REFERENCES

Ermakov D.M., Raev M.D., Suslov A.I. and Sharkov E.A., 2007. Electronic long-standing database for the global radiothermal field of the Earth in context of multi-scale investigation of the atmosphere-ocean system. *Issledovanie Zemli iz Kosmosa*, 1, pp. 7-13. (in Russian)

Ermakov, D.M., Chernushich, A.P., Sharkov, E.A. and Shramkov, Ya.N., 2011. Stream Handler system: an experience of application to investigation of global tropical cyclogenesis. Proceedings of ISRSE-34, Sydney, Australia http://www.isprs.org/proceedings/2011/ISRSE-34/211104015Final00456.pdf (24 Feb. 2015)

Ermakov, D., Chernushich A. and Sharkov E., 2013a. A closed algorithm to create detailed animated water vapor fields over the oceans from polar-orbiting satellites' data. Proceedings of ESA Living Planet Symposium, SP-722, December 2013, ESA Communications, Noordwijk, The Netherlands.

Ermakov, D.M., Chernushich A.P., Sharkov E.A. and Pokrovskaya I.V., 2013b. Searching for an energy source of the intensification of tropical cyclone Katrina using microwave satellite sensing data. *Izvestiya, Atmospheric and Oceanic Physics*, 49(9), pp. 963-973.

Ermakov, D.M., Raev M.D., Chernushich A.P. and Sharkov E.A., 2013c. An algorithm for construction global ocean-atmosphere radiothermal fields with high spatiotemporal sampling by satellite microwave measurements. *Issledovanie Zemli iz Kosmosa, (Earth Research from Space)*, 4, pp. 72-82. (in Russian)

Ermakov, D.M., Sharkov E.A., Pokrovskaya I.V. and Chernushich A.P., 2013d. Revealing the energy sources of alternating intensity regimes of the evolving Alberto tropical cyclone using microwave satellite sensing data *Izvestiya, Atmospheric and Oceanic Physics*, 49(9), pp. 974-985.

Ermakov, D.M., Sharkov E.A. and Chernushich A.P., 2014. The role of tropospheric advection of latent heat in the investigation of tropical cyclones. *Issledovanie Zemli iz Kosmosa, (Earth Research from Space)*, 4, pp. 3-15. (in Russian)

Lin, I.-I., Pun, I.-F. and Lien, C.-C., 2014. "Category-6" supertyphoon Haiyan in global warming hiatus: Contribution from subsurface ocean warming. *Geophysical Research Letters*, 41(23), pp. 8547-8553.

Mori, N., Kato, M., Kim, S., Mase H., Shibutani, Y., Takemi, T., Tsuboki, K. and Yasuda, T., 2014. Local amplification of storm surge by Super Typhoon Haiyan in Leyte Gulf. *Geophysical Research Letters*, 41(14), pp. 5106-5113.

Nerushev, A.F. and Kramchaninova, E.K., 2011. Method for determining atmospheric motion characteristics using measurements on geostationary meteorological satellites. *Izvestiya, Atmospheric and Oceanic Physics*, 47(9), pp. 1104-1113.

Pokrovskaya I.V. and Sharkov E.A., 2006. *Tropical cyclones and tropical disturbances of the World Ocean: chronology and evolution. Version 3.1 (1983 – 2005)*. Poligraph servis, Moscow, 728 p.

Richardson, I.E.G., 2003. *H.264 and MPEG-4 video compression*. John Wiley & Sons Ltd, The Atrium, Southern Gate, Chichester, West Sussex PO19 SQ, England. 306 p.

Ruprecht, E., 1996. Atmospheric water vapor and cloud water: an overview. *Advances in Space Research*, 18(7), pp. 5-16.

Wentz, F.J., 1997. A well-calibrated ocean algorithm for SSM/I. *Journal of Geophysical Research*, 102(C4), pp. 8703-8718.

Wimmers, A.J. and Velden C.S., 2011. Seamless advective blending of total precipitable water retrievals from polar orbiting satellites. *Journal of Applied Meteorology and Climatology*, 50(5), pp. 1024-1036.

6

QUALIFICATION OF POINT CLOUDS MEASURED BY SFM SOFTWARE

Kazuo Oda, Satoko Hattori, Hiroyuki Saeki, Toko Takayama, Ryohei Honma

Asia Air Survey Co., Ltd. (kz.oda, stk.hattori, jan.saeki, tk.takayama, ryh.honma)@ajiko.co.jp

Commission V

KEY WORDS: Five Point Algorithm, Relative Orientation, UAV

ABSTRACT:

This paper proposes a qualification method of a point cloud created by SfM (Structure-from-Motion) software. Recently, SfM software is popular for creating point clouds. Point clouds created by SfM Software seems to be correct, but in many cases, the result does not have correct scale, or does not have correct coordinates in reference coordinate system, and in these cases it is hard to evaluate the quality of the point clouds. To evaluate this correctness of the point clouds, we propose to use the difference between point clouds with different source of images. If the shape of the point clouds with different source of images is correct, two shapes of different source might be almost same. To compare the two or more shapes of point cloud, iterative-closest-point (ICP) is implemented. Transformation parameters (rotation and translation) are iteratively calculated so as to minimize sum of squares of distances. This paper describes the procedure of the evaluation and some test results.

1. INTRODUCTION

Recently, SfM (Structure-from-Motion) software is popular for 3D reconstruction and point cloud generation. SfM applications, such as Smart3DCaputure, PhotoScan, and Pix4D, are convenient for non-professional operator of photogrammetry, because these systems only require sequence of photos to generate point clouds with colour index which corresponds to the colour of original image pixel where the each point is projected. If the condition of capturing image is well-done, the result seems to be quite accurate. However, in many cases, the result is not constructed with correct scale or correct coordinates in reference coordinate system.

Basically, the quality of the point clouds created by dense image matching of SfM software should estimate by comparing true point cloud or more precise point cloud. In some cases, point cloud data measured by a laser scanner are adopted. But in many cases, laser scanner data do not have sufficient precision for, such as small objects. If the objects are on a cliff and hard to be accessed, measuring by TLS (Terrestrial Laser Scanner), might be impossible.

We focused on the evaluation of correctness of shape of the point cloud created by SfM. Even if it does not have correct scale, the shape of the object might be correct. To evaluate this correctness, we propose the difference of point clouds with different source of images. It the shapes of the point clouds with different source of images is correct, two shapes of different source might be almost same.

To compare the two or more shapes of point cloud, iterative-closest-point (ICP) is implemented by Besl et al., 1992 and Takai et al., 2013. Transformation parameters (scale, rotation, and translation) are iteratively calculated so as to minimize sum of squares of distances. If the shape of the point cloud is correct, the distances of corresponding points between the two point clouds are expected to be small. Distance in ICP should be determined as the distance between a point of one point cloud and its nearest face of the other point cloud. The distances of the

two point clouds reflect the error of the two point clouds' shapes. This method can be applied for point clouds without correct scale.

This evaluation cannot be applied for some cases. One is the case that the object is isotropic like a sphere, or the case that the object is planar. This means that the method should check the anisotropy of the object by statistical analysis of distribution of normal vectors.

This ICP optimization and error estimation can also be used for extracting the part of deformation of the shape.

2. QUALIFIATION METHOD

The qualification method follows the procedure shown in Figure 1.

2.1 Capturing Two Groups of Images

Two groups of images are collected: Image set A and Image set B. Two groups of images should be captured under almost same condition, but it should not be same, because the errors in the point clouds created from these image groups should not have same tendency in systematic error caused by aerial triangulation and dense image matching.

2.2 Creating point clouds

Two point clouds of each image groups are generated by SfM software. The two of these point clouds should be almost the same coordinate system. This can be attained by creating GCPs in the point clouds of one group of images and execute bundle adjustment in the other group of images. To avoid the deformation caused by systematic error of bundle adjustment, the errors of GCP coordinates should be large in bundle adjustment.

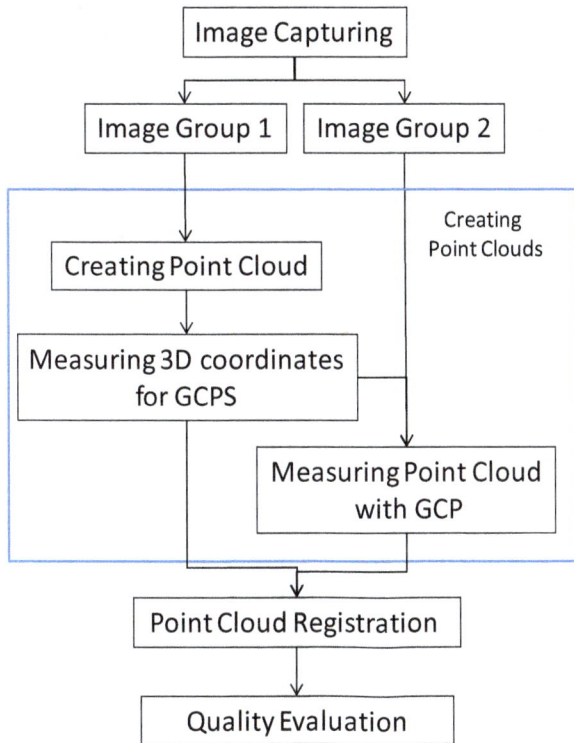

Figure 1. Procedure of the Qualification

2.3 Point Cloud Registration

Registration of two point clouds is executed with ICP algorithm. ICP iteratively try to minimize the distance of two point clouds. One point cloud, the source, is moved to fit the other point cloud, the target, by rigid transformation which is the combination of translation and rotation.

We adopt CCICP (Classification and Combined ICP) algorithm. The CCICP algorithm minimizes point-to-plane, point-to-point distances, simultaneously, and also reject incorrect correspondences based on point classification by PCA (Principle Component Analysis) (Takai et al 2013). The points in the local point clouds are classified into linear points, planar points and scatter points depending on the results of the PCA which is shown Figure 2 (Demanke et al 2011).

Linear Point   Planar Point   Scatter Point
Figure 2. Point classification by PCA

Point-to-plane and point-to-point distances are minimized simultaneously in CCICP. Point-to-plane distance minimization is applied to planar-planar correspondences and point-to-point distance minimization is applied to the other correspondences. Point-to-plane and point-to-point distance minimization problem is solved using the method of Low (Low, 2004). Point-to-point distance (Dpt_pt) and point-to-plane distance (Dpt_pl) is defined in following equation:

$$Dpt\_pt = \sqrt{|(T \cdot P_s - P_t)|^2}$$

$$Dpt\_pl = \sqrt{|(T \cdot P_s - P_t) \cdot n_t|^2}$$

where T is a transformation matrix in homogeneous coordinate system P_s is a point in source point cloud, P_t is the matching point in target point cloud, and n_t is the normal vector of point P_t calculated by PCA. T is described as following

$$T = \begin{pmatrix} 1 & -\gamma & \beta & t_x \\ \gamma & 1 & -\alpha & t_y \\ -\beta & \alpha & 1 & t_z \\ 0 & 0 & 0 & 1 \end{pmatrix}$$

where α, β, γ are rotation angles about x, y, z axis (≈ 0) and t_x, t_y, t_z are translation.

CCICP minimizes sum of square difference of corresponding coordinates of points. Detail of CCICP algorithm is described in the paper of Takai et al (2013).

2.4 Quality Evaluation

CCICP outputs the mean value of square distances E. Root of E value includes systematic error of both point cloud. Therefore, root mean square error of each point cloud is smaller than root of E.

3. TESTS AND RESULT

3.1 Test Object

The test object is a lava stone of Izu Oshima Island. The size of the lava stone is about 10 cm width and 5 cm height. Surface of the lava stone is partly rough with small holes, and partly smooth. The colour of the lava stone is almost black. Also, the stone object is set on a map-printed cloth (Figure 3).

Figure 3. The test object: lava stone of Izu Oshima Island

3.2 Image capturing

The cloth and the stone had been set on a rotary chair (Figure 4) and images had been captured with SONY Cyber-shot DSC-WX200 (Figure 5). Two set of images, (image set A and image set B) had been captured. For each set of images, more than hundred of images had been captured all around the test object, with two different angles of depression (Table 1).

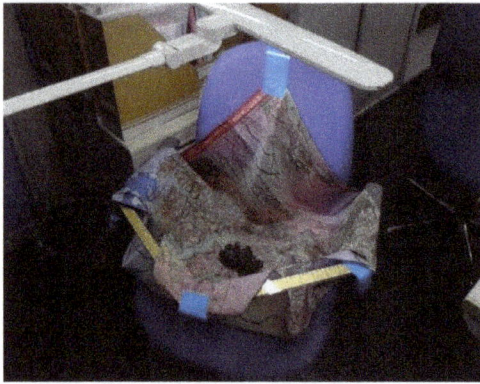

Figure 4. Image capturing stage

Figure 5. SONY Cyber-shot DSC-WX200

Table 1. Two set of images

	Image Set A	Image Set B
Number of Images	111	125
Camera	SONY Cyber-shot DSC-WX200	SONY Cyber-shot DSC-WX200
Image Resolution	About 0.03mm	About 0.03mm

3.3   Generation of Point Clouds

Two set of images had been processed with SfM software, named Pix4Dmapper of Pix4D, for the generation of point cloud.

Image set A had been processed without GCP, therefore the image coordinates of point cloud of image set A is arbitrary and the unit of the coordinate system is approximately 1 mm (Figure 6). 111 images had been processed and all of the images had been adopted for point cloud generation. The point cloud is dense around at stone and sparse at surrounding (Figure 7.). In this paper, we refer to 1 unit length as 1mm.

Image set B had been processed with 125 images (Figure 8). Six GCPs measured in image set A (Figure 9) had been used in bundle adjustment procedure in Pix4D. To avoid the systematic deformation with bundle adjustment, the accuracy of GCPs had been set to 20 mm.

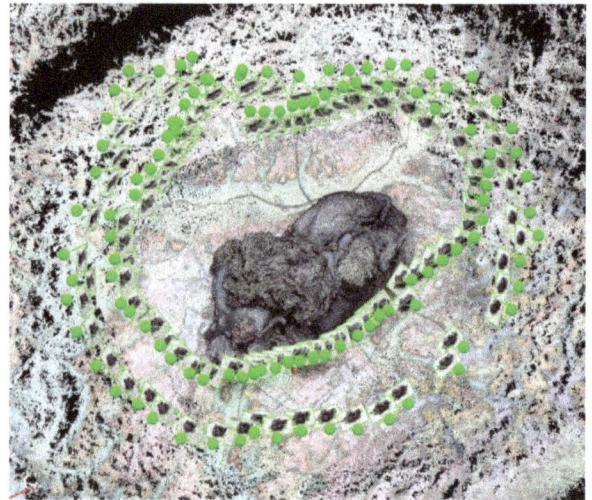

Figure 6. Point cloud A generated from image set A and estimated camera position and rotation

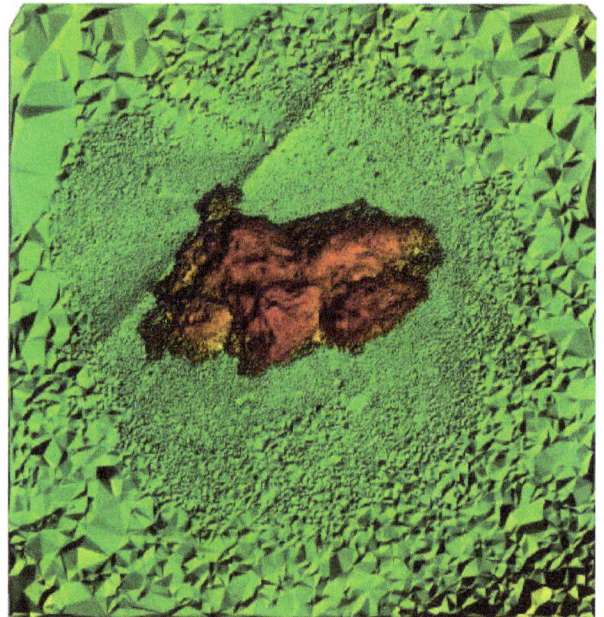

Figure 7. DSM created from the point cloud

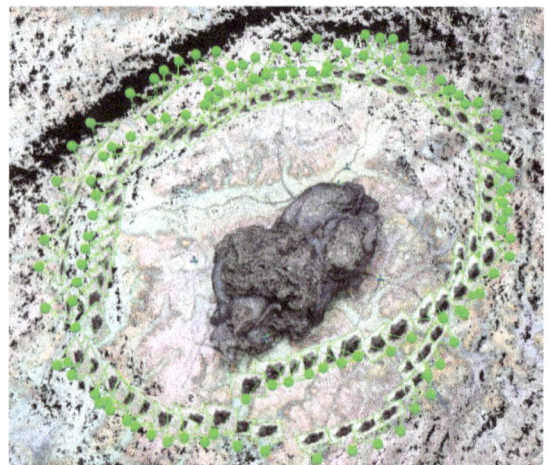

Figure 8. Point cloud B generated from image set B and estimated camera position and rotation

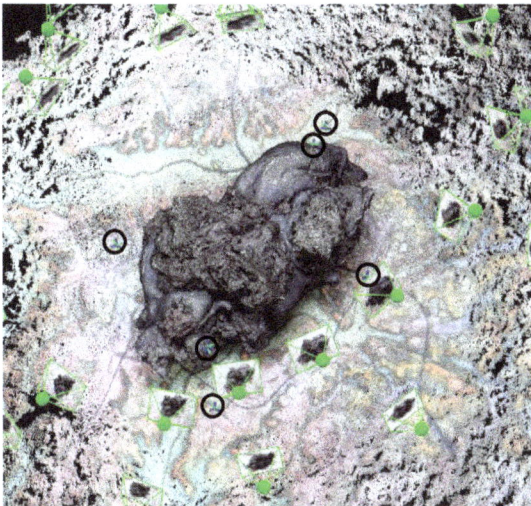

(1) Alignment of 6 GCPs

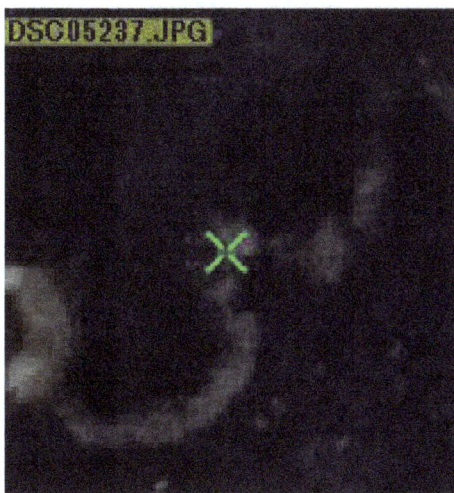

(2) A GCP on the stone

(3) A GCP on the cloth
Figure 9. GCPs in point cloud

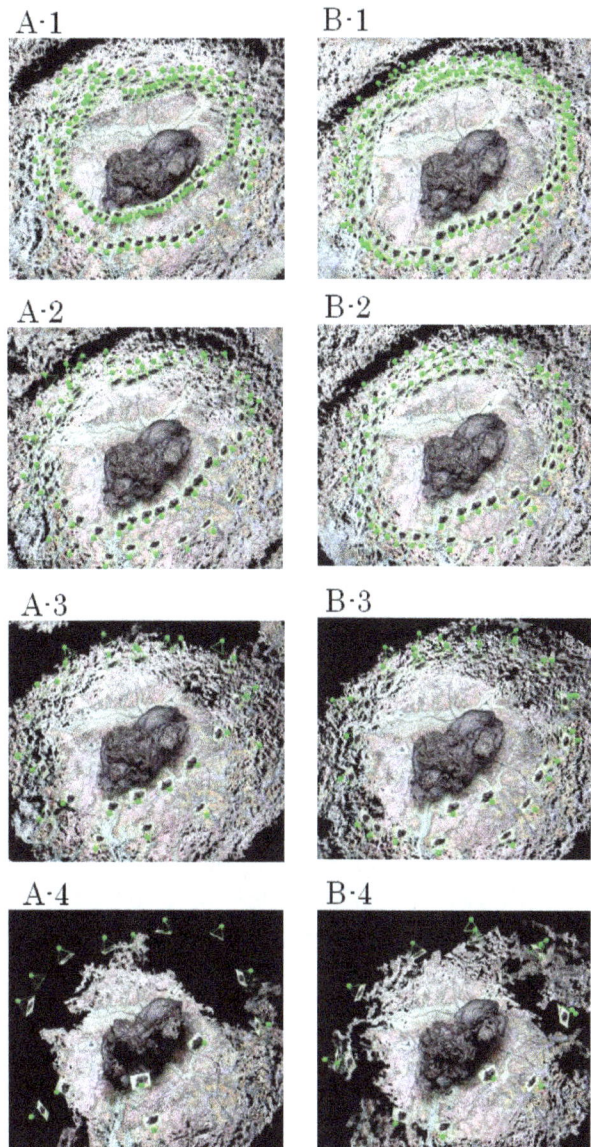

Figure 10. Point Clouds with reduced images

Table 2. Data set name list of point clouds

Point clouds with image set A		Point clouds with image set B	
Data set name	Number of image	Data set name	Number of images
A-1	111	B-1	125
A-2	56	B-2	63
A-3	28	B-3	32
A-4	14	B-4	16

Point clouds with reduced images had been generated with image set A and B respectively. We refer to point clouds generated from 1/1, 1/2, 1/4, 1/8 of image set A as A-1, A-2, A-3 and A-4 respectively, and similarly, B-1, B-2, B-3 and B-4 respectively. The number of points tends to decrease in both point cloud data set of both image set (Figure 10, Table 2).

3.4 Image Registration

Image registration by CCICP had been executed for four pairs of point clouds: A-1 and B-1, A-2 and B-2, A-3 and B-3, A-4 and B-4. Source point cloud generated from image set A had been fitted to target point cloud generated from image set B. Search distance of matching point was set to 1mm and search distance for PCA was set to 0.8mm. Number of sampling points for CCICP matching had been decreased to 1/10 of full data points of point clouds. Matching points were limited to the object stone and points of the cloth had been eliminated.

Table 3 shows the number of matching pairs for CCICP. Matching points were more than 90% of sample points of CCICP. It is considered that eliminated pairs in CCICP include points with big errors, pairs with different classification, or sparse point cloud, but their numbers were relatively small. Table 3 also shows that more than 90% of CCICP pairs were planar points. Figure 11 shows result of PCA classification of A-1.

Table 3. Numbers of matching pairs for CCICP

Source	Target	number of points	sample points for CCICP	pairs	planar	other
A-1	B-1	1326160	132616	126858	128109	3263
				96%	97%	2%
A-2	B-2	876770	87677	86888	85241	1647
				99%	97%	2%
A-3	B-3	534260	53426	53044	52522	522
				99%	98%	1%
A-4	B-4	171620	17162	16026	15739	287
				93%	92%	2%

(1) Point cloud with real colour

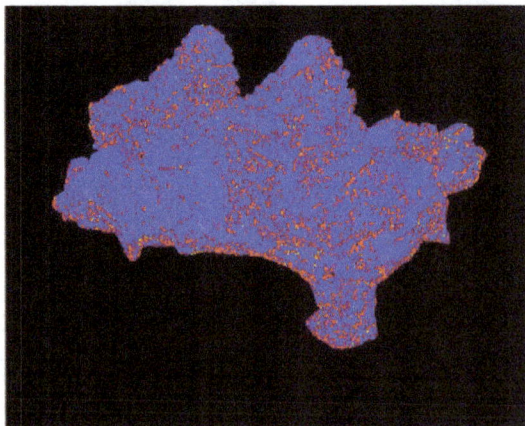

Blue: planar points Yellow: scatter points Red: Linear points
(2) Classification result

Figure 11. PCA classification of A-1

Figure 12 shows a sample profile of point clouds before and after CCICP registration, and Table 4 shows mean distances of point clouds before and after CCICP registration. In all cases, mean distances are about pixel size (0.03mm) or less. This shows that these setting of images and number of images do not affect precision of measurement so much, while it greatly affects number of measured points.

Table 4. Mean distance before/after CCICP registration

Source	Target	Number of points	Sample points for ICP	Before optimization (mm)	After optimization (mm)
A-1	B-1	1326160	132616	0.123	0.032
A-2	B-2	876770	87677	0.141	0.016
A-3	B-3	534260	53426	0.186	0.028
A-4	B-4	171620	17162	0.535	0.042

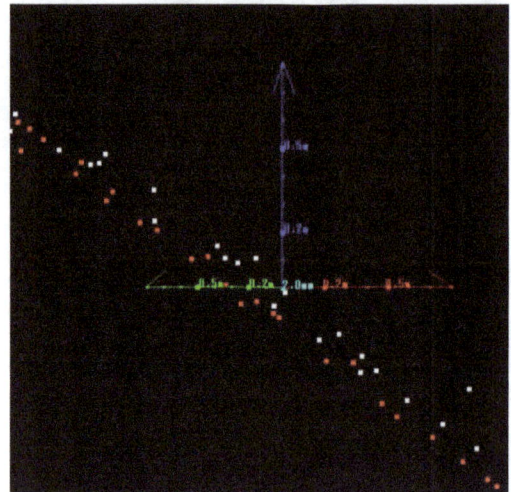

(1) point cloud profile before CCICP

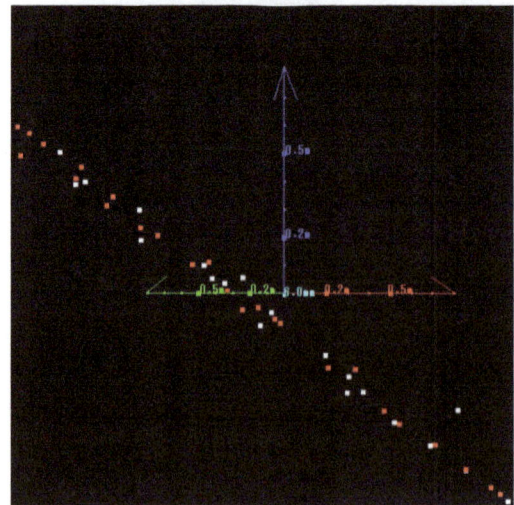

(2) Point cloud profile after CCICP

White: Source points, Red: Target points
Figure 12. Result of CCICP point cloud registration

Figure 13 shows 3D error (distance of two point clouds) for registration between of A-1 and A-2. This shows that big errors (more than 0.09mm) cluster to some parts, while error distribution of other part have no eminent tendency. This means this qualification method can access local matching quality in a point cloud.

(1) Point cloud with real colour

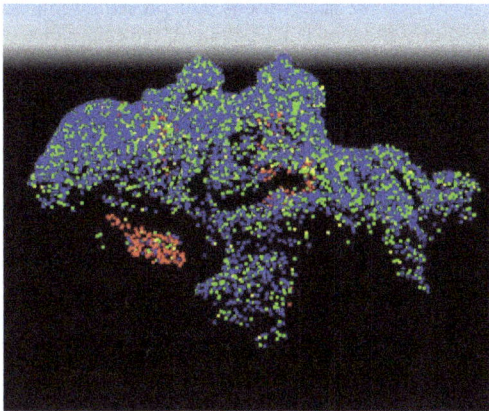

(2) Point cloud coloured by error
Blue: < 0.03mm Green: < 0.06mm
Yellow: < 0.09mm Red: >0.09mm
Figure 13. 3D error distribution for registration between A-1
and A-2

4. CONCLUSION AND FUTURE WORKS

Qualification method of point cloud generated by SfM has been proposed. With CCICP registration, point cloud quality can be estimated, as well as local matching quality in a point cloud.
The error level of the point cloud is estimated by mean distance of two point clouds.

This method requires redundant image capturing, but it is easy for small objects with large overlapping configuration.
Numerical relationship between mean distance of CCICP and errors in point cloud generation has not been theoretically discussed in this paper. We are planning to clarify the relationship by analyzing the result of CCICP with simulated errors of point clouds.

5. ACKNOWLEDGEMENT

We wish to thank Prof. Satoshi Kanai and Assoc. Prof. Hiroaki Date, Hokkaido University, for help for CCICP programming and inspiring suggestion.

REFERENCES

Besl, Paul J.; N.D. McKay (1992). A Method for Registration of 3-D Shapes. IEEE Trans. on Pattern Analysis and Machine Intelligence (Los Alamitos, CA, USA: IEEE Computer Society) 14 (2): 239–256. doi:10.1109/34.121791.

S. Takai, H. Date, S. Kanai, Y. Niina, K. Oda, and T. Ikeda(2013). Accurate registration of MMS point clouds of urban areas using trajectory, ISPRS Workshop Laser Scanning 2013, 277-282, Nov. 13th, 2013, Antalya, Turkey.

Demantke, J., Mallet, C., David. N., Vallet, B.(2011). Dimensionality based scale selection in 3D LiDAR Point Cloud. The International Archives of the Photogrammetry Remote Sensing and Spatial Information Sciences, 38 (Part 5/W12) (on CDROM).

7

EARLY VALIDATION OF PROBA-V GEOV1 LAI, FAPAR AND FCOVER PRODUCTS FOR THE CONTINUITY OF THE COPERNICUS GLOBAL LAND SERVICE

J. Sánchez [a, *], F. Camacho [a], R. Lacaze [b], B. Smets [c]

[a] EOLAB, Parc Científic Universitat de València, Catedrático José Beltrán, 2. 46980 Paterna (Valencia), Spain
jorge.sanchez@eolab.es
[b] HYGEOS, Toulouse, France - rl@hygeos.com
[c] VITO, Belgium - bruno.smets@vito.be

KEY WORDS: Copernicus Global Land, PROBA-V, LAI, FAPAR, FCover, validation

ABSTRACT:

This study investigates the scientific quality of the GEOV1 Leaf Area Index (LAI), Fraction of Absorbed Photosynthetically Active Radiation (FAPAR) and Fraction of Vegetation Cover (FCover) products based on PROBA-V observations. The procedure follows, as much as possible, the guidelines, protocols and metrics defined by the Land Product Validation (LPV) group of the Committee on Earth Observation Satellite (CEOS) for the validation of satellite-derived land products. This study is focused on the consistency of SPOT/VGT and PROBA-V GEOV1 products developed in the framework of the Copernicus Global Land Services, providing an early validation of PROBA-V GEOV1 products using data from November 2013 to May 2014, during the overlap period (November 2013-May 2014). The first natural year of PROBA-V GEOV1 products (2014) was considered for the rest of the quality assessment including comparisons with MODIS C5. Several criteria of performance were evaluated including product completeness, spatial consistency, temporal consistency, intra-annual precision and accuracy. Firstly, and inter-comparison with both spatial and temporal consistency were evaluated with reference satellite products (SPOT/VGT GEOV1 and MODIS C5) are presented over a network of sites (BELMANIP2.1). Secondly, the accuracy of PROBA-V GEOV1 products was evaluated against a number of concomitant agricultural sites is presented. The ground data was collected and up-scaled using high resolution imagery in the context of the FP7 ImagineS project in support of the evolution of Copernicus Land Service. Our results demonstrate that GEOV1 PROBA-V products were found spatially and temporally consistent with similar products (SPOT/VGT, MODISC5), and good agreement with limited ground truth data with an accuracy (RMSE) of 0.52 for LAI, 0.11 for FAPAR and 0.14 for FCover, showing a slight bias for FCover for higher values.

1. INTRODUCTION

From 1st January 2013, the Copernicus Global Land Service is operational, providing in near real time a set of biophysical variables over the whole globe. Leaf Area Index (LAI), Fraction of Absorbed Photosynthetically Active Radiation (FAPAR) and Fraction of Vegetation Cover (FCover) are delivered at 1 km resolution and 10-days frequency. These vegetation biophysical variables play a key role in several surface processes, including photosynthesis, respiration and transpiration. The first version of LAI, FAPAR and FCover variables, called GEOV1, was based on SPOT/VGT observation until the end of the mission in May 2014 (Baret et al., 2013). A SPOT/VGT GEOV1 archive of 15 years is now available (1999-2014) at the Global Land service. The continuity of the GEOV1 products is based on PROBA-V observations at 1 km. To provide continuity of the variables at 1km, the processing chains are to be updated towards the new PROBA-V mission which started its lifetime in November, 2013. Although the sensor of the latter mission is very compatible with the former mission, a spectral correction is to be performed to continue the 15-year time series at 1km resolution in a consistent manner. Therefore a pre-processing module is developed that performs next to the spectral correction a transformation of the new PROBA-V input data into SPOT-VGT compatible input data. Validation of PROBA-V GEOV1 is thus mandatory before delivering the products to the users.

This paper describes the main results of the early scientific quality assessment of PROBA-V GEOV1 LAI, FAPAR, FCover products. This preliminary validation is focused on the consistency with SPOT/VGT GEOV1 products during the overlap period (November 2013 - May 2014). MODIS C5 LAI/FPAR products are also considered for the intercomparison

The procedure follows the guidelines and metrics defined by the Land Product Validation (LPV) group of the Committee on Earth Observation Satellite (CEOS) for the validation of satellite-derived land products. Several criteria of performance were evaluated including product completeness, spatial consistency, temporal consistency, intra-annual precision and accuracy.

The accuracy of PROBA-V GEOV1 products was evaluated against a number of agricultural sites. The ground data was collected in the context of the FP7 ImagineS project (http://fp7-imagines.eu) in support of the evolution of Copernicus Land Service (Camacho et al. 2014).

The quality assessment method is briefly described in Section 2. Section 3 shows the main results; conclusions are provided en Section 4.

2. QUALITY ASSESSMENT METHOD

Several criteria of performance were assessed in agreement with previous global LAI validation exercises (Camacho et al., 2013), the OLIVE (On Line Validation Exercise) tool hosted by CEOS cal/val portal (Weiss et al., 2014), and with the recent CEOS LPV Global LAI product validation good practices (Fernandes et al., 2014). First and intercomparison with the existing global products was conducted to examine the spatial and temporal consistency of GEOV1 PROBA-V products. Second, a direct validation approach was conducted using

---

* Corresponding author. This is useful to know for communication with the appropriate person in cases with more than one author.

ground reference maps to quantify the overall uncertainties of the products.

2.1 Intercomparison Approach

The reference global satellite products used are: GEOV1 based on SPOT/VGT observations (Baret et al., 2013) and Terra MODIS LAI/FAPAR (MOD15A2) collection 5 (Knyazikhin et al., 1998).

The products were intercompared over the BELMANIP2.1 (Weiss et al., 2014) network of sites that was designed to represent globally the variability of land surface types. Furthermore, the products are analysed for 6 generic classes, namely: Evergreen Broadleaf Forest, Evergreen Deciduous Forest, Needle-leaf Forest, Croplands, Herbaceous and Shrub/Sparse/Bare Areas. The different products must be compared over a similar spatial support area and temporal support period. The intercomparison was conducted using an average value over 3x3 pixels to reduce coregistration errors between products and differences in their sensor Point Spread Function (PSF) which determines the actual footprint of the data. The temporal support period for the quantitative assessment is 10-days with monthly composites. The original temporal sampling was used for compute missing values, histograms and the smoothness.

The following criteria of performance and metrics are assessed:

• Product Completeness: corresponds to the absence of spatial and temporal gaps in the data. Temporal variations of GEOV1 missing values for SPOT/VGT and PROBA-V LAI products have been computed over the whole images.

• Spatial Consistency: can be quantitatively assessed by comparing the spatial distribution of a reference validated product with the product under study. Global histograms of residuals at a monthly basis were analyzed. This analysis is complemented by the analysis of Probability Density Function (PDFs) of retrievals per biomes.

• Temporal Consistency: The consistency of temporal variations of the vegetation variables are qualitatively analyzed as compared to reference validated products.

• Intra-annual precision (smoothness): corresponds to temporal noise assumed to have no serial correlation within a season. In this case, the anomaly of a product LAI value from the linear estimate based on its neighbors can be used as an indication of intra-annual precision or smoothness. It can be characterized as suggested by Weiss et al., (2014): for each triplet of consecutive observations, the absolute value of the difference between the center P(dn+1) and the corresponding linear interpolation between the two extremes P(dn) and P(dn+2) was computed:

$$\delta = \left| P(d_{n+1}) - P(d_n) - \frac{P(d_n) - P(d_{n+2})}{d_n - d_{n+2}}(d_n - d_{n+1}) \right| \quad (1)$$

Histograms of the smoothness are presented adjusted to a negative exponential function. The exponential decay constant is used as quantitative indicator of the typical smoothness value.

• Relative Uncertainties: The inter-comparison of products offers an indirect means of assessing uncertainties (systematic or random) between products. The global statistical analysis is performed over BELMANIP2.1 sites considering all the dates available.

2.2 Direct Validation

The accuracy assessment was performed against ground truth data processed according to CEOS LPV guidelines for validation of LAI products. The data set used to validate is the ground data collected in the framework of the ImagineS project over agricultural sites for the period under study. Up-scaling of ground data was achieved with high-resolution satellite image using an empirical transfer function (Camacho et al., 2014). Eight sites were made available for the accuracy assessment for 2014 coming from FP7 ImagineS (Table 1).

Site Country	Lat Lon (deg)	Land Cover	Dates (mm/yyyy)	LAI	FAPAR	FCover
25Mayo_1 Argentina	-37.906 -67.746	Crops	02/2014	1.30	0.39	0.32
25Mayo_2 Argentina	-37.939 -67.789	Shrub	02/2014	0.42	0.19	0.16
LaReina_1 Spain	37.819 -4.862	Crops	05/2014	1.08	0.30	0.29
LaReina_2 Spain	37.793 -4.827	Crops	05/2014	1.59	0.42	0.41
Merguellil Tunisia	35.5662 9.912	Crops	01/2014	0.18*	N/A	N/A
			04/2014	0.93*	N/A	N/A
LaAlbufera Spain	39.2743 -0.316	Crops (Rice)	06/2014	0.58	0.21	0.18
			06/2014	1.51	0.46	N/A
			07/2014	3.77	0.73	N/A
			08/2014	5.78	0.85	N/A
			08/2014	5.9	0.85	N/A
Rosasco Italy	45.253 8.562	Crops (Rice)	07/2014	4.2	0.85	N/A
Pshenichne Ukraine	50.07 30.23	Crops	06/2014	2.26	0.65	0.54

(*) LAIeff

Table 1: Characteristics of the validation sites from ImagineS project for 2014 and associated ground biophysical values.

Due to the limited number of concomitant ground measurements, the number of ground reference maps was increased by using data from a different year from Camacho et al., 2013, and available at CEOS OLIVE Cal/Val portal (http://calvalportal.ceos.org/). These sites have been filtered by analyzing the inter-annual stability of the MODIS C5 FAPAR products, as MODIS time series expands from 2000 till the most recent dates. Only stable forest and grassland sites have been used: a maximum difference of ±0.05 in the MODIS FAPAR value between the concomitant date and the equivalent day of the current year was allowed. A total of 20 additional sites were finally considered, their main characteristics are presented in the Appendix.

3. RESULTS

3.1 Product Completeness

Figure 1 shows the temporal evolution of the fraction of missing values for SPOT/VGT and PROBA-V GEOV1 products. Over the six months overlap period, SPOT/VGT and PROBA-V provided consistent results, with SPOT/VGT showing a slightly better fraction of valid observations (around 5%), which could be partly explained due to the different overpass time between PROBA-V (10:45 am) and SPOT/VGT satellites (currently around 45 minutes before, GIOGL1-ATBD-PROBA2VGT).

The length of the missing values, evaluated over BELMANIP2.1 sites (Figure 2), shows very similar distributions for PROBA-V and SPOT/VGT GEOV1 products, with around 50% of the gaps shorter than 30 days. On the other hand, the length of the gaps in MODIS C5 is shorter, with around 60% of gaps corresponding to one missing observation. This could be partly explained by the richer spectral information of MODIS as compared to SPOT/VGT or PROBA-V which is more suitable for cloud screening.

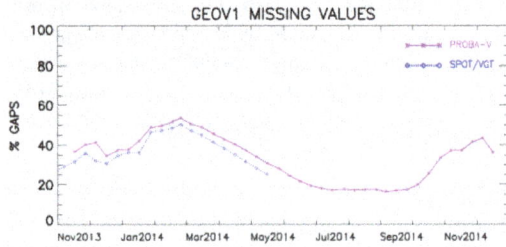

Figure 1: Temporal variations of missing values for SPOT/VGT (blue dashed line) and PROBA-V (purple solid line) GEOV1 products during the November2013-December2014 period.

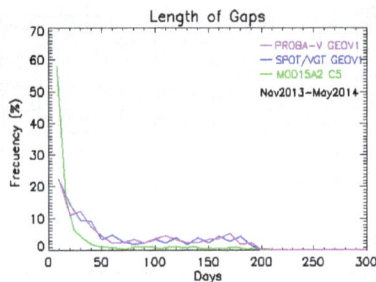

Figure 2: Distribution of the temporal lenght of the missing values over BELMANIP2.1 sites during the Nov13-May14 period for PROBA-V GEOV1, SPOT/VGT GEOV1 and MODIS C5 products.

3.2 Histograms of residuals

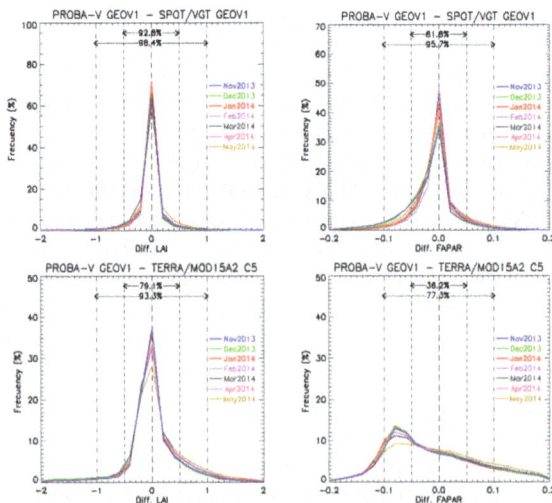

Figure 3: Distribution of differences between PROBA-V and SPOT/VGT GEOV1 (Top) and between PROBA-V GEOV1 and MODIS C5 (Bottom) for LAI (Left side) and FAPAR (right side) products.

Histograms of diferences among the global products under study were analyzed monthly (Figure 3). For LAI differences, above 93% of values (in average) are within ±0.5 LAI units for all the dates evaluated, which corresponds with a good spatial consistency, and 98% of difference values lies between ±1 LAI units between both GEOV1 products (PROBA-V and SPOT/VGT). PROBA-V GEOV1 and MODIS C5 LAI

differences are lager with a 79% of values between ±0.5 and 93% between ±1 LAI units. For FAPAR, a good agreement is found between PROBA-V and SPOT/VGT GEOV1 products with a percentage around 82% of differences within ±0.05 and up to 96% for the ±0.1 interval, which means also quite good consistency, but showing an asymmetric histogram with a slight negative bias (i.e. larger FAPAR values from SPOT/VGT observations). The histogram of residuals for FAPAR between PROBA-V and MODIS shows the inconsistency between both products. Only 36% of the pixels are within 0.05, with the peak located around -0.08 (higher values of MODIS). These results are similar to that found between SPOT/VGT GEOV1 and MODIS C5 FAPAR products for the 2003-2005 period (Camacho et al., 2013), but differences seems to be even larger between PROBA-V and MODIS mainly over DBF areas for the studied period.

3.3 Distributions of retrievals

Figure 4 shows the statistical distributions of LAI retrieved values per biomes for the several satellite products under study. Very similar distributions were found for both GEOV1 products (PROBA-V and SPOT/VGT) for all biome type. As compared with MODIS C5, the main discrepancies were found for evergreen broadleaf forest. Larger values for MODIS C5 over herbaceous and shrubs and bare areas are found, as well as a negative bias over needle-leaf forest.

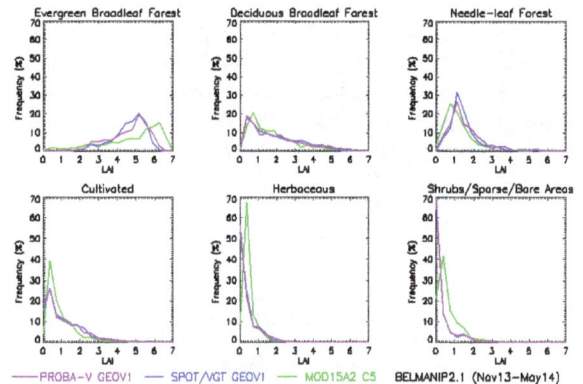

Figure 4: Distribution of LAI values of each product for the BELMANIP2.1 sites during Nov13-May14 period for each biome type.

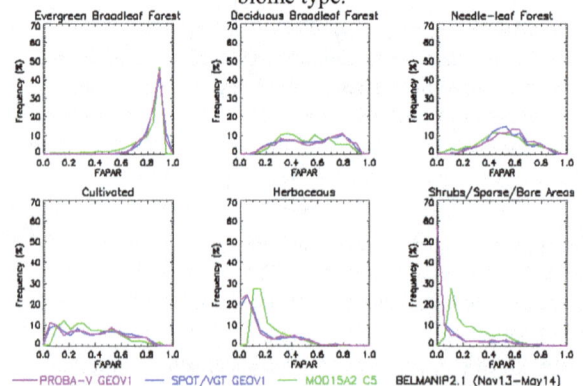

Figure 5: Distribution of FAPAR values of each product for the BELMANIP2.1 sites during Nov13-May14 period for each biome type.

Figure 5 shows the FAPAR histograms. A good agreement is found between MODIS and GEOV1 distributions for forest and cultivated areas. MODIS shows larger values for herbaceous, shrubs and bare areas, in agreement with previous validation exersices (Camacho et al., 2013).

Figure 6 shows the FCover histograms, where PROBA-V and SPOT/VGT GEOV1 retrievals display consistent results for all biome type except for evergreen broadlef forest and needle-leaf forest, with slight larger values of PROBA-V products.

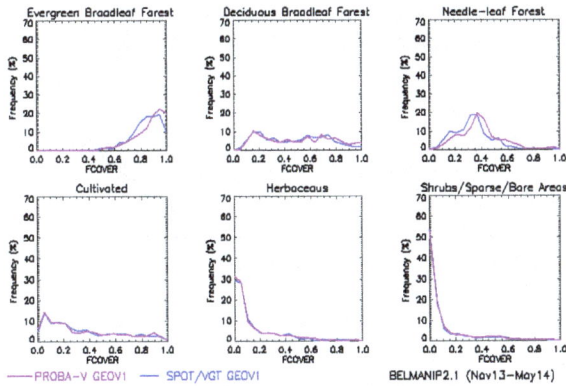

Figure 6: Distribution of FCover values of each product for the BELMANIP2.1 sites during Nov13-May14 period for each biome type.

3.4 Intra-Annual Precision

The smoothness of PROBA-V GEOV1 products is almost identical to that of SPOT/VGT similar products (Figure 7). The GEOV1 products are very smooth, indicating a high intra-annual precision. MODIS products are noisier, as can be observed in the distribution of the smoothness values, which quantifies what it is clearly visible on the temporal profiles (see section 3.5).

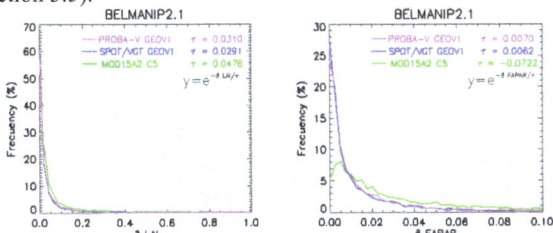

Figure 7: Histograms of the delta function (smoothness) for LAI, and FAPAR products for BELMANIP2.1 sites during the November2013-May2014 period. The curves are adjusted to an exponential function and the exponential decay constant is presented in the figure.

3.5 Temporal Consistency

The consistency of PROBA-V GEOV1 temporal variations, as compared to SPOT/VGT GEOV1 and MODIS C5 products, was assessed over the BELMANIP2.1 network of sites. A good consistency of PROBA-V GEOV1 temporal variations for this time period was found for all the different biomes over the globe (see Figures 8 and 9). It is noticeable the good continuity achieved in GEOV1 products using PROBA-V data.

Note, however, that the SPOT/VGT GEOV1 FCover product showed over some desertic sites unexpected seasonal variations with a maximum up to 0.2.

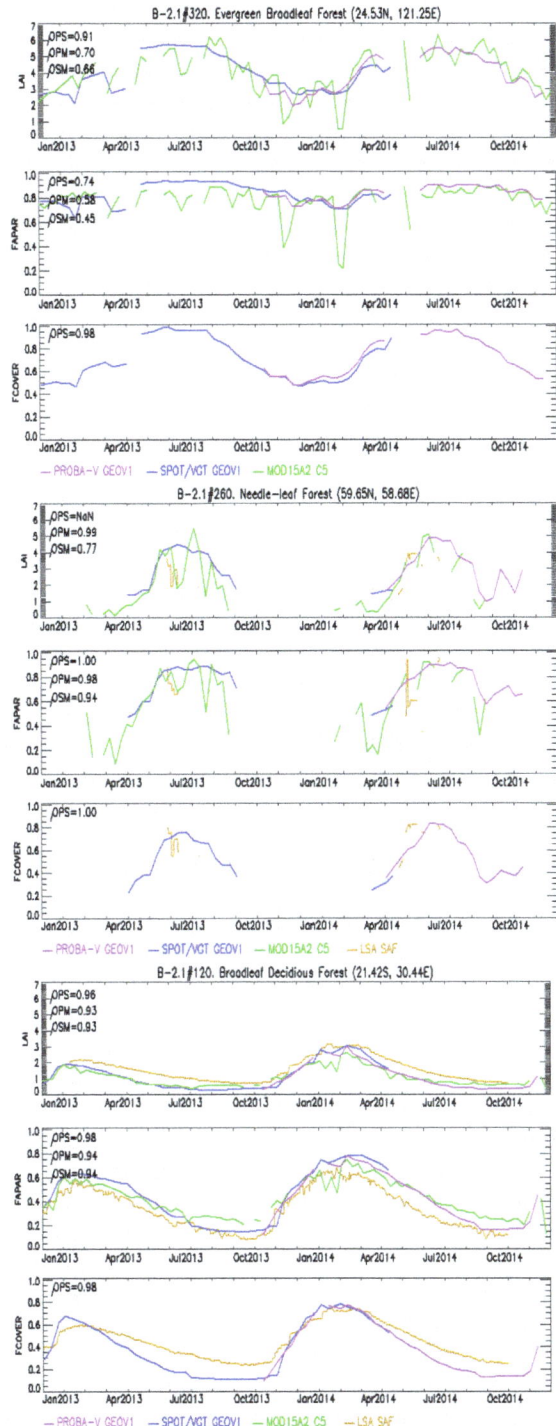

Figure 8: Temporal profiles of LAI (top), FAPAR (middle) and FCover (bottom) values of each product (PROBA-V GEOV1, SPOT/VGT GEOV1, MODIS C5 and LSA SAF) for selected sites of each biome type.

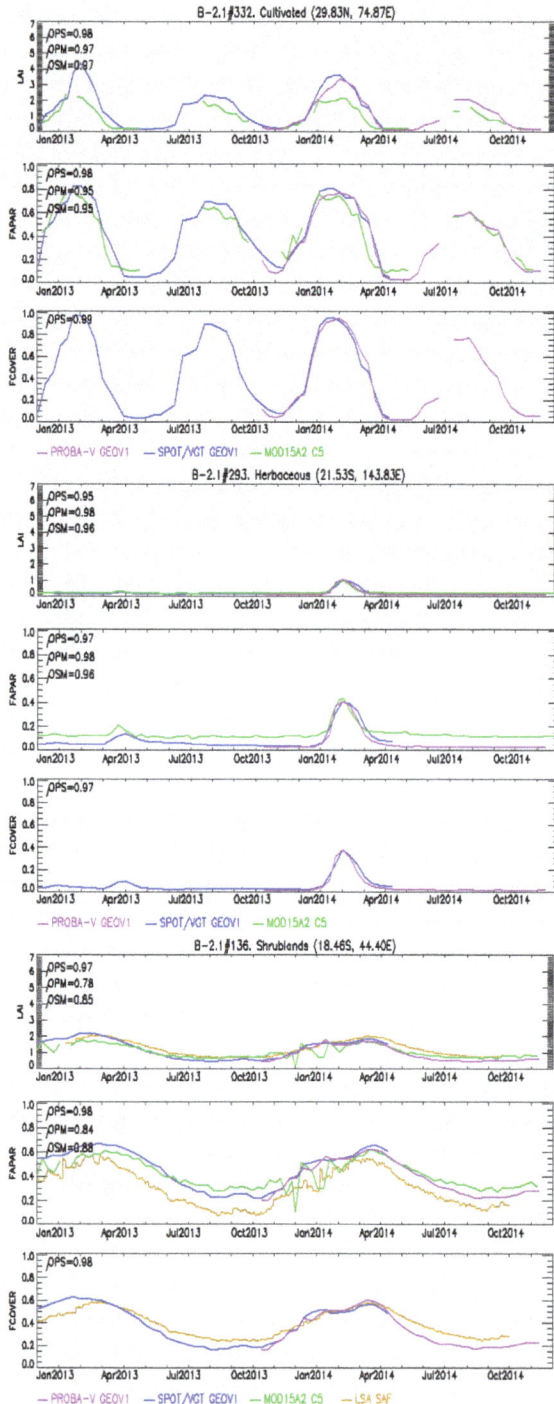

Figure 9: Temporal profiles of LAI (top), FAPAR (middle) and FCover (bottom) values of each product (PROBA-V GEOV1, SPOT/VGT GEOV1, MODIS C5 and LSA SAF) for selected sites of each biome type.

3.6 Relative Uncertainties

The consistency of PROBA-V GEOV1 with the reference global products was evaluated over BELMANIP2.1 sites at 10-days frequency during the November 2013 to May 2014 period.

Scatter-plots between GEOV1 (PROBA-V vs SPOT/VGT) products (Figure 10) show the optimal consistency between both products, with overall correlations higher than 0.96, no bias (note the perfect fit with slope 1.0 and offset 0.0), and

overall discrepancies (RMSE) of around 0.3, 0.03 and 0.04 for LAI, FAPAR and FCover respectively.

Overall discrepancies between PROBA-V and MODIS for the overlap period (November 2013 to May 2014) are larger (Figure 11), with some scattering observed for LAI products, and both scattering and bias (mainly for low values) for the FAPAR. The overall discrepancies between PROBA-V and MODIS are around 0.7 and 0.09 for LAI and FAPAR respectively.

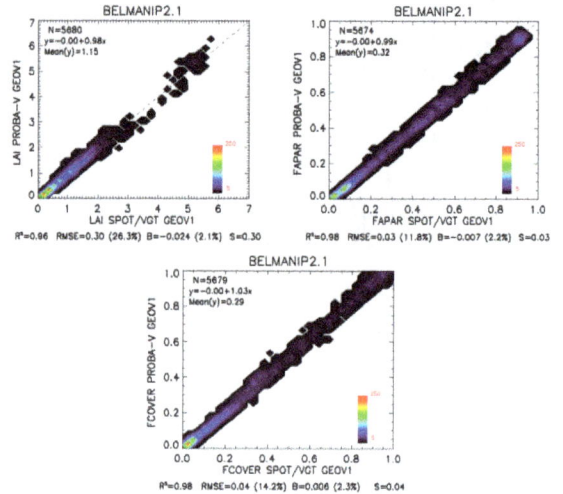

Figure 10: LAI, FAPAR and FCover PROBA-V GEOV1 versus SPOT/VGT GEOV1 products scatter-plots over all BELMANIP-2.1 sites for the November 2103 - May 2014 period. The terms B and S represent the mean and the standard deviation of the difference.

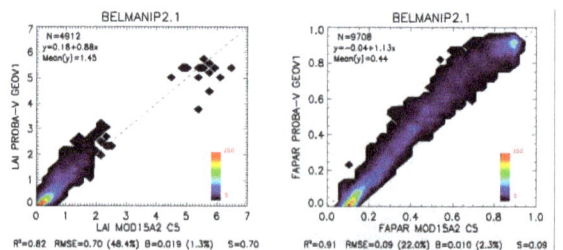

Figure 11: LAI and FAPAR PROBA-V GEOV1 versus MODIS C5 products scatter-plots over all BELMANIP-2.1 sites for the November 2103 - May 2014 period. The terms B and S represent the mean and the standard deviation of the difference.

3.7 Direct Validation

A robust regression method was used to compute the accuracy (RMSE_W) in addition to the ordinary least square algorithm to reduce the outlier effects. The algorithm uses an iteratively reweighted least square with a bisquare weighting function and provides a weight associated to each sample. Samples with associated weight lower than 0.3 were considered outliers and not considered for accuracy estimation (RMSE). The closest satellite date to the ground data is used for the accuracy assessment.

PROBA-V GEOV1 LAI product shows a good accuracy (RMSE=0.52, RMSE_W=0.51), and shows also very low mean bias (0.10) using all data but a positive bias for concomitant data over crops (Figure 12, Table 2). GEOV1 LAI provides a good agreement across the whole range of LAI values, with however two outliers identified corresponding to paddy rice fields in La Albufera site (rows #8 #11, Table 1) and a non-concomitant forest sample. The large discrepancies over La Albufera were observed end of June, when the rice crops were growing very rapidly, and at the end of August, before to start the harvest, which indicates that discrepancies could be partly attributed to the impact of the compositing period (30 days) of the satellite estimates. MODIS presents also similar accuracy (RMSE= 0.69) for the different biome types, but slightly degraded precision probably due to the reported instability over short time periods (Figure 9). Opposite to GEOV1, MODIS tends to underestimate the concomitant data provided by ImagineS over cropland sites, except in La Albufera site for the growing period. This can be explained as the composite period of MODIS is 8-days after the date of the product (future observations).

For the FAPAR, PROBA-V GEOV1 product shows also a quite good accuracy (RMSE=0.11, RMSE_W=0.1) with a slight positive bias (0.05) and very good correlations (Figure 13). For concomitant ground values over cropland sites (Table 2), no outliers were removed but a clear overestimation was observed again over La Albufera, (rows #7, #8, #9 in Table1) during the rapid growing stage of the rice. MODIS shows also quite good accuracy (RMSE=0.1) except for low values where a positive bias was observed. The agreement with ImagineS concomitant sites is better than GEOV1, with almost all points within the uncertainty level of ±0.1.

Finally, for FCover the accuracy (RMSE) obtained is 0.14 (0.1 for the weighted RMSE) with a positive bias of 0.1. The larger discrepancies are observed over forest sites, which was not observed for SPOT/VGT GEOV1 products, and should be confirmed with concomitant data when available. For concomitant data over some croplands sites of 25Mayo (#1), LaReina (#3) and Psenichne (#13) a large overestimation was observed (samples not identified as outliers). These results seems to indicate that the FCover product tend to overestimate the ground references, even if the dataset considered is still very limited.

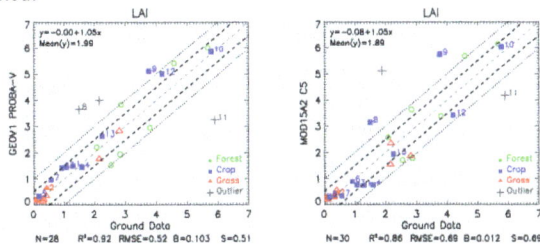

Figure 12: Comparison of satellite LAI product with the ground-based maps. Left side: PROBA-V GEOV1. Right side: MODIS C5. Filled symbols correspond to concomitant values of 2014 from ImagineS and unfilled symbols to a different year. Forest stands for Broadleaf Evergreen, Broadleaf Deciduous and Needle-leaf Forests, Crops stands for Cultivated and Grass refers to Herbaceous, Shrubs, Sparse and Bare Areas. Numbers identify the ground data (Table 1). Crosses identify outliers (weight<0.3).

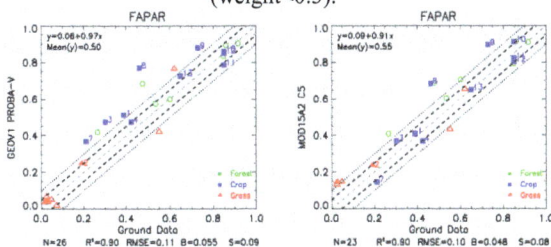

Figure 13: As in Figure 12 for FAPAR.

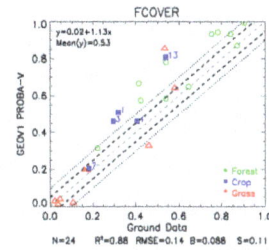

Figure 14: As in Figure 54 for FCover

		Concomitant data (2014)		
		LAI	FAPAR	FCover
PROBA-V GEOV1 vs Ground Data	N	11	11	7
	Correlation	0.96	0.84	0.95
	Bias	0.38	0.09	0.15
	RMSE	0.54	0.13	0.18
	RMSE_W	N/A	N/A	N/A
	Offset	0.22	0.2	-0.04
	Slope	1.07	0.79	1.52
MOD15A2 C5 vs Ground Data	N	12	11	N/A
	Correlation	0.82	0.88	N/A
	Bias	0.067	0.034	N/A
	RMSE	0.84	0.09	N/A
	RMSE_W	N/A	N/A	N/A
	Offset	-0.12	0.02	N/A
	Slope	1.09	1	N/A
		All data (2000-2014)		
		LAI	FAPAR	FCover
PROBA-V GEOV1 vs Ground Data	N	28	26	25
	Correlation	0.92	0.9	0.88
	Bias	0.10	0.06	0.09
	RMSE	0.52	0.11	0.14
	RMSE_W	0.51	0.10	0.1
	Offset	0	0.06	0.02
	Slope	1.05	0.97	1.13
MOD15A2 C5 vs Ground Data	N	30	23	N/A
	Correlation	0.86	0.9	N/A
	Bias	0.012	0.05	N/A
	RMSE	0.69	0.1	N/A
	RMSE_W	0.71	0.1	N/A
	Offset	-0.08	0.09	N/A
	Slope	1.05	0.91	N/A

Table 2: Performance of PROBA-V GEOV1 and MOD15A2 C5 products against reference ground based maps. RMSE_W stands for weighted RMSE.

4. CONCLUSIONS

In this study, an early scientific validation of GEOV1 (LAI, FAPAR and FCover) products based on PROBA-V observation was performed for the overlap period between SPOT/VGT and PROBA-V. The methodology used follows the guidelines proposed by the CEOS LPV group for validation of remote sensing vegetation products. First, an intercomparison with existing global products (SPOT/VGT GEOV1 and MODIS C5) was performed. The BELMANIP2.1 network of sites was used to perform the global statistical analysis at 3x3 km^2 and at a 10-days time step. Second, the uncertainties were quantified by direct comparison with ground-based reference maps.

GEOV1 LAI, FAPAR and FCover estimates from PROBA-V data were found consistent with that of GEOV1 based on SPOT/VGT observations. The completeness of PROBA-V GEOV1 products is similar to that of SPOT/VGT GEOV1 products (around 5% lower).

Both PROBA-V and SPOT/VGT shows consistent spatial distribution of retrievals, whereas discrepancies between PROBA-V and MODIS are larger, mainly for the FAPAR product (large discrepancies). The temporal variations of PROBA-V were found also consistent with that of SPOT/VGT GEOV1 products.

The overall consistency between SPOT/VGT and PROBA-V evaluated in term of RMSE over BELMANIP-2 is better than the CEOS requirements on accuracy (0.5 for LAI, 0.05 for FAPAR), which demonstrates the good consistency between both products.

The direct validation shows that the accuracy of PROBA-V GEOV1 LAI product was very close to CEOS requirement on accuracy using limited concomitant data (RMSE=0.54) or using additional non-concomitant references (RMSE=0.51). For the FAPAR the accuracy is also quite good (0.11 for all data), but a slight overestimation was observed mainly as compared to concomitant data (bias=0.09). The FCover shows the worst performance, with a systematic positive bias observed for mainly for forest and cropland sites (up to 0.15 for concomitant data) and overall error of 0.14. PROBA-V shows similar performances than MODIS for both LAI and FAPAR products. In summary, these validation results of the PROBA-V GEOV1 products over one year (2014) period shows very good spatial and consistency with the SPOT/VGT GEOV1 products for the overlap period. However, a positive bias as compared to SPOT/VGT has been detected for the FCover mainly for values larger than 0.5. This bias seems to be confirmed by the limited ground observations available. All the criteria evaluated including accuracy assessment shows positive results. The main drawback of the product is the completeness which is slightly lower than in SPOT/VGT GEOV1 products.

REFERENCES

Baret, F., M. Weiss, R. Lacaze, F. Camacho, H. Makhmara, P. Pacholcyzk, B. Smets, 2013. GEOV1: LAI and FAPAR essential climate variables and FCOVER global time series capitalizing over existing products. Part1: Principles of development and production. *Remote Sensing of Environment* 137: 299–309.

Camacho, F., Cernicharo, J., Lacaze, R., Baret, F., and Weiss, M.,2013. GEOV1: LAI, FAPAR Essential Climate Variables and FCover global time series capitalizing over existing products. Part 2: Validation and intercomparison with reference products. *Remote Sensing of Environment* 137: 310–329.

Camacho, F., C. Latorre, R. Lacaze, F. Baret, F. De la Cruz, V. Demarez, C. Di Bella, H. Fang, J. García-Haro, M. P. Gonzalez, N. Kussul, E. López-Baeza, C. Mattar, E. Nestola, E. Pattey, I. Piccard, C. Rudiger, I. Savin, A. Sanchez - Azofeifa, M. Boschetti, D. Bossio, M. Weiss, A. Castrignano, M. Zribi (2014). "A Network of Sites for Ground Biophysical Measurements in support of Copernicus Global Land Product Validation". Proceedings of the IV RAQRS conference, Valencia, Spain, 22-26 September 2014. http://ipl.uv.es/raqrs/?q=content/4th-international-symposium-raqrs.

Fernandes, R., Plummer, S., Nightingale, J., et al., 2014. "Global Leaf Area Index Product Validation Good Practices". CEOS Working Group on Calibration and Validation - Land Product Validation Sub-Group.Version 2.0: Public version made available on LPV website. http://lpvs.gsfc.nasa.gov/.

Knyazikhin, Y., Martonchik, J. V., Myneni, R. B., Diner, D. J., & Running, S. W. (1998). Synergetic algorithm for estimating vegetation canopy leaf area index and fractionof absorbed photosynthetically active radiation from MODIS and MISR data. Journalof Geophysical Research, 103(D24), 32,257–32,275.

Weiss, M., Baret, F., Block, T. et al., (2014). On Line Validation Exercise (OLIVE): A Web Based Service for the Validation of Medium Resolution Land Products. Application to FAPAR products.*Remote Sensing: 2014. 6(5):4190-4216.*

APPENDIX

Characteristics of the validation sites and associated ground biophysical maps used in the direct validation for non concomitant dates.

Site Country	Lat Lon (deg)	Land Cover	Dates (mm/yyyy)	LAI	FAPAR	FCover
KONZ USA	39.09 -96.57	Herb.	06/2000	2.17	N/A	N/A
			08/2000	2.16	N/A	N/A
SEVI USA	34.35 -106.69	Shrubs	07/2002 - 11/2003	0.05 - 0.4	N/A	N/A
Larose2 Canada	45.38 -75.17	Needle-leaf F.	08/2003	2.86	N/A	N/A
Appomattox USA	37.22 -78.88	Needle-leaf F.	08/2000	1.89	N/A	N/A
Camerons Australia	-32.6 116.25	Evergreen F.	03/2004	2.08	0.47	0.41
GN/Agara Australia	-31.53 115.88	Deciduous F.	03/2004	1.0*	0.27	0.22
Hiriskangas Finland	62.64 27.01	Needle-leaf F.	08/2003	N/A	N/A	0.64
Jarvaselja Estonia	58.3 27.26	Needle-leaf F.	07/2000	N/A	N/A	0.75
			06/2001	N/A	N/A	0.78
			06/2002	N/A	N/A	0.79
			06/2005	N/A	N/A	0.84
Laprida Argentina	-36.99 -60.55	Crops	10/2002	2.8	0.62	0.53
Nezer France	44.57 -1.04	Needle-leaf F.	07/2000	N/A	N/A	0.54
			06/2001	N/A	N/A	0.87
			04/2002	2.54	0.53	N/A
Puechabon France	43.72 3.65	Needle-leaf F.	06/2001	2.84	0.6	0.54
Rovaniemi Finland	66.46 26.35	Needle-leaf F.	06/2004	N/A	N/A	0.42
Sonian Belgium	50.77 4.41	Needle-leaf F.	06/2004	5.66	0.91	0.9
Turco Bolivia	-18.24 -68.18	Sparse	07/2001	0.3	N/A	0.11
			08/2001	0.04	0.03	0.02
			04/2003	N/A	0.05	0.04
Wankama Niger	16.65 2.64	Herb.	06/2005	N/A	0.07	0.04
Mongu Zambie	-15.44 23.25	Shrubs	02/2000- 05/2000	N/A	0.55 - 0.59	0.46 - 0.58

Dahra Norht Senegal	15.43 -15.4	Shrubs	07/2001- 08/2001	N/A	0.02- 0.03	N/A
Tessekre South Kenya	15.819 -15.06	Herb.	07/2001	N/A	0.03	N/A
Budongo8 Uganda	1.77 31.61	Evergreen F.	11/2007	0.87	N/A	0.26
Harth Forest France	47.81 7.45	Deciduous F.	06/2013- 09/2013	3.8- 4.58	0.85- 0.86	N/A

More information and full list of validation sites can be found on the CEOS cal/val site (http://calvalportal.ceos.org/cvp/web/olive/descriptions)

8

CLIMATE ABSOLUTE RADIANCE AND REFRACTIVITY OBSERVATORY (CLARREO)

J. Leckey [a], *

[a] National Aeronautics and Space Agency (NASA) Langley Research Center (LARC), Bldg. 1202 MS 468, Hampton, Virginia 23681 USA – john.p.leckey@nasa.gov

THEME: Atmosphere, weather and climate (ATMC)

KEY WORDS: Infrared, Far Infrared, Radiance, Climate, Atmosphere

ABSTRACT:

The Climate Absolute Radiance and Refractivity Observatory (CLARREO) is a mission, led and developed by NASA, that will measure a variety of climate variables with an unprecedented accuracy to quantify and attribute climate change. CLARREO consists of three separate instruments: an infrared (IR) spectrometer, a reflected solar (RS) spectrometer, and a radio occultation (RO) instrument. The mission will contain orbiting radiometers with sufficient accuracy, including on orbit verification, to calibrate other space-based instrumentation, increasing their respective accuracy by as much as an order of magnitude. The IR spectrometer is a Fourier Transform spectrometer (FTS) working in the 5 to 50 μm wavelength region with a goal of 0.1 K ($k = 3$) accuracy. The FTS will achieve this accuracy using phase change cells to verify thermistor accuracy and heated halos to verify blackbody emissivity, both on orbit. The RS spectrometer will measure the reflectance of the atmosphere in the 0.32 to 2.3 μm wavelength region with an accuracy of 0.3% ($k = 2$). The status of the instrumentation packages and potential mission options will be presented.

1. INTRODUCTION

According to the Fifth Assessment Report of the Intergovernmental Panel on Climate Change (IPCC) – "Each of the last three decades has been successively warmer at the Earth's surface than any preceding decade since 1850. The period from 1983 to 2012 was *likely* the warmest 30-year period of the last 1400 years in the Northern Hemisphere" (IPCC 2013). Measuring climate change on a decadal time scale is vital to both assessing model accuracy and attributing climate change to its source (IPCC 2012). While critically important, the small decadal scale signals of climate change are dwarfed by short-term natural variability and remain out of reach for the majority of current global satellites. The Climate Absolute Radiance and Refractivity Observatory (CLARREO) mission seeks to increase the absolute accuracy of climate change measurements using SI traceable standards on orbit.

Figure 1 shows the National Climate Data Center (NCDC) Merged Land-Ocean Surface Temperature Analysis (MLOST) temperature trend data broken into 3 distinct time periods: 1911-1940, 1951-1980, and 1981-2012 (IPCC 2013). It is clear from these results that the last period, 1981-2012, shows the strongest warming per decade of the entire data set while the period 1951-1980 was the least warming. During the period 1979-2012, the NCDC MLOST global mean surface temperature (GMST) trend was 0.151 $\pm$ 0.037 °C/10y compared to only 0.081 $\pm$ 0.013 °C/10y from 1901-2012 (Vose et al., 2012b). In addition to global surface warming, these data show large decadal and interannual variability (IPCC 2013).

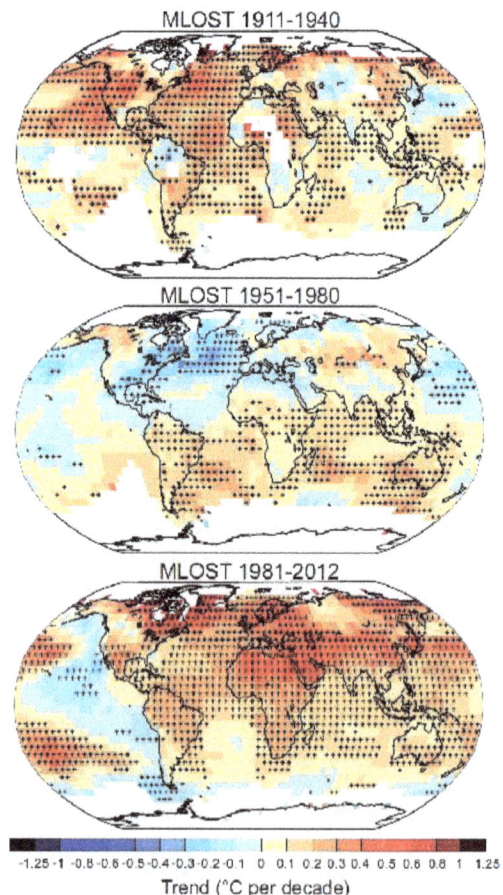

Figure 1. NCDC MLOST GMST trends (°C/10y) over three separate time periods: 1911-1940, 1951-1980, and 1981-2012 (IPCC 2013) where white signifies incomplete data

---

* Corresponding author.

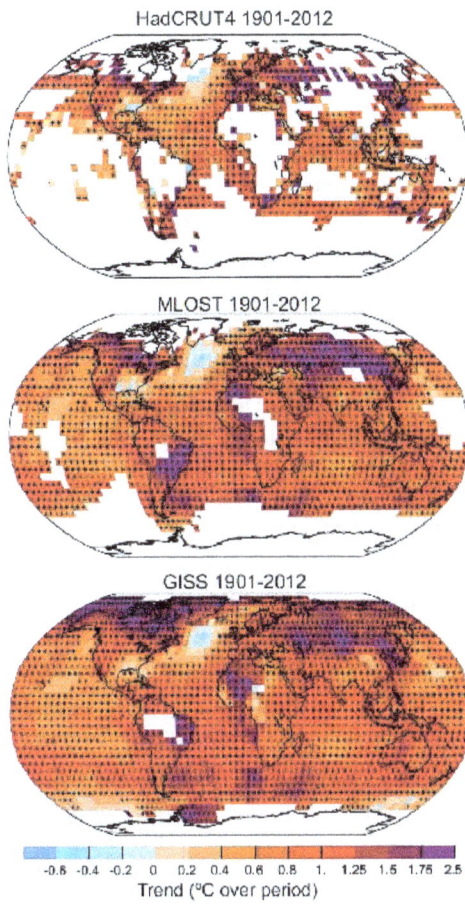

Figure 2. Surface Temperature Trends from three separate climate model data sets (HadCRUT4, MLOST, and GISS) for the period 1901 – 2012 (IPCC 2013) where white signifies incomplete data

The surface temperature trend data sets from the Hadley Centre/Climate Research Unit (HadCRUT4), MLOST, and the Goddard Institute for Space Studies (GISS) are shown in Figure 2. The GMST trends for the period 1901-2012 are 0.075 $\pm$ 0.013 °C/10y, 0.081 $\pm$ 0.013 °C/10y, and 0.083 $\pm$ 0.013 °C/10y from HadCRUT4, MLOST, and GISS, respectively. It is clear that all of the data sets show surface temperature warming and also closely agree with each other for most geographic regions; however, a 15% uncertainty remains on the absolute accuracy of the warming trend due to uncertainties in climate models and a need for further model validation with observations. From the plots we can also see discrepancies in the warming trend by as much as a factor of two (south central South America) between the climate models. With the increase in accuracy from a true climate observing system, these uncertainties will be reduced and the precise amount of warming can be confirmed.

In 2007, the National Research Council Decadal Survey on Earth Science and Applications from Space identified the level of uncertainties present in climate variables as a problem for both future policy making and risk management strategies. The Decadal Survey also identified the need to make measurements with high enough accuracy to be able to resolve small climate change signals over decadal time scales (NRC 2007). One of the major challenges in Earth observing satellites is the uncertainty in the calibration over long time scales. This is a challenge that CLARREO will address by carrying an absolute calibration standard onto orbit. With the goal of making

measurements to resolve decadal scale changes in climate, the requirements of CLARREO are driven by long term absolute accuracy rather than typical instantaneous noise requirements. The CLARREO benchmark measurements as defined in the NRC Decadal Survey are:

1. Spectrally resolved infrared (IR) radiance emitted from Earth to space measured to an accuracy of 0.07 K ($k = 2$).
2. Spectrally resolved nadir reflectance of reflected solar (RS) radiation measured to an accuracy of 0.3% ($k = 2$).
3. Global Navigation Satellite Systems Radio Occultation (RO) measurement of the phase delay of the transmitted GNSS signal occulted by the atmosphere from low Earth orbit for attitudes (5 km to 20 km) and with an accuracy of 0.06% ($k = 2$).

The IR, RS, and RO measurements were strategically chosen to target the least understood climate forcings. The primary climate variable targets (listed with the corresponding instrument noted) are: the atmospheric distribution of temperature and water vapor (IR, RS, and RO), broadband reflected (RS) and emitted (IR) radiative fluxes, surface albedo (RS), temperature (IR), emissivity (IR), and cloud properties (RS and IR). In combination, the CLARREO data will resolve with a high degree of confidence the rate at which climate change is occurring, improve future climate modeling, and improve the absolute accuracy of currently orbiting weather and climate satellites by providing an absolute reference calibration on orbit.

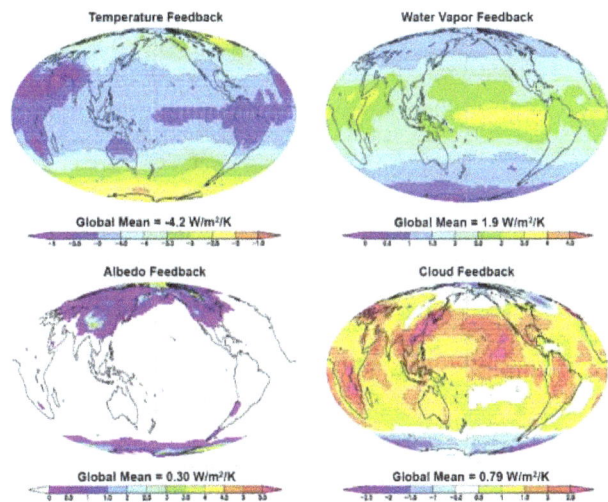

Figure 3. IPCC AR4 Climate Model Feedback Maps showing temperature, water vapor, Albedo, and Cloud feedbacks (Soden 2008)

In order to resolve the inherently small climate scale changes over decadal time scales absolute accuracy is important, not instantaneous noise levels. This difference in methodology relaxes some instrumentation requirements while introducing new requirements. CLARREO will make measurements over a large spatial area and over long time scales with a focus on accuracy over precision. The large spatial scale over which CLARREO will operate will enable the cross-calibration of numerous current and future Earth observing missions with previously unachievable accuracy.

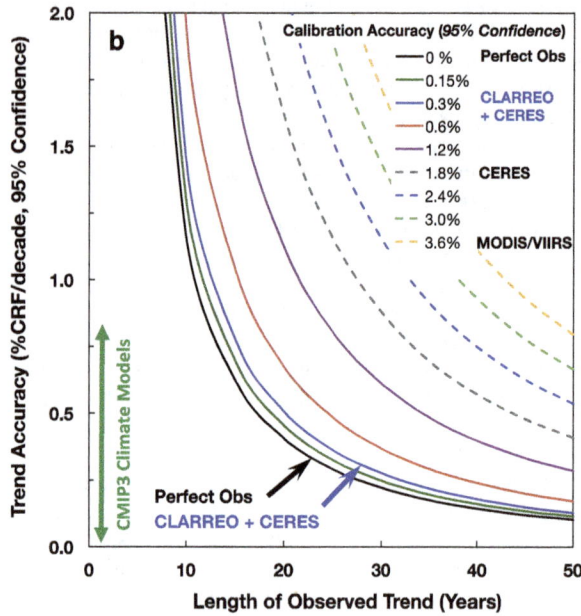

Figure 4. Accuracy of decadal trends of current earth observing missions and the potential improvement CLARREO will yield (Wielicki 2013)

Figure 4 shows the accuracy trend of cloud radiative forcing per decade as a function of the length of observed trend in years. Current missions are shown in Figure 4, including CERES and MODIS/VIIRS, where it is clear that the length of time necessary to reach 1% accuracy in cloud radiative forcing is 30+ years. This length of time is reduced by nearly 20 years with the addition of an instrument with CLARREO's absolute accuracy (Wielicki 2013). Similar analyses can be made for a variety of climate variables where the length of time necessary to achieve a specific absolute accuracy is dramatically reduced with the introduction of CLARREO's instruments.

2. INFRARED SPECTROMETER

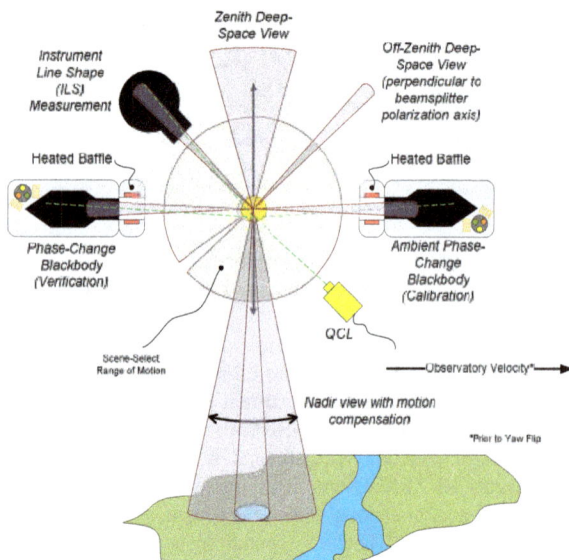

Figure 5. IR Spectrometer concept showing the variety of validation sources that are routinely viewed

The CLARREO Infrared Spectrometer is a Fourier transform spectrometer (FTS) over the spectral range $5\,\mu m - 50\,\mu m$. The significance of the IR spectrometer is its collection of SI-traceable verification standards that are launched with the spectrometer onto orbit. Traditional earth observing missions carefully calibrate their sensors before launch and make the assumption that nothing substantial changes during launch and deployment of the instrument. Alternatively, CLARREO will not only undergo careful calibration before launch, but will undergo regular on-orbit re-calibrations using the onboard standards.

The IR instrument is shown in Figure 6 in an exploded view consisting of the two detector assemblies, the optical and Fourier transform spectrometer hardware, and the scene select mechanism. The scene select mechanism looks at 5 different scenes iteratively: the nadir Earth view (up in Figure 6), 45° off nadir view, deep space view (cold), verification blackbody, and ambient blackbody. A rotating mirror is inside the center of the scene select mechanism to rapidly (second timescale) switch between the different possible views. The system is calibrated using the deep space view and the ambient blackbody and then the calibration is verified using the verification blackbody.

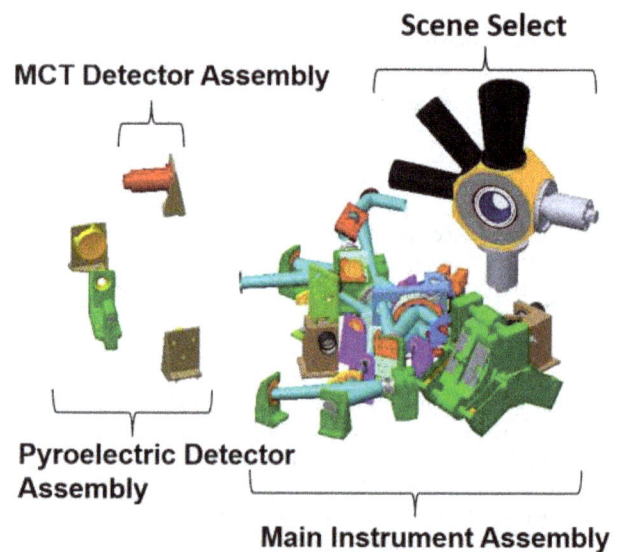

Figure 6. IR Spectrometer exploded view with major subsystems labelled

The blackbodies are unique due to two features: a heated baffle surrounding the entrances of the blackbodies and phase change cells. When working properly, the view of a blackbody as seen by the scene select mechanism is of radiation purely from deep within the blackbody and not from the edges or other external blackbody components. The heated baffle emits radiation into the blackbody and as long as the view of the blackbody is pure, there will be no additional signal from the increase in radiation entering the blackbody. If additional signal is seen, a correction or adjustment is necessary. The phase change cells contain three different pure materials that melt/solidify at temperature that span the operating points of the blackbodies. The cells are warmed as the blackbody is warmed and as the material melts, the temperature change versus time curve flattens while the material is melting and then resumes heating once the material has melted. The three material melt points provide three absolute, SI-traceable, temperature measurements to fix the temperature of the blackbody to an absolute scale. Although a temperature probe's calibration may drift with time, the material melt points do not drift and thus the temperature probes will be recalibrated with every material melt.

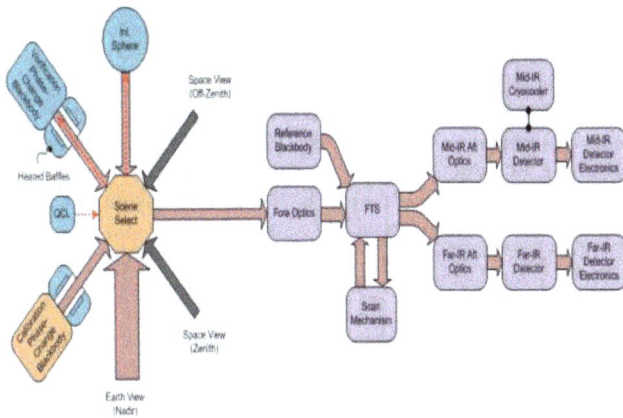

Figure 7. Optical functional diagram for the IR CLARREO instrument

The optical functional diagram is shown in Figure 7. The quantum cascade laser can be used to illuminate the integrating sphere (to verify instrument lineshape) or to act as a high intensity source to send into the blackbodies to check reflection. The orientation of the scene select is passed through the fore optics and into the FTS. The core of the FTS is a Caesium Iodide beam splitter and a scan mechanism to create an interferogram. The output of the FTS is then split to a pair of detectors. The Mercury Cadmium Telluride (MCT) detector is used for mid-IR detection and a pyroelectric detector is for far-IR detection. Overall, the IR FTS spectrometer has enough verification standards to act as a SI-traceable calibration standard on orbit.

3. REFLECTED SOLAR SPECTROMETER

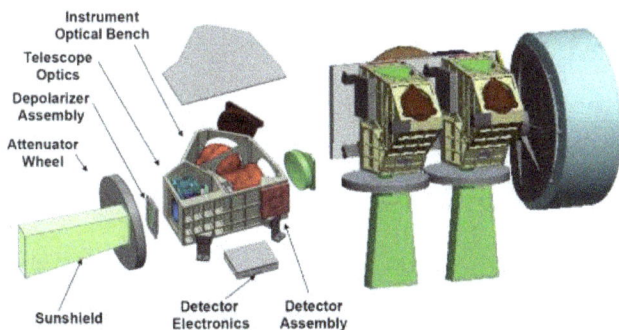

Figure 8. RS Spectrometer exploded view

The Reflected Solar (RS) spectrometer consists of two separate spectrometers as may be seen in Figure 8. The first operates over the spectral range 320 nm – 640 nm and the second operates from 600 nm – 2300 nm, where both have 4 nm spectral sampling and have the goal of less than 0.3% calibration accuracy. The aggressive absolute accuracy

requirement is nearly an order of magnitude better than any comparable instrument. The spectrometers observe the Earth's surface and measure radiance that can be converted to reflectance by making similar lunar and solar measurements and taking the ratio of the measurements with geometrical effects removed. This variety of measurements is especially difficult to make because the signals can vary by an order of magnitude between the brightest and dimmest signals. The overall RS spectral range was chosen so that CERES and VIIRS can be cross calibrated by the CLARREO instrument. This objective will require that the RS instrument have sufficient pointing accuracy as to definitively know that the same part of the Earth will be measured by each of the different missions to allow for a proper intercalibration.

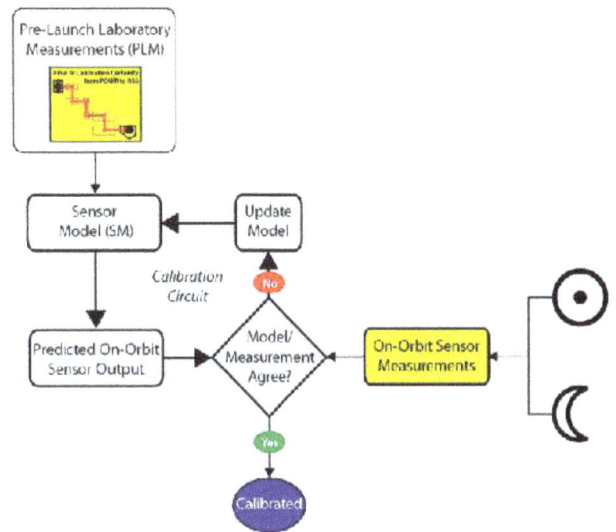

Figure 9. RS Calibration approach.

In order to meet these required accuracy standards, a rigorous calibration approach must be created. The RS instrument calibration starts with a series laboratory measurements that will all be inputs to a sensor model. Prior to launch, radiance and irradiance modes will be thoroughly calibrated including all of the necessary geometrical measurements for a conversion to reflectance. After the instrument is launched, pre-launch measurements incorporated into the sensor model will be compared to solar absolute irradiance measurements to demonstrate post launch accuracy and prove that nothing significantly changed during launch. Stellar and lunar measurements post-launch will all be incorporated into the sensor model so that the model can evolve with time and still retain absolute accuracy. If discrepancies arise between the pre-launch model predictions and measurements post-launch, the model will need to evolve based on current measurements. Once the model prediction matches the on orbit measurements, the sensor will be fully calibrated. The solar and lunar measurements will yield information on temporal changes in the sensor over time once an initial model agreement takes place.

4. RADIO OCCULTATION

The Radio Occultation (RO) instrument is a standard GPS receiver system used for global navigation with the added ability to perform radio occultation using data from multiple satellites. The general concept is that the GPS signals are refracted as they pass through the atmosphere and undergo Doppler shifts that can be measured. The frequency shifts are a function the density of the atmosphere and the distance the signal traverses through the atmosphere. Figure 10 illustrates the full RO concept beginning with a Global Navigation Satellite System (GNSS) satellite signal that is transmitted through the atmosphere as it rises or sets behind the Earth's limb. The Doppler shift is related to how the signal is bent and is converted to an angle, α in Figure 10. The angle can be used to extract a refractivity profile that is then translated into the temperature, pressure, and humidity in the atmosphere.

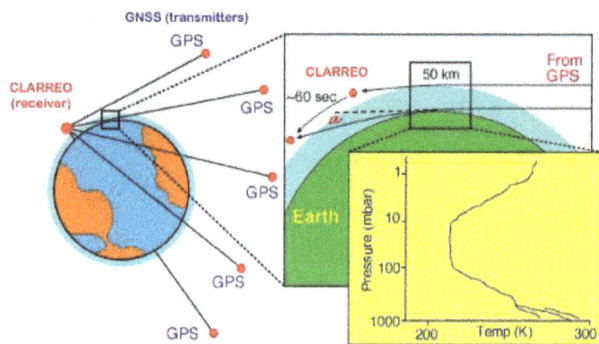

Figure 10. Radio Occultation Instrument Concept

Currently, there are only two frequencies, L1 and L2, at which the GPS satellites operate. There are more frequencies that future GPS systems (2016+) will use for operation that CLARREO would benefit from; however, the minimum scientific goals can be accomplished using ~2000 occultation measurements per day of the L1 and L2 frequencies. These occultation signals will be collected using a pair of array antennas, one in the ram and one in the wake on orbit. A variety of RO options are currently under investigation including laser ranging and other techniques that will provide the RO measurements combined with precise on orbit location measurements necessary to meet the CLARREO science objectives.

5. MISSION OPTIONS

A variety of launch configurations have been explored ranging from multiple free-flying independent satellites, a single satellite containing all the instruments, and an option to put versions of the instruments on the international space station (ISS). In the President's FY16 budget, a version of the ISS option has been specifically identified as a budget line item. If the budget is passed in its current form, the construction of an ISS CLARREO mission could begin in FY16 and launch several years after that. Once CLARREO is on orbit the climate record will be measured with sufficient accuracy to resolve climate change over decadal time scales.

ACKNOWLEDGEMENTS (OPTIONAL)

I would like to acknowledge the work of the CLARREO team. I would like to thank Bruce, Costy, Dave, and Rich for helpful discussions and suggestions.

REFERENCES

IPCC, 2007a: Climate Change 2007: "The Physical Science Basis," Contribution of Working Group I to the Fourth Assessment Report of the Intergovernmental Panel on Climate Change [Solomon, S., D. Qin, M. Manning, Z. Chen, M. Marquis, K.B. Averyt, M. Tignor and H.L. Miller (eds.)]. Cambridge University Press, Cambridge, United Kingdom and New York, NY, USA

IPCC, 2013: Climate Change 2013: "The Physical Science Basis", Contribution of Working Group I to the Fifth Assessment Report of the Intergovernmental Panel on Climate Change [Stocker, T.F., D. Qin, G.-K. Plattner, M. Tignor, S.K. Allen, J. Boschung, A. Nauels, Y. Xia, V. Bex and P.M. Midgley (eds.)]. Cambridge University Press, Cambridge, United Kingdom and New York, NY, USA, 1535

NRC, 2007: "Earth Science and Applications from Space: National Imperatives for the Next Decade and Beyond," The National Academy Press, 428

Soden, B. J., I. M. Held, R. Colman, K. M. Shell, J. T. Kiehl and C. A. Shields, 2008: "Quantifying climate feedbacks using radiative kernels," Journal of Climate, 21, pp. 3504 - 3520

Vose, R. S., et al., 2012b: "NOAA's Merged Land-Ocean Surface Temperature" Analysis. Bull. Am. Meteor. Soc., 93, 1677–1685

Wielicki, B.A., D.F. Young, M.G. Mlynczak, K.J. Thome, S. Leroy, J. Corliss, J.G. Anderson, C.O. Ao, R. Bantges, F. Best, K. Bowman, H. Brindley, J. Butler, W. Collins, J.A. Dykema, D.R. Doelling, D.R. Feldman, N. Fox, X. Huang, R. Holz, Y. Huang, Z. Jin, D. Jennings, D.G. Johnson, K. Jucks, S. Kato, D.B. Kirk-Davido , R. Knuteson, G. Kopp, D.P. Kratz, X. Liu, C. Lukashin, A.J. Mannucci, N. Phojanamongkolkij, P. Pilewskie, V. Ramaswamy, H. Revercomb, J. Rice, Y. Roberts, C.M. Roithmayr, F. Rose, S. Sandford, E.L. Shirley, W.L. Smith, Sr., B. Soden, P.W. Speth, W. Sun, P.C. Taylor, D. Tobin, X. Xiong, 2013: "Achieving Climate Change Absolute Accuracy in Orbit" Bull. Amer. Meteor. Soc., pp 1519 - 1539

9

COMPARISONS OF AEROSOL OPTICAL DEPTH PROVIDED BY SEVIRI SATELLITE OBSERVATIONS AND CAMx AIR QUALITY MODELLING.

A. P. Fernandes[a*], M. Riffler[b,c], J. Ferreira[a], S. Wunderle[c], C. Borrego[a] and O. Tchepel[d]

[a]CESAM & Department of Environment and Planning, University of Aveiro, 3810-193 Aveiro, Portugal - (apsfernandes*, jferreira, cborrego)@ua.pt
[b]Oeschger Centre for Climate Change Research, University of Bern, Switzerland - michael.riffler@oeschger.unibe.ch
[c]Remote Sensing Research Group, Department of Geography, University of Bern, Switzerland - stefan.wunderle@giub.unibe.ch
[d]CITTA, Department of Civil Engineering, University of Coimbra, 3030 - 788 Coimbra, Portugal – oxana@uc.pt

KEY WORDS: Aerosol Optical Depth (AOD), Satellite data, Chemical Transport Model, dust outbreak.

ABSTRACT:

Satellite data provide high spatial coverage and characterization of atmospheric components for vertical column. Additionally, the use of air pollution modelling in combination with satellite data opens the challenging perspective to analyse the contribution of different pollution sources and transport processes. The main objective of this work is to study the AOD over Portugal using satellite observations in combination with air pollution modelling. For this purpose, satellite data provided by Spinning Enhanced Visible and Infra-Red Imager (SEVIRI) on-board the geostationary Meteosat-9 satellite on AOD at 550 nm and modelling results from the Chemical Transport Model (CAMx - Comprehensive Air quality Model) were analysed. The study period was May 2011 and the aim was to analyse the spatial variations of AOD over Portugal. In this study, a multi-temporal technique to retrieve AOD over land from SEVIRI was used. The proposed method takes advantage of SEVIRI's high temporal resolution of 15 minutes and high spatial resolution.

CAMx provides the size distribution of each aerosol constituent among a number of fixed size sections. For post processing, CAMx output species per size bin have been grouped into total particulate sulphate (PSO4), total primary and secondary organic aerosols (POA + SOA), total primary elemental carbon (PEC) and primary inert material per size bin (CRST_1 to CRST_4) to be used in AOD quantification. The AOD was calculated by integration of aerosol extinction coefficient (Qext) on the vertical column.

The results were analysed in terms of temporal and spatial variations. The analysis points out that the implemented methodology provides a good spatial agreement between modelling results and satellite observation for dust outbreak studied (10th -17th of May 2011). A correlation coefficient of r=0.79 was found between the two datasets. This work provides relevant background to start the integration of these two different types of the data in order to improve air pollution assessment.

1. INTRODUCTION

Over the last decade, air pollution has become a major problem in Portugal due to high concentration of particulate matter (PM) in the atmosphere, being the exceedance of daily limit values one of the main issues for air pollution management. Therefore, better characterisation of the emission sources and understanding the atmospheric processes involved in the aerosol formation, transport and deposition are of prime concern. Atmospheric aerosols are associated with various environmental impacts from local to global scales. Thus, aerosols cause detrimental health effects in humans and an increase of fine particles concentration is associated with rising morbidity and mortality (e.g. Pope and Dockery, 2006; WHO, 2006a,b). At global scale, atmospheric aerosols have direct and indirect effects on the climate system and affect both temperature and precipitation patterns on the earth's surface (IPCC, 2007).

The concentrations and compositions of aerosols vary strongly in space and time (Dentener et al., 2006; Kaufman and Koren, 2006; Tsigaridis et al., 2006; van der Werf et al., 2006; Koch et al., 2007) because the residence time of particles in the atmosphere is only in the order of hours to weeks, depending mainly on the particle size and meteorological conditions. Therefore, adequate techniques should be identified to characterise these variations and to improve our understanding on the pollution sources and their possible effects. In this context, the use of satellite data in combination with air pollution modelling opens a challenging perspective. An increasing interest to use satellite observations in

air pollution modelling is mainly related with two important properties of the data in comparison with surface measurements: more complete spatial coverage and characterization of atmospheric components for vertical column (Tchepel et al., 2013; Vijayaraghavan et al., 2008; Engel-Cox et al., 2004). Satellite data may be used to evaluate, initialize, constrain, and improve the performance of air pollution models (Vijayaraghavan et al., 2008). On the other side, chemical transport models provide essential information on aerosol composition, size distribution and vertical profiles that may be used to improve satellite aerosol retrievals (Hu et al., 2009; Randall, 2008).

Long-range transport of atmospheric pollution, including mineral dust from natural sources, is one of the research topics where an integration of the satellite data with air pollution modelling may provide promising results. In Europe, and particularly in Mediterranean countries, desert dust particles transported from arid and semi-arid regions of North Africa have a strong impact on air quality (Monteiro et al., 2015; Basart et al., 2012; Pay et al., 2012; Querol et al., 2009, 2004; Rodríguez et al., 2001). It has been estimated that this natural contribution to PM may range from 5% to 50% in different European Countries (Marelli, 2007). Therefore, the development of a harmonized methodology that contributes to a better understanding of the natural aerosol burden is an important issue.

In this study, we present a combined analysis of satellite data and air pollution modelling to assess the contribution of PM to the air pollution levels in Portugal. The Comprehensive Air Quality

---

Model (CAMx) model was used to provide 3D fields on aerosol concentration, size distribution and chemical composition. The modelling results were evaluated against ground-based observations and analysed in combination with aerosol satellite observations provided by Spinning Enhanced Visible and Infra-Red Imager (SEVIRI).

2. METHODOLOGY

2.1 Modelling approach

To investigate the spatial variations of aerosol optical depth (AOD) over Portugal, May 2011 was chosen as a study period for this study. During this period Portugal was influenced by African dust outbreaks (Monteiro et al., 2015). The air quality modelling system WRF-CAMx, constituted by the Weather Research & Forecasting (WRF) model (Skamarock et al., 2008) and the CAMx model (ENVIRON, 2013) weas considered as a suitable tool for the purpose of this study. The input/output structure of WRF-CAMx is presented in Figure 1. CAMx is a 3D chemistry-transport Eulerian photochemical model that allows for an integrated assessment of gaseous and particulate air pollution over many scales, ranging from sub-urban to continental. CAMx is well-known and has been extensively applied for Portugal and worldwide (Tchepel et al., 2013; Ferreira et al., 2012, 2010; Huang et al., 2010; Borrego et al., 2008).

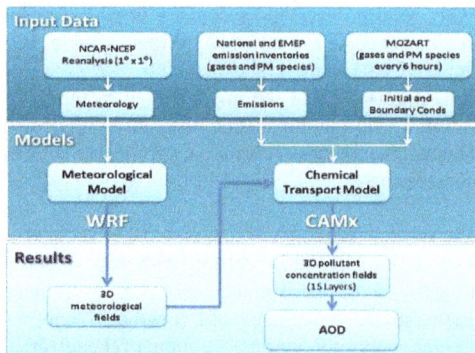

Figure 1. Air quality modelling system – input/output structure

The WRF-CAMx air quality modelling system has been applied for May 2011 to simulate 3D pollutant concentration fields. The modelling setup included 2 nesting domains covering Europe (D1) and Portugal (D2) with 27 and 9 km horizontal resolution, respectively (Figure 2), both with about 15 km vertical column (non-regularly subdivided on 15 levels considering higher details near ground).

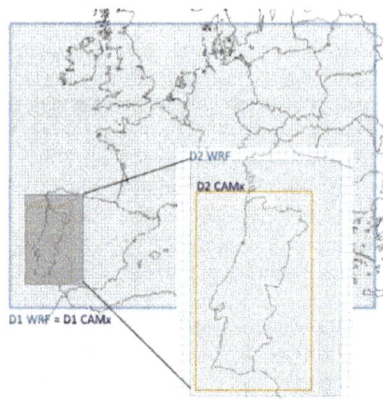

Figure 2. Modelling domains for WRF and CAMx.

For this WRF-CAMx application, initial and boundary conditions for Europe were taken from the Model for OZone and Related chemical Tracers (MOZART), an offline global chemical transport model (Emmons et al., 2010). MOZART outputs at every 6 hours were downloaded for May 2011 (http://www.acd.ucar.edu/wrf-chem/mozart.shtml), at 1.9°x2.5° horizontal resolution and 56 vertical levels. A pre-processing tool allowed for the conversion of MOZART gaseous and aerosol species into CAMx species according to the chemical mechanism in use – CB05. Emission inputs were prepared for the two simulation domains. The EMEP-EU27 gridded emissions (http://webdab1.umweltbundesamt.at/scaled_country_year.html) by SNAP sector available for the pollutants CO, NH3, NMVOC, SOx, PM2.5 and PMcoarse, were disaggregated by area to EU grid. This emission inventory was subject to a comparative analysis with other emission inventories available for Europe (Ferreira et al., 2013). Despite the fact that it is not the highest resolution inventory, it is suitable for the purpose considering that the EU simulation is only used to get initial and boundary conditions for the domain of interest. For the Portuguese domain, the national emission inventory (http://www.apambiente.pt/index.php?ref=17&subref=150) developed for regulation purposes were used. Emissions of CO, NOx, NH3, NMVOC, SO2, PM10 and PM2.5 from anthropogenic sources are available by municipality for the whole territory of Portugal, and were disaggregated to the 9x9 km2 grid cell domain. Biogenic emissions were provided as well. The aerosol chemistry module in CAMx performs the following three processes: 1) Aqueous sulphate and nitrate formation in resolved cloudwater using the RADM aqueous chemistry algorithm (Chang et al., 1987); 2) partitioning of condensable organic gases to secondary organic aerosols to form a condensed "organic solution phase" using a semi-volatile, equilibrium scheme called SOAP (Strader et al., 1999); and 3) partitioning of inorganic aerosol constituents between the gas and aerosol phases using the ISORROPIA thermodynamic module (Nenes et al., 1998, 1999).

For post processing, CAMx output species per size bin have been grouped into total particulate sulphate (PSO4), total primary and secondary organic aerosols (POA + SOA), total primary elemental carbon (PEC) and primary inert material per size bin (CRST_1 to CRST_4: 0.1-1.0µm, 1.0 -2.5 µm, 2.5 -5.0 µm, 5.0 - 10.0 µm) to be used in AOD quantification and compared with the aerosol data provided by the SEVIRI satellite sensor.

The comparison of modelled aerosol load with satellite retrieval is done in terms of AOD which is a measure of the attenuation of the incoming solar radiation by particle scattering and absorption. AOD is calculated by integration the aerosol extinction coefficient (Qext) over the vertical column from the surface level to the top of the modelling domain (Eq. (1)):

$$AOD\ (\lambda) = \int_{0}^{z_{max}} Q_{ext}\ (\lambda, z),\, dz \qquad \text{(Eq. 1)}$$

Both AOD and Qext depend on the wavelength (λ) and are related with particle chemical composition, size distribution and shape (Martin et al., 2003; Tegen and Lacis, 1996). These data are used in combination with aerosol column mass loading provided by the modelling system to estimate AOD at 550 nm.

2.2 SEVIRI aerosol retrievals

The MSG SEVIRI sensor measures the reflected and emitted electromagnetic radiation of the Earth's atmosphere and surface utilizing 11 spectral channels between 0.6 μm and 14 μm and one broad band visible channel with higher spatial resolution. MSG is in a geostationary orbit enabling SEVIRI to capture the entire observed disc every 15 minutes. To estimate AOD from the satellite data, the channel centred at 0.6 μm with a spatial resolution of 3 km at the sub-satellite point (approximately 5 km for central Europe) is used. A detailed description of the retrieval algorithm can be found in Popp et al. (2007).

A short summary of the algorithm is given in the following lines. To retrieve AOD from SEVIRI, the top-of-atmosphere (TOA) signal (converted to reflectance units) measured by the satellite sensor needs to be decomposed into the Earth surface's and atmospheric contribution. First, the surface reflectance for every pixel is estimated selecting the lowest observed reflectance (corrected for ozone, water vapor, and background aerosol concentration) from a temporal window of 15 days. A fixed aerosol model (continental aerosol type, single scattering albedo ω0=0.89 at λ=0.55 μm) and meteorological data (ozone and water vapour concentration) from the European Centre for Medium Range Wether Forecasts (ECMW) operational analysis are then fed into a radiative transfer model to invert the AOD from the (estimated) surface and (measured) TOA reflectance. The Simplified Model for Atmospheric Correction (SMAC, Rahman and Dedieu, 1994) is used for the radiative transfer simulations which is a parameterized version of 6S (Vermote et al., 1997). As a last step, a spatial averaging filter (moving 5×5 pixel box) is applied to the retrieved AOD to reduce noise such that, finally, each pixel represents the AOD of an area of approximately 25×25 km^2.

To compare the AOD from SEVIRI with the modelling results, the spatial join tool from a geographic information system was used to combine both grids. Then, we selected the maximum value in each pixel for each hour. The result of this methodology is a grid with SEVIRI data adequate to compare with CAMx results.

3. RESULTS AND DISCUSSION

In this section, we present the results from the analysis and comparison of the air quality model outputs, SEVIRIand AERONET observations in order to verify their agreement. During May 2011, over mainland Portugal, two dust outbreaks from North of Africa were identified by Monteiro et al. (2015). One of them occurred during several days (10-17 May 201) was selected to investigate the long-range transport of mineral dust.

3.1 SEVIRI observations

The AOD data provided by SERIVI over mainland Portugal were analysed in terms of average of the daily values, in order to assess its spatial and temporal distribution. SERIVI obtains observations with a high temporal resolution of 15 minutes and a spatial resolution of approximately 9 km^2 (Figure 3a). In the first step, it was necessary get a single SEVERI observation per hour. Therefore, the maximum value for each pixel and each hour was selected. The last step was to apply the spatial join tool from the geographic information system to combine both grids. The result is a grid with a maximum value for each pixel and each hour with SEVIRI data adequate to compare with CAMx results (Figure 3b).

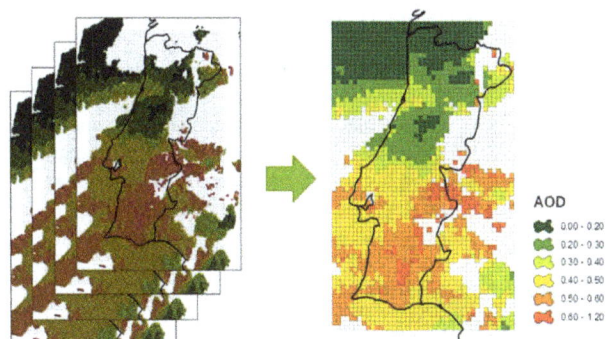

Figure 3. SEVIRI data for 14 May 2011 at 14 UTC: a) original pixel size (3x3 km^2) and b) pixel size of the modelling results (9x9 km^2).

3.2 SEVIRI observations and AERONET data

SEVIRI AOD retrievals were compared with Aerosol Robotic Network (AERONET) level 2.0 AOD measurements from 4 stations (Figure 4a) in the modelling domain for May 2011 (http://aeronet.gsfc.nasa.gov/)(Figure 2). For that purpose, AERONET observations were averaged to the 15 min resolution of SEVIRI observations. For both data, SEVIRI and AERONET, the daily mean AOD value was calculated. Figure 4b shows the scatterplot of daily mean AOD values provided by SEVIRI and AERONET for all stations located within the modelling domain at May 2011.

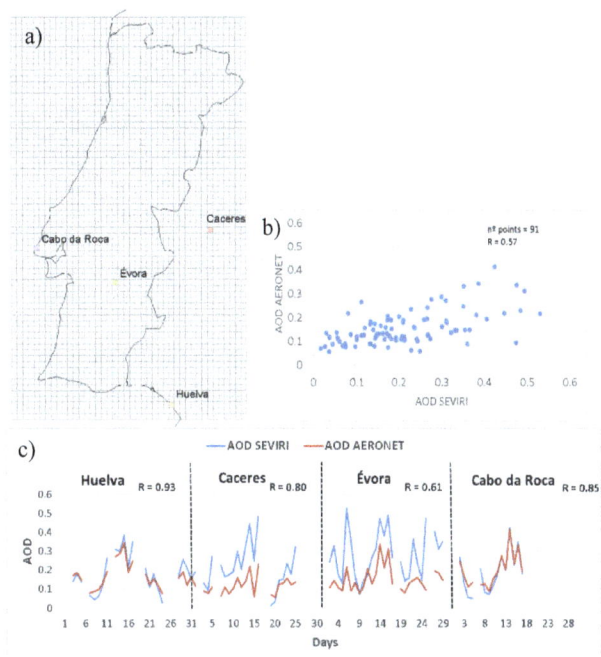

Figure 4. a) Localition of AERONET Stations, b) Scatterplot of daily mean AOD values from AERONET versus SEVIRI and c) trendlines of the daily mean AOD value from AERONET and SEVIRI for May 2011.

SEVIRI AODs agree well with AERONET observations with a correlation coefficient value of 0.57 (Figure 4b). The trendlines of the daily mean AOD value from AERONET and SEVIRI for all stations at May 2011 are presented in Figure 4c. The lowest correlation coefficient (0.61) was found for Évora station, located in the centre of the domain.

3.3 Air quality modelling results

Figure 5 presents a particulate matter mass field for Portugal, for a specific hour of day as an example. These values result from the sum of organic carbon, black carbon, sulphate and mineral dust from all the vertical levels. As one can see, the southern part of the domain presents higher particulate matter (PM) mass than the other regions of the study area influenced mainly by mineral dust.

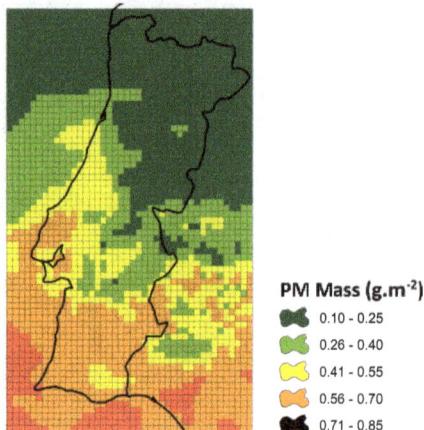

Figure 5. PM mass (g.m^{-2}) are shown for 14th May 2011 at 14 UTC.

The analysis of the contribution of each PM component by vertical layers is important to better understand the atmospheric processes that influence aerosol loading and to identify possible long-range transport of mineral dust. For this propose, the 3D model outputs were analysed and an example for 2 random points is presented (Figure 6a), one in the north and another in the south of the domain. The plots (Figure 6b and 6c) display the vertical profile of the concentrations obtained for each PM component. At both points, the highest PM concentration was simulated at 2000 meters. The results in the South point (Figure 6b) showed higher PM concentration values in all layers comparing to the North point (Figure 6c). The component with largest contribution is the dust bin 2 (1.0-2.5μm).

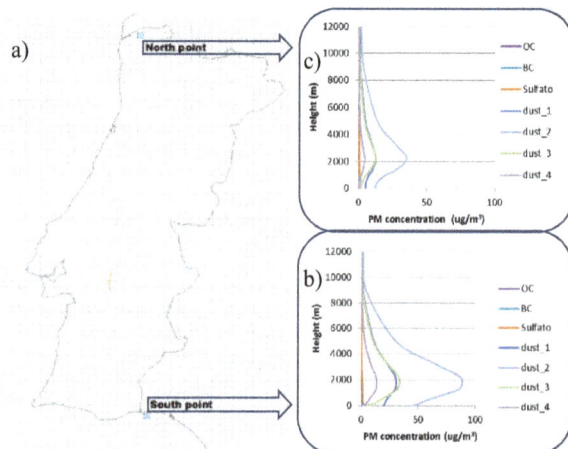

Figure 6. Vertical distribution of PM concentration: a) localization of the two points, b) results for South point and c) results for North point.

3.4 Air quality model results and SEVIRI observations

A spatial analysis of the monthly results obtained from modelling application and their comparison with remotely sensed aerosol data are presented in Figure 7 a) and b) respectively. The monthly results were obtained by computing the average of the daily AOD values between 6 -17 UTC (diurnal time). The southern part of the domain presents higher concentrations, showing a clear influence of the mineral dust.

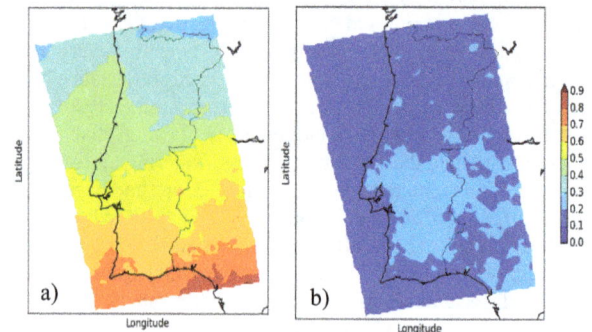

Figure 7. Monthly mean Aerosol Optical Depth (AOD) for May 2011. a) Spatial distribution obtained with CAMx and b) Spatial distribution of SEVIRI data.

The scatter plot of the modelling results (Figure 7a) versus satellite data (Figure 7b) is presented in Figure 8 for each cell for May 2011. The modelling data are in agreement with the observations showing a correlation coefficient of 0.38 (Figure 8). Moreover, simulated and observed values exhibit different spatial patterns and distinct magnitudes (Figures 7 and 8).

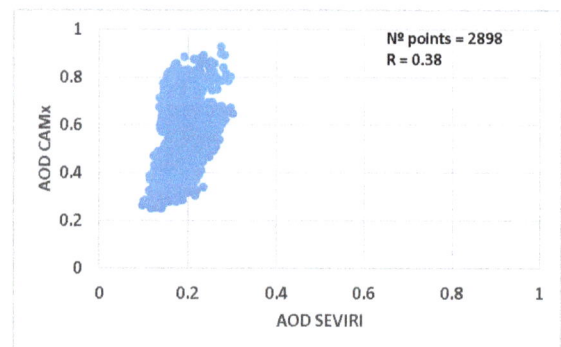

Figure 8. Scatterplot of monthly mean AOD for each grid cell from CAMx and SEVIRI for May 2011.

3.5 Analysis of the dust outbreak

During May 2011, over mainland Portugal, two dust outbreaks from North of Africa were identified by Monteiro et al. (2015). One of them occurred during several days (10-17 May 2011) and was the biggest dust outbreak during 2011. This dust outbreak was selected to analyse the spatial variation of daily mean AOD value from CAMx (Figure 9a) and SEVIRI (Figure 9b).

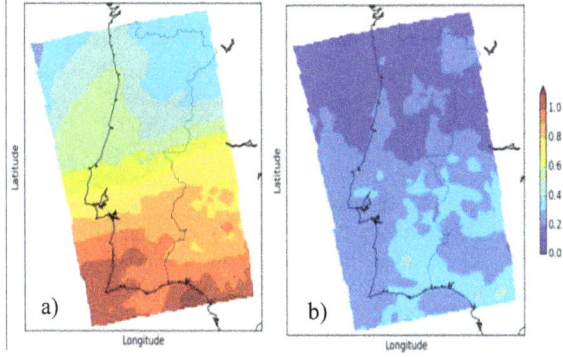

Figure 9. Mean Aerosol Optical Depth (AOD) shown for 10th-17th May 2011. a) Spatial distribution obtained from CAMx data and b) Spatial distribution of SEVIRI data.

The scatter plot of the modelling results versus satellite data is presented in Figure 10 for each cell, for the dust outbreak period. The modelling data are in agreement with the observations showing a correlation of r=0.79 between the two datasets for the studied period (10th-17th of May 2011) (Figure 10).

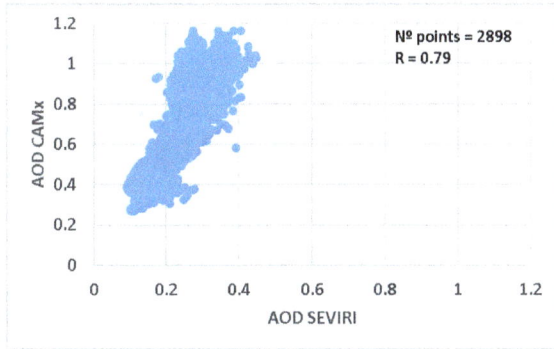

Figure 10. Scatterplot of mean AOD spatial distribution between CAMx and SEVIRI for 10th -17th of May 2011.

CAMx results of mean AOD for 10th -17th of May 2011 were evaluated against SEVIRI observations with standard statistical techniques to determine mean bias (MB) (Eq. 2) and root mean square error (RMSE) (Eq. 3):

$$MB = \frac{1}{N}\sum_1^N(CAMx - SEVIRI) \qquad \text{(Eq. 2)}$$

$$RMSE = \left[\frac{\sum_1^N(CAMx-SEVIRI)^2}{N}\right]^{1/2} \qquad \text{(Eq. 3)}$$

For each daily mean spatial plot, statistics were generated if the number of grid cells with successfully retrieved SEVIRI AOD values was greater than 1500 for each day over all study domain (total grid cells = 2898).

Figure 11 shows the mean bias and root-mean square error derived using the data at each grid cell over the 8-day time period. It can be observed that, in the south part of the domain, MB and RMSE are higher, reaching values between 0.5 and 0.9 this could be related to the fact that CAMx simulated higher particulate matter concentrations, mainly influenced by mineral dust (Figure 6).

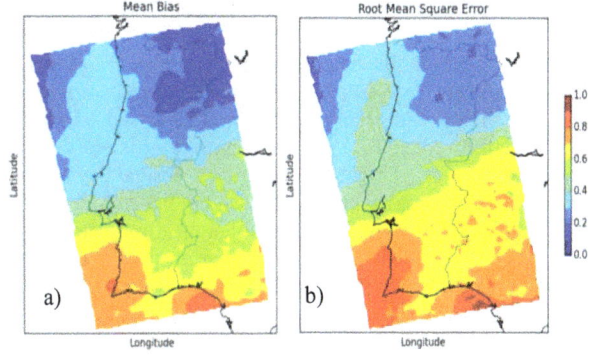

Figure 11. Statistical analysis between CAMx results and SEVIRI observations for 10th -17th of May 2011. a) Mean bias and b) Root mean square error.

Spatial analysis of AOD over three regions of the study domain is given in Figure 12.

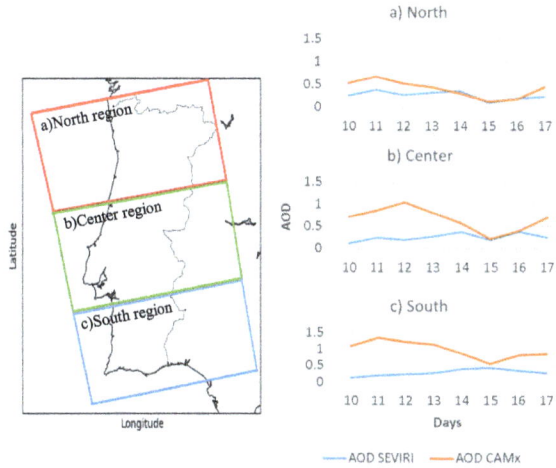

Figure 12. Daily mean AOD values obtained from CAMx results and SEVIRI observation for each region a) North (red box), b) Center (green box) and c) South (blue box) for the period 10th -17th May 2011.

The analysis of AOD over the regions shows significant spatial variations, more evident in CAMx results. However, AOD over the south region is observed to be high compared to other regions. There is a better agreement between SEVIRI and CAMx in the North region probably because of the less important contribution of mineral dust in this region. Moreover, the higher difference between CAMx and SEVIRI averages in all regions was identified during the first four days of the study period.

4. CONCLUSION

Atmospheric aerosols play an important role in the energy budget of the Earth climate system by interacting with solar and terrestrial radiation. Therefore, the development of a harmonized methodology that contributes to a better understanding of the aerosol burden is an important issue to determine their impacts on the global climate and human health.

The chemical transport model CAMx was applied to characterise the 3D distribution of aerosols over Portugal. Modelling results were analysed in combination with SEVIRI observations in terms of AOD at 550 nm.

The results show that the implemented methodology provides a reasonable agreement between the modelling outputs and satellite observations. For the selected and analysed dust outbreak (10th-17th May 2011) a correlation coefficient of r=0.79 was found between the two datasets. Spatially, the differences are bigger over the region where mineral dust exhibited higher concentration, with mean bias and root mean square error values between 0.5 and 0.9. This paper presents relevant background to start the integration of these two different types of the data in order to improve air pollution assessment.

ACKNOWLEDGEMENTS

The authors acknowledge the financial support of the FCT under the projects CLICURB (EXCL/AAG-MAA/0383/2012) and MAPLIA (PTDC/AAG-MAA/4077/2012), the Ph.D grants of A. Fernandes (SFRH/BD/86307/2012) and the post-doc grants of J. Ferreira (SFRH/BPD/40620/2007).

REFERENCES

Basart, S., Pay, M.T., Jorba, O., Pérez, C., Jiménez-Guerrero, P., Schulz, M., Baldasano, J.M., 2012. Aerosol in the CALIOPE air quality modelling system: validation and analysis of PM levels, optical depths and chemical composition over Europe. Atmospheric Chemistry and Physics 12, 3363e3392. http://dx.doi.org/10.5194/acp-12-3363-2012.

Borrego, C., Lopes, M., Valente, J., Tchepel, O., Miranda, A.I. and Ferreira, J., 2008. The role of PM10 in air quality and exposure in urban areas. In Air Pollution 2008 Conference, 22-24 Setembro, Skyathos, Grécia. Eds. C.A. Brebbia, J.W.S. Longhurst W.I.T. Transactions on Ecology and the Environment, Vol 116, WIT Press, Southampton, UK, pp 511-520. doi:10.2495/AIR08521.

Chang, J.S., Brost, R.A., Isaksen, I.S.A., Madronich, S., Middleton, P., Stockwell, W.R., Walcek, C.J., 1987. A three-dimensional Eulerian acid deposition model: physical concepts and formulation. Journal of Geophysical Research 92, 14,681e14,700. http://dx.doi.org/10.1029/JD092iD12p14681.

Dentener, F., Kinne, S., Bond, T., Boucher, O., Cofala, J., Generoso, S., Ginoux, P., Gong, S., Hoelzemann, J.J., Ito, A., Marelli, L., Penner, J.E., Putaud, J.P., Textor, C., Schulz, M., van der Werf, G.R., Wilson, J., 2006. Emissions of primary aerosol and precursor gases in the years 2000 and 1750 prescribed data-sets for AeroCom. Atmospheric Chemistry and Physics 6, 4321e4344. http://dx.doi.org/10.5194/acp-6-4321-2006.

Emmons, L. K., Apel, E. C., Lamarque, J.-F., Hess, P. G., Avery, M., Blake, D., Brune, W., Campos, T., Crawford, J., DeCarlo, P. F., Hall, S., Heikes, B., Holloway, J., Jimenez, J. L., Knapp, D. J., Kok, G., Mena-Carrasco, M., Olson, J., O'Sullivan, D., Sachse, G., Walega, J., Weibring, P., Weinheimer, A., and Wiedinmyer, C., 2010. Impact of Mexico City emissions on regional air quality from MOZART-4 simulations, Atmos. Chem. Phys., 10, 6195-6212, doi:10.5194/acp-10-6195-2010.

Engel-Cox, J.A., Hoff, R.A., Haymet, A.D.J., 2004. Recommendations on the use of satellite remote-sensing data for urban air quality. Journal of the Air & Waste Management Association 54, 1360e1371.

ENVIRON, 2013. User's guide to the Comprehensive Air Quality model with extensions (CAMx) version 6.00 (May, 2013), http://www.camx.com

Ferreira, J., Guevara, M., Baldasano, J.M., Tchepel, O., Shaap, M., Miranda, A.I., Borrego, C., 2013. A comparative analysis of two highly spatially resolved European atmospheric emission inventories. Atmos. Environ., 75, 43-57. Doi: 10.1016/j.atmosenv.2013.03.052.

Ferreira, J., Reeves, C.E., Murphy, J.G., Garcia-Carreras, L., Parker, D.J., Oram, D.E., 2010. Isoprene emissions modelling for West Africa: MEGAN model evaluation and sensitivity analysis. Atmos. Chem. Phys., 10, 8453-8467. doi:10.5194/acp-10-8453-2010

Ferreira, J., Rodriguez, A., Monteiro, A., Miranda, A.I., Dios, M., Souto, J.A., Yarwood, G., Nopmongcol, U., Borrego, C., 2012. Air quality simulations for North America - MM5-CAMx modelling performance for main gaseous pollutants. Atmospheric Environment. 53, 212-224. doi: 10.1016/j.atmosenv.2011.10.020

Hu, R.M., Sokhi, R.S., Fisher, B.E.A., 2009. New algorithms and their application for satellite remote sensing of surface PM2.5 and aerosol absorption. Journal of Aerosol Science 40 (5), 394-402.

Huang, Q., Cheng, S.Y., Li, Y.P., Li, J.B., Chen, D.S., Wang, H.Y., 2010. An integrated MM5-CAMx modelling approach for assessing PM10 contribution from different sources in Beijing, China. Journal of Environmental Information, 15 (2), 47-61.

IPCC, 2007. Climate change 2007: the physical science basis. In: Solomon, S., Qin, D., Manning, M., Chen, Z., Marquis, M., Averyt, K.B., Tignor, M., Miller, H.L. (Eds.), Contribution of Working Group I to the Fourth Assessment Report of the Intergovernmental Panel on Climate Change. Cambridge University Press, Cambridge, United Kingdom and New York, USA. 996 p.

Kaufman, Y.J., Koren, I., 2006. Smoke and pollution aerosol effect on cloud cover. Science 313, 655e658. http://dx.doi.org/10.1126/science.1126232.

Kautzman K. E., 2014. Reflective Aerosols and the Greenhouse Effe. Global Environmental Change, Handbook of Global Environmental Pollution. Volume 1 , pp 23-30

Koch, D., Bond, T.C., Streets, D., Unger, N., van derWerf, G.R., 2007. Global impacts of aerosols from particular source regions and sectors. Journal of Geophysical Research 112, D02205. http://dx.doi.org/10.1029/2005JD007024.

Marelli L., 2007. Contribution of Natural Sources to Air Pollution Levels in the EU e a Technical Basis for the Development of Guidance for the Member States (Post Workshop Report from 'Contribution of Natural Sources to PM Levels in Europe' Workshop Organized by JRC, Ispra, October 2006. EUR 22779 EN).

COMPARISONS OF AEROSOL OPTICAL DEPTH PROVIDED BY SEVIRI SATELLITE OBSERVATIONS...   55

Martin, R.V., D.J. Jacob, R.M. Yantosca, M. Chin, and P. Ginoux, 2003 Global and regional decreases in tropospheric oxidants from photochemical effects of aerosols, J. Geophys. Res., 108, 4097, doi:10.1029/2002JD002622.

Monteiro, A., Fernandes, A.P., Gama, C., Borrego, C., Tchepel, O., 2015. Assessing the mineral dust from North Africa over Portugal region using BSC-DREAM8b model. Atmospheric Pollution Research, doi: 10.5094/APR.2015.009.

Nenes, A., Pilinis, C., Pandis, S.N., 1998. ISORROPIA: a new thermodynamic model for multiphase multicomponent inorganic aerosols. Aquatic Geochemistry 4, 123e152.

Nenes, A., Pilinis, C., Pandis, S.N., 1999. Continued development and testing of a new thermodynamic aerosol module for urban and regional air quality models. Atmospheric Environment 33, 1553e1560.

Pay, M.T., Jiménez-Guerrero, P., Jorba, O., Basart, S., Querol, X., Pandolfi, M., Baldasano, J.M., 2012. Spatio-temporal variability of concentrations and speciation of particulate matter across Spain in the CALIOPE modeling 240 O. Tchepel et al. / Atmospheric Environment 64 (2013) 229e241 system. Atmospheric Environment 46, 376e396. http://dx.doi.org/10.1016/ j.atmosenv.2011.09.049

Pope, C.A., Dockery, D.W., 2006. Health effects of fine particulate air pollution: lines that connect. Journal of the Air & Waste Management Association 56, 709e742

Popp, C., Hauser, A., Foppa, N., & Wunderle, S., 2007. Remote sensing of aerosol optical depth over central Europe from MSG-SEVIRI data and accuracy assessment with ground-based AERONET measurements. Journal of Geophysical Research, 112, D24S11.

Querol, X., Alastuey, A., Ruiz, C.R., Artiñano, B., Hansson, H.C., Harrison, R.M., Buringh, E., ten Brink, H.M., Lutz, M., Bruckmann, P., Straehl, P., Schneider, J., 2004. Speciation and origin of PM10 and PM2.5 in selected European cities. Atmospheric Environment 38, 6547e6555. http://dx.doi.org/10.1016/ j.atmosenv.2004.08.037.

Querol, X., Pey, J., Pandolfi, M., Alastuey, A., Cusack, M., Pérez, N., Moreno, T., Viana, M., Mihalopoulos, N., Kallos, G., Kleanthous, S., 2009. African dust contributions to mean ambient PM10 mass-levels across the Mediterranean Basin. Atmospheric Environment 43, 4266e4277. http://dx.doi.org/10.1016/ j.atmosenv.2009.06.013.

Rahman, H. and Dedieu, G, 1994. SMAC: a simplified method for the atmospheric correction of satellite measurements in the solar spectrum, Int. J. Remote Sens., 15(1), 123–143

Randall, V.M., 2008. Satellite remote sensing of surface air quality. Atmospheric Environment 42, 7823e8784. http://dx.doi.org/10.1016/j.atmosenv.2008.07.01.

Rodriguez, S., Querol, X., Alastues, A., Kallos, G., Kakaliagou, O., 2001. Saharan dust contribution to PM10 and TSP levels in southern and eastern Spain. Atmospheric Environment 35, 2433e2447. http://dx.doi.org/10.1016/S1352-2310(00)00496-9.

Skamarock, W.C., Klemp, J.B., Dudhia, J., Gill, D.O., Barker, D.M., Huang, X.Y., Wang, W., Powers, J.G., 2008. A Description of the Advanced Research WRF Version 3. NCAR Technical Note NCAR/TN-475+STR, pp. 113, doi:10.5065/D68S4MVH.

Strader, R., Lurmann, F., Pandis, S.N., 1999. Evaluation of secondary organic aerosol formation in winter. Atmospheric Environment 33, 4849e4863. http://dx.doi.org/10.1016/S1352-2310(99)00310-6.

Tchepel O., Ferreira J., Fernandes A.P., Basart S., Baldasano J.M., Borrego C., 2013. Analysis of long-range transport of aerosols for Portugal using 3D Chemical Transport Model and satellite measurements. Atmospheric Environment. 64, 229-241.

Tegen, I., and A. A. Lacis, 1996. Modeling of particle size distribution and its influence on the radiative properties of mineral dust aerosol, Journal of Geophysical Research-Atmospheres, 101(D14), 19237- 19244.

Tsigaridis, K., Krol, M., Dentener, F.J., Balkanski, Y., Lathiere, J., Metzger, S., Hauglustaine, D.A., Kanakidou, M., 2006. Change in global aerosol composition since preindustrial times. Atmospheric Chemistry and Physics 6, 5143e5162. http://dx.doi.org/10.5194/acp-6-5143-2006.

van der Werf, G.R., Randerson, J.T., Giglio, L., Collatz, G.J., Kasibhatla, P.S., Arellano, A.F., 2006. Interannual variability in global biomass burning emissions from 1997 to 2004. Atmospheric Chemistry and Physics 6, 3423e3441. http://dx.doi.org/10.5194/acp-6-3423-2006.

Vermote, E., Tanr´e, D., and Morcrette, J.-J., 1997. Second simulation of the satellite signal in the solar spectrum, 6S: an overview, IEEE T. Geosci. Remote, 35(3), 675–686

Vijayaraghavan, K., Snell, H.E., Seigneu, C., 2008. Practical aspects of using satellite data in air quality modeling. Environmental Science & Technology 42 (22), 8187-8192.

WHO, 2006a. Air Quality Guidelines. Global Update 2005. Particulate Matter, Ozone, Nitrogen Dioxide and Sulfur Dioxide. WHO Regional Office for Europe, Copenhagen. ISBN 9289021926, 484 p.

WHO, 2006b. Health Risks of Particulate Matter from Long-range Transboundary Air Pollution. Joint WHO/Convention Task Force on the Health Aspects of Air Pollution. WHO Regional Office for Europe, Copenhagen. 99 p.

10

REGIONAL SCALE CROP MAPPING USING MULTI-TEMPORAL SATELLITE IMAGERY

N. Kussul [a, *], S. Skakun [a], A. Shelestov [a, b], M. Lavreniuk [c], B. Yailymov [a], O. Kussul [d]

[a] Space Research Institute NAS Ukraine and SSA Ukraine, Department of Space Information Technologies and Systems, Kyiv, Ukraine – nataliia.kussul@gmail.com; serhiy.skakun@gmail.com
[b] National University of Life and Environmental Sciences of Ukraine, Kyiv, Ukraine – andrii.shelestov@gmail.com
[c] Taras Shevchenko National University of Kyiv, Kyiv, Ukraine – nick_93@ukr.net
[d] National Technical University of Ukraine "Kyiv Polytechnic Institute", Kyiv, Ukraine – olgakussul@gmail.com

KEY WORDS: Crop classification, Missing data, Landsat-8, Neural networks, Ensemble, Ukraine

ABSTRACT:

One of the problems in dealing with optical images for large territories (more than 10,000 sq. km) is the presence of clouds and shadows that result in having missing values in data sets. In this paper, a new approach to classification of multi-temporal optical satellite imagery with missing data due to clouds and shadows is proposed. First, self-organizing Kohonen maps (SOMs) are used to restore missing pixel values in a time series of satellite imagery. SOMs are trained for each spectral band separately using non-missing values. Missing values are restored through a special procedure that substitutes input sample's missing components with neuron's weight coefficients. After missing data restoration, a supervised classification is performed for multi-temporal satellite images. An ensemble of neural networks, in particular multilayer perceptrons (MLPs), is proposed. Ensembling of neural networks is done by the technique of average committee, i.e. to calculate the average class probability over classifiers and select the class with the highest average posterior probability for the given input sample. The proposed approach is applied for regional scale crop classification using multi temporal Landsat-8 images for the JECAM test site in Ukraine in 2013. It is shown that ensemble of MLPs provides better performance than a single neural network in terms of overall classification accuracy, kappa coefficient, and producer's and user's accuracies for separate classes. The overall accuracy more than 85% is achieved. The obtained classification map is also validated through estimated crop areas and comparison to official statistics.

1. INTRODUCTION

Geographical location and distribution of crops at global, national and regional scale is an extremely valuable source of information for many applications. Reliable crop maps can be used for more accurate agriculture statistics estimation (Gallego et al., 2010, 2013, 2014), stratification purposes (Boryan and Zhengwei, 2013), better crop yield prediction (Becker-Reshef et al., 2010; Kogan et al., 2013a, 2013b).

Remote sensing images from space have always been an obvious and promising source of information for deriving crop maps. This is mainly due capabilities to timely acquire images and provide repeatable, continuous, human independent measurements for large territories. Yet, there are no globally available satellite-derived crop specific maps at present moment. Only coarse-resolution imagery (at least 250 m spatial resolution) has been utilized to derive global cropland extent (e.g. GlobCover, MODIS). Nevertheless, even these maps provide variable quality and reliability in capturing cropland (Fritz et al., 2013). With availability of Landsat-8 and Sentinel-2 images and their synergic exploitation (Roy et al., 2014), it becomes possible to generate crop specific maps at high spatial resolution scale for main agriculture regions.

It should be however noted that most studies on crop mapping using high and medium resolution satellite imagery (e.g. Landsat-5/7, SPOT, AWiFS) have been carried out at local scale (Conrad et al., 2010; Peña-Barragán et al., 2011; Yang et al., 2011). One of the exceptions is the creation of the Cropland Data Layer (CDL) of the US Department of Agriculture (USDA) National Agricultural Statistics Service (NASS) (Boryan et al., 2011). The CDL product provides crop maps for 47 states at 56 m spatial resolution. Another effort to create a global cropland product based on Landsat TM and ETM+ is performed by Yu et al. (2013a, 2013b). Yu et al. (2013b) created a 30 m global land cover product called FROM-GLC (Fine Resolution Observation and Monitoring of Global Land Cover). Producer's accuracy (PA) and user's accuracy (UA) for cropland class were 75.25% and 55.62%, respectively, which is below the target of 85% for agriculture applications (McNairn et al., 2009).

One of the main issues in utilizing optical imagery is the presence of clouds and shadows that introduce missing values. At local scale, it is usually possible to acquire cloud-free images in the crucial period of vegetation cycle. However, this is not the case for large territories. That is why, most of the existing studies on large scale crop mapping use high- and medium-resolution cloud-free optical images coupled with weather-independent synthetic-aperture radar (SAR) (McNairn et al., 2009) or use coarse-resolution imagery at high temporal resolution (Pittman et al., 2010; Wardlow and Egbert, 2008). In order to deal with missing data in optical satellite imagery, a number of approaches have been proposed. On of the most popular approach is compositing. Yan and Roy (2014) utilize a 30 m Web Enabled Landsat data (WELD) time series to derive cropland and agriculture crop field boundaries. The WELD is based on compositing Landsat ETM+ images with cloud cover <80% within 150 × 150 km tiles on weekly, monthly, seasonal,

---

and annual basis. However, missing value can still happen in composite products. Another popular approach is related to fill in missing data with different techniques such as multi-spectral and multi-temporal. Roy et al. (2014) utilize course resolution MODIS data for filling gaps and predicting Landsat data. Yu et al. (2013b) improve the 30 m FROM-GLC global land cover map based on Landsat TM and ETM+ imagery by adding coarse resolution MODIS imagery. It allows them to increase overall accuracy from 64.89% to 67.08%. Chen et al. (2011) propose a neighbourhood similar pixel interpolator (NSPI) for filling gaps in Landsat ETM+ SLC-off images. Latif et al. (2008) propose self-organizing Kohonen maps (SOMs) for reconstructing missing values in a time-series of low-resolution satellite imagery. It should be however noted that only few studies on filling in techniques assessed their efficiency on generating dedicated products, for example land cover maps (Chen et al., 2011). Moré et al. (2006) propose a hybrid classifier to dealing with missing data in a time-series of Landsat imagery. First, unsupervised classification for different combinations of input data is performed based on clustering algorithm IsoMM. Then, an algorithm called ClsMix is run to assign every spectral class to a thematic class through training areas defined by the user. The proposed approach achieves overall accuracy of 88.6% comparing to 67.2% obtained by the maximum likelihood (ML) classifier.

No previous studies used restored missing data from high- and medium resolution satellite imagery (such as Landsat-8) to provide crop classification and mapping for large areas. In this paper, a new approach to classification of multi-temporal Landsat-8 imagery with missing data due to clouds and shadows is presented. The approach combines different neural networks (NNs) architectures to restore missing values in a time-series of satellite imagery and provide supervised classification for crop discrimination. Results are presented for the Joint Experiment of Crop Assessment and Monitoring (JECAM) test site in Ukraine with the area of more than 28,000 km^2 (Gallego et al., 2014; Shelestov et al., 2013). The resulting classification map from Landsat-8 imagery is produced, and derived crop area estimates are compared to official statistics. To our best knowledge, the obtained crop map is one of the first ones produced at regional scale using new Landsat-8 images.

2. METHODOLOGY

2.1 Restoration of missing data in satellite images

SOM is a type of artificial neural network that is trained using unsupervised learning to produce a discretised representation of the input space of the training samples, called a map (Kohonen, 1995). The map seeks to preserve the topological properties of the input space. SOM is formed of the neurons located on a regular, usually one- or two-dimensional grid. Neurons compete with each other in order to pass to the "excited" state. The output of the map is, so called, neuron-winner or best-matching unit (BMU) whose weight vector has the greatest similarity with the input sample $\mathbf{x}$

$$i(\mathbf{x}) = \arg\min_{l=1,L}\|\mathbf{x} - \mathbf{w}_l\| \qquad (1)$$

where
$i(\mathbf{x})$ = SOM output, i.e. the number of BMU
$\mathbf{x}$ = an input vector
L = a number of neurons in the output grid
$\mathbf{w}_l$ is a vector of weight coefficients for neuron l

$\|\bullet\|$ means metric (e.g. Euclidean)

It should be noted that dimension of weight vectors $\mathbf{w}_l$ is identical to dimension of the input vectors $\mathbf{x}$. Figure 1 shows a general procedure for restoration of missing values in a time-series of data sets. The reconstruction of satellite images is performed for each spectral band separately, i.e. a separate SOM is trained for each spectral band. Pixels that have no missing values in the time-series are selected for training. Selecting the number of training pixels represents a trade-off, in particular increasing the number of training samples will lead to the increased time of SOM training while increasing the quality of restoration. Also, training data sets should be selected automatically. As such, we propose to select training samples on a regular grid of pixels. Therefore, the SOM seeks to project a large number of non-missing data to the subspace vectors in the map.

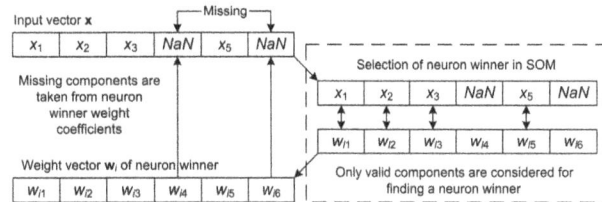

Figure 1. A procedure to restore missing values in input data using SOM

Restoration of missing values is performed in the following way (Figure 1). The multi-temporal pixel values with missing components are input to the SOM. A neuron-winner in the SOM is selected following Eq. (1). It is worth noting, however, that missing values are omitted from metric estimation when selecting BMU, i.e. only components with valid values in the input vector are used. When the BMU is selected, missing values are substituted by corresponding components of the BMU weight values. Detailed description of the algorithm and its performance evaluation is described in (Skakun and Basarab, 2014).

2.2 Committee of neural networks for image classification

Support vector machine (SVM), decision tree (DT) and RF classifiers have been probably the most popular ones for remote sensing image classification in the past years (Boryan et al., 2011; McNairn et al., 2009; Pittman et al., 2010; Shao and Lunetta, 2012; Wardlow and Egbert, 2008). Many papers report better performance of SVM, DT and RF comparing to other techniques, including MLP (McNairn et al., 2009). However, some other studies show MLP to outperform SVM and DT (Gallego et al., 2012, 2014). Though the MLP training phase might be resource and time consuming (but this is becoming less problematic with the use of high-performance computations (Kravchenko et al., 2008; Kussul et al., 2009, 2010a, 2010b, 2012; Shelestov et al., 2006; Shelestov and Kussul, 2008)), and might require experience from the user, it has several advantages over SVM and DT. In particular, MLP is fast at processing new data which can be critical to the processing of large volumes of satellite data, and can produce probabilistic outputs which can be used for indicating reliability of the map. In many cases, in our opinion, not a full potential of MLP has been explored. In particular, cost function for MLP training is usually considered square (e.g. root mean square error – RMSE) in remote sensing literature while it had been shown that cross-entropy (CE) error function provides better performance in terms of speed of training and classification accuracy (Bishop, 2006; Meier et al., 2011; Simard et al. 2003). Another potential is to explore a committee of neural networks since the committee of classifiers tends to outperform the single classifier (Zhang and Xie, 2014).

Therefore, an MLP classifier is used as a basic one in this study for classification of restored multi-temporal satellite imagery. The MLP classifier has a hyperbolic tangent activation function for neurons in the hidden layer and logistic activation function in the output layer. The CE error function is defined using the following equation (Bishop, 2006)

$$E(\mathbf{w}) = -\ln p(\mathbf{T} \mid \mathbf{w}) = -\sum_{n=1}^{N}\sum_{k=1}^{K} t_{nk}\ln y_{nk} \qquad (2)$$

where $E(\mathbf{w})$ = CE error function that depends on the neurons' weight coefficients $\mathbf{w}$

$\mathbf{T}$ = set of vectors of target outputs in the training set composed of N samples

K = number of classes

t_{nk} and y_{nk} = target and MLP outputs, respectively

In the target output for class k, all components of vector t_n are set to 0, except for the k-th component which is set to 1. The CE error $E(\mathbf{w})$ is minimized by means of the scaled conjugate gradient algorithm by varying weight coefficients $\mathbf{w}$ (Bishop, 2006).

A committee of MLPs is used to increase performance of individual classifiers. Two approaches to forming the committee are evaluated in this study. Both these approaches are modifications of the bagging technique (Bishop, 2006). Within the first approach, committee is formed using MLPs trained on different data sets. Within the second approach, committee is formed using MLPs with different parameters trained on the same training data. These approaches are quite simple, non-computation intensive and proved to be efficient for other applications (Meier et al., 2011).

Outputs from different MLPs are integrated using the technique of average committee (Meier et al., 2011). Under this technique the average class probability over classifiers is calculated, and the class with the highest average posterior probability for the given input sample is selected (Figure 3). The following equation formalizes this procedure

$$k* = \arg\max_{k=1,K} p_k^e, \quad p_i^e = \frac{1}{L}\sum_{l=1}^{L} p_i^l \qquad (3)$$

where $k*$ = class to which the committee of classifiers assigns the input sample

p_i^e = resulting posterior probability of the committee

p_i^l = posterior probability of each MLP

L = number of classifiers in the committee, and K is the number of classes

The average committee procedure has advantage over majority voting technique in two aspects: (i) it gives probabilistic output which can be used as an indicator of reliability for mapping particular pixel or area; (ii) it does not have ambiguity when two or more classes give the same number of "votes".

Classification of satellite images is performed on a per-pixel basis. Though, it was previously reported that per-field classification often outperforms per-pixel classification, it requires availability of accurate field boundaries. Unfortunately, field boundaries for most regions of Ukraine are not available at present moment, and therefore, it complicates the use of per-field classification in the operational context (McNairn et al., 2009).

3. STUDY AREA DESCRIPTION

The proposed methodology is evaluated for the JECAM test site in Ukraine. The JECAM test site in Ukraine was established in 2011 and covers the administrative region of Kyiv oblast with the geographic area of 28,100 km^2 and almost 1.0 M ha of cropland. Northern part of the region is dominated by forests and grasslands, while central and southern parts are agriculture intensive areas. Land cover classes are quite heterogeneous including croplands, forests, grassland, rivers, lakes and wetlands. The climate in the region is humid continental with approximately 709 mm of annual precipitations. Landscape is mostly flat terrain with slopes ranging from 0% to 2%; near 10% of the territory is hilly with slopes about 2-5%. The crop calendar is September-July for winter crops, and April-October for spring and summer crops. Major crop types include maize (25.1% of total cropland area in 2013), winter wheat (16.1%), soybeans (12.6%), vegetables (10.3%), sunflower (9.3%), spring barley (6.8%), winter rapeseed (4.0%), and sugar beet (1.3%). A remark should be made considering vegetables. In the region, vegetables are mainly (approximately 96%) produced by small farmers and people living in villages for self-consumption purposes (so called family gardens (Gallego et al., 2014)). The fields are mainly located next to the houses and, as a rule, are very small in size (less than 0.1 ha). This requires special techniques and the use of very high-resolution satellite data that were not available for the test site at large scale. Therefore, vegetables are not considered among major crops types within this study. Due to relatively large number of major crops and other factors there is no a typical simple crop rotation scheme in this region. Most farmers use different crop rotations depending on specialization. Fields in the region are quite large (except family gardens) with size generally ranging up to 250 ha.

4. MATERIALS DESCRIPTION

4.1 Ground measurements

Ground surveys were conducted in June 2013 to collect data on crop types and other land cover classes. European Land Use and Cover Area frame Survey (LUCAS) nomenclature is used in this study as a basis for land cover / land use types. In total, 386 polygons are collected covering the area of 22,700 ha (Table 1). Data are collected along the roads using mobile devices with built-in GPS.

4.2 Landsat-8 satellite imagery

Remote sensing images acquired by Operational Land Imager (OLI) sensor aboard Landsat-8 satellite are used for crop mapping over the study region. Landsat-8/OLI acquires images in eight spectral bands (bands 1-7, 9) at 30 m spatial resolution and in panchromatic band 8 at 15 m resolution (Roy et al., 2014). Three scenes with path/row coordinates 181/24, 181/25 and 181/26 cover the test site region. Dates of acquisition are April 16, May 02, May 18, June 19, July 05, and August 06.

N	Class	Polygons		Area	
		No.	%	ha	%
1	Artificial	6	1.6	23.0	0.1
2	Winter wheat	51	13.2	3960.8	17.4
3	Winter rapeseed	12	3.1	937.3	4.1
4	Spring crops	9	2.3	455.9	2.0
5	Maize	87	22.5	7253.3	31.9
6	Sugar beet	8	2.1	632.5	2.8
7	Sunflower	30	7.8	2549.0	11.2
8	Soybeans	60	15.5	3252.3	14.3
9	Other cereals	32	8.3	1364.0	6.0
10	Forest	17	4.4	1014.3	4.5
11	Grassland	48	12.4	747.5	3.3
12	Bare land	10	2.6	67.2	0.3
13	Water	16	4.1	448.3	2.0
	Total	386	100	22705.3	100

Table 1. Number of polygons and total area of crops and land cover types collected during the ground survey

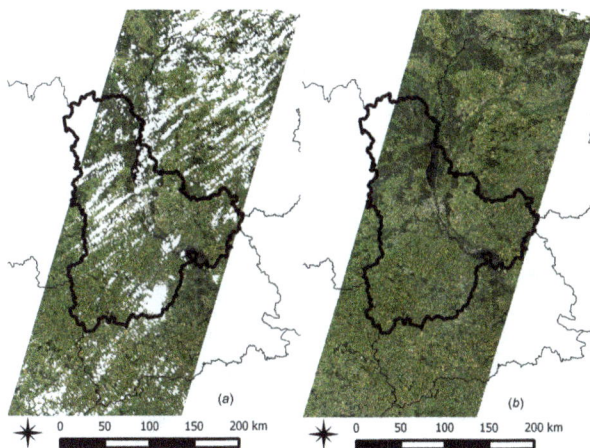

Figure 2. Example of restoration of missing data in Landsat-8 images acquired on the 5th of July 2013. Original image with identified clouds and shadows as missing data is show in (a). Result of restoration is shown in (b). For both images true colour composite of bands 4-3-2 is shown. SR reflectance values are scaled from 0 to 0.15

The following pre-processing steps are applied for all Landsat-8 images: (1). Conversion of digital numbers (DNs) values to the top-of-atmosphere (TOA) reflectance values using conversion coefficients in the metadata file (Roy et al., 2014). (2). Conversion from the TOA reflectance to the surface reflectance (SR) using the Simplified Model for Atmospheric Correction (SMAC) (Rahman and Dedieu, 1994). The source code for the model is acquired from http://www.cesbio.ups-tlse.fr/multitemp/?p=2956. Parameters of the atmosphere to run the model (in particular, aerosol optical depth) are acquired from the Aeronet network's station in Kyiv (geographic coordinates +50.374N and +30.497E). (3). Detection of clouds and shadows using Fmask algorithm proposed by Zhu and Woodcock (2012).

4.3 Preparation of satellite images for classification

Multi-temporal Landsat-8 images acquired in bands 2 through 7 are reconstructed using SOMs and used for classification of satellite imagery. Bands 1 and 9 are not used due to the strong atmospheric influence. Panchromatic band and thermal bands by Thermal Infrared Sensor (TIRS) are not utilized as well. Multi-temporal SR values in six spectral bands form a feature vector that is input to the classifier. Therefore, a total amount of 36 variables have been introduced in the classification. All variables are normalized to have mean 0 and standard deviation 1. Feature vectors of SR values are derived for fields collected during ground survey. All surveyed fields are randomly divided into training set (50%) to train the classifier and testing set (50%) for testing purposes. Fields are selected in such a way so there is no overlap between training and testing sets. All classification results, in particular overall accuracy (OA), kappa coefficient, user's (UA) and producer's (PA) accuracies are reported for testing set. The input features are classified into one of the 13 classes (Table 1).

5. RESULTS

5.1 Restoration of missing values in time-series of Landsat-8 images

The results of restoration show that relative root mean square error (*RRMSE*) are dependent on the number of missing data, and increase when the number of missing values increases (Skakun and Basarab, 2014). RRMSE values are dependant on the Landsat-8 spectral bands with minimum value being for Band 5 (11.4%) and maximum value being for Band 4 (19.7%). Quality of reconstruction of vegetated areas is higher than for artificial surface. The example of missing data restoration for images acquired on the 5th of July 2013 is shown in Figure 2.

5.2 Landsat-8 images classification

Three different classification schemes are compared in the study. The first scheme (Scheme 1) utilizes a single MLP classifier that is trained on all training data. For this, the number of hidden neurons in MLP is varied (from 20 to 80) in order to select the MLP classifier that yields the largest OA. The second scheme (Scheme 2) utilizes a committee of MLPs that are trained on different training data sets that are randomly divided into five disjoint subsets. For each subset, a number of MLPs are trained and the best MLP in terms of OA is selected into the committee. Therefore, the committee is composed of five MLP classifiers. The third scheme (Scheme 3) utilizes a committee of seven MLPs that are trained on all training data and have different number of hidden neurons, in particular 20, 30, 40, 50, 60, 70, and 80. The obtained classification metrics, in particular OA, Kappa, PA and UA, are summarized in Table 2. The use of multi-temporal Landsat-8 imagery and a committee of MLP classifiers allow us to achieve overall accuracy of slightly over 85% which is considered as target accuracy for agriculture applications (McNairn et al., 2009). The use of committee of MLP classifiers comparing to the single MLP classifier is essential, and it is statistically confirmed by using z-test (Foody, 2004). In particular, z value is equal to 5.36 when comparing Scheme 3 to Scheme 1 which is larger than the threshold value of $|z|>1.96$. It means that hypothesis of no significant difference between two classifiers would be rejected at the widely used 5 percent level of significance.

		Scheme 1: Best single MLP		Scheme 2: Committee of MLPs		Scheme 3: Committee of MLPs	
	OA, %	84.60		85.11		85.32	
	Kappa	0.8144		0.8211		0.8235	
		PA, %	UA, %	PA, %	UA, %	PA, %	UA, %
1	Artificial	74.5	93.2	100.0	97.9	100.0	97.9
2	Winter wheat	95.6	90.6	96.0	91.9	95.7	91.8
3	Winter rapeseed	94.5	96.1	93.3	99.2	93.5	99.4
4	Spring crops	12.1	15.3	46.2	38.8	40.6	34.6
5	Maize	92.6	86.6	90.3	86.8	90.5	86.8
6	Sugar beet	83.0	93.7	94.4	88.0	94.9	89.6
7	Sunflower	86.1	82.1	83.6	84.2	84.1	85.4
8	Soybeans	66.6	77.1	68.8	76.6	69.7	77.1
9	Other cereals	71.8	76.9	70.2	78.1	70.9	78.0
10	Forest	96.7	91.9	96.9	91.9	96.9	92.9
11	Grassland	84.2	88.9	90.7	88.0	91.0	89.0
12	Bare land	86.7	88.8	86.7	98.5	86.7	99.0
13	Water	99.3	98.1	100.0	98.0	100.0	98.1

Table 2. Classification results of using different neural network approaches

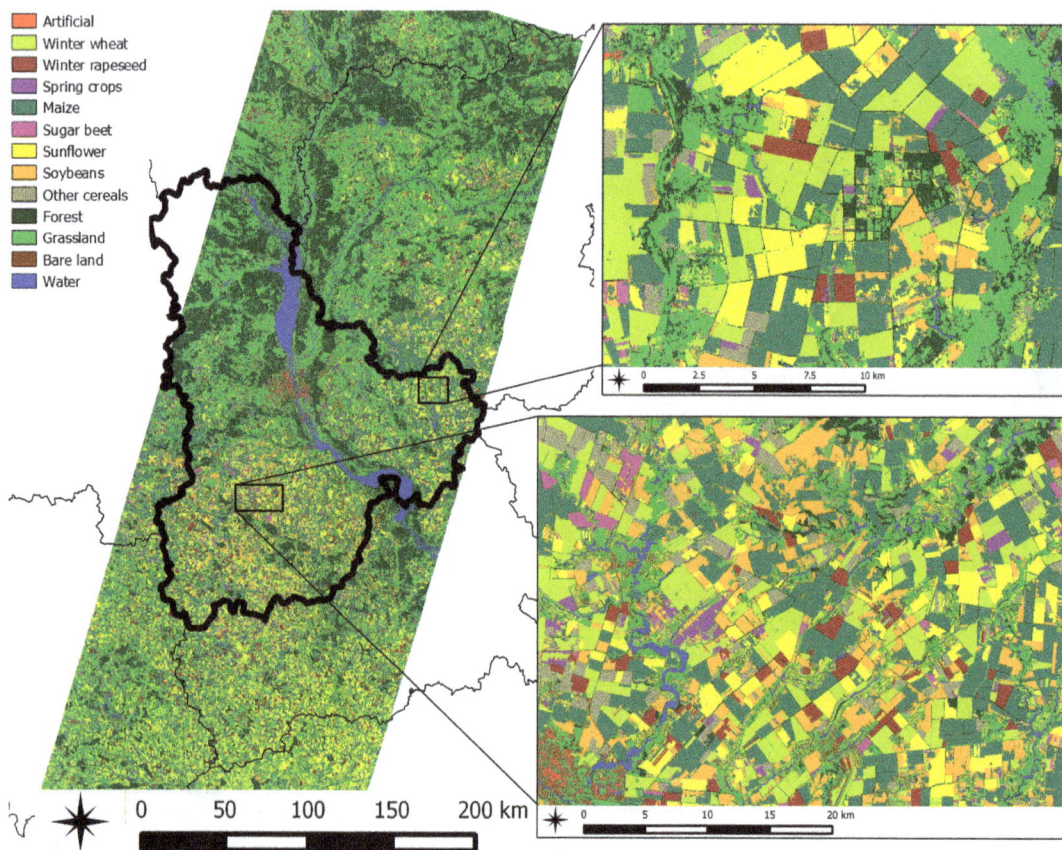

Figure 3. Final map obtained by classifying multi-temporal Landsat-8 imagery using a committee of MLP classifiers

Target accuracy of 85% (in terms of both producer's and user's accuracies) is also achieved for the following agriculture classes:

- *winter wheat* (class 2, PA=95.6%, UA=90.6%): main confusion with other cereals (class 9) and spring crops (class 4).
- *winter rapeseed* (class 3, PA=93.5, UA=99.4%): main confusion with other cereals (class 9).

- *maize* (class 5, PA=90.5%, UA=86.8%): main confusion with soybeans (class 8); in particular almost 88% of commission error and 75% of omission error for maize is due to confusion with soybeans.
- *sugar beet* (class 6, PA=94.9%, UA=89.6%): main confusion with soybeans (class 8) and maize (class 5); in particular, almost 55% of commission error is due to confusion with maize, and almost 95% of omission error is due to confusion with soybeans.

For the following agriculture classes the accuracy of 85% is not obtained:

- *spring crops* (class 4, PA=40.6%, UA=34.6%): classification using available set of satellite imagery fail to produce reasonable performance for spring crops. The main confusion of this class is with winter wheat (class 2) and other cereals (class 9). The reasons for this are as follows. When collecting ground data, it was impossible to discriminate winter crops from spring crops in the fields. Therefore, all wheat samples are assigned winter wheat class (since proportion of spring wheat is small), and all barley samples are assigned spring crops class (since proportion of winter barley is small). Unfortunately, reliable satellite data (including coarse resolution MODIS) for the autumn period of 2012 are not available due to strong cloud contamination. Confusion with other cereals can be explained by almost identical vegetation cycle of spring barley and other cereals produced in the region, namely with rye and oats. Combining spring crops and other cereals classes would improve accuracies for both these classes to PA=79.93% and UA=83.53%.
- *sunflower* (class 7, PA=84.1%, UA=85.4%): main confusion with soybeans; in particular, almost 74% of commission error and 41% of omission error is due to confusion with soybeans.
- *soybeans* (class 8, PA=69.7%, UA=77.1%): this is the least discriminated summer crop with main confusion with maize; in particular, almost 61% of commission error and 71% of omission error is due to confusion with maize.

All non-agriculture classes including forest and grassland yield PA and UA of more than 85%. The final classification map is shown in Figure 3.

5.3 Comparison to official statistics

The derived crop map over the Kyiv oblast is used to estimate crop statistics and compare it to the official one. The official statistics on crops for the region was released only in January 2014, while the crop map was produced using the remote sensing images acquired until the 6th of August 2013. Therefore, within operational context, the map could be potentially produced within August-September 2013 which is 4-5 months in advance of the official statistics report.

A simple pixel counting procedure is applied for crop area estimation. Pixel counting is known to be biased (Gallego et al., 2010), and the bias can be approximated as

$$Bias = Commission\ error - omission\ error. \quad (4)$$

Using commission and omission errors from the confusion matrix, this bias is used to correct pixel counting estimates and provide final crop area values. The results are given in Table 3.

In general, there is a good correspondence between satellite derived crop area estimates and official statistics except winter rapeseed and sugar beet. The former crop class is overestimated +28% while the latter crop is underestimated -28%.

Class no.	Class	Crop area: official statistics, x 1000, ha	Crop area: Landsat-8 derived, x 1000, ha	Relative error, %
2	Winter wheat	187.3	184.5	-1.5
3	Winter rapeseed	46.7	59.9	28.3
5	Maize	291.7	342.4	17.4
6	Sugar beet	15.5	11.2	-27.9
7	Sunflower	108.2	117.6	8.7
8	Soybeans	145.9	168.5	15.5

Table 3. Comparison of official statistics and crop area estimates derived from Landsat-8 imagery for Kyiv region

5.4 Discussion of results

The results achieved in this study show the efficiency of different neural networks architectures for classification of multi-temporal satellite imagery with missing data. The use of SOMs makes possible to restore missing data by training the neural network in an unsupervised fashion. Only data with all valid components are used for SOMs training. In such a way, the neural network projects data from training set into the subspace of neurons weight coefficients which are further used for restoration of missing values. The restoration is not perfect and introduces the error. It is found that the error is dependent on the spectral band: in particular, the relative error of restoration is 11.4% to 19.7% for Landsat-8 bands 2-7. However, the error shows small variations when varying training data size for SOM training. It should be also noted that data for SOM training and SOM size are selected automatically. It is very important when processing large volumes of data, and is one of the advantages of the proposed approach.

After all missing values are restored a supervised classification procedure is performed. For this, a committee of MLP classifiers is used. Two approaches to compose a committee are evaluated with both showing better performance over a single MLP classifier. The use of MLPs committee allows us to achieve overall accuracy of 85.32% and Kappa coefficient of 0.8235 when classifying multi-temporal Landsat-8 images over the JECAM test site in Ukraine. Accuracy of 85% is usually considered as a target for space-based agriculture applications (McNairn et al., 2009). Analysis of user's and producer's accuracies shows that some crop-specific classes achieve the target accuracy (such as winter wheat, winter rapeseed, maize and sugar beet) while others do not (spring crops, sunflower and soybeans). Spring crops class (mostly, barley) is the least discriminated class due to difficulties in discriminating winter and spring classes in the field during summer ground surveys, and confusion with other spring and summer cereals such as rye and oats. If spring crops and other cereals classes are combined together, accuracies considerably increase: from 40.6% to 79.93% of PA and from 34.6% to 83.53% of UA. Winter crops (wheat and rapeseed) yield very good performance with PA and UA both more that 91%. There is a mixed performance for summer crops. In particular, maize and sugar beet exceeded the threshold of 85% while sunflower (almost exceeded with 84.1% of PA and 85.4% of UA) and soybeans did not. Soybeans class is least discriminated summer crop far below the 85% threshold: PA=69.7%, UA=77.1%. Main confusion of soybeans is with other summer crops, namely maize, sunflower, and sugar beet. This is due to similar vegetation cycle of summer crops which requires much better temporal resolution. Another way to improve discrimination of summer crops is to utilize SAR imagery. These activities are ongoing and will be reported in future papers.

The derived crop map is used for crop area estimation for the Kyiv oblast. The estimates are compared to the official statistics and show good correspondence. Relative error for major crops is within ±28%. It should be emphasised that the latest image that is used to produce a crop map was acquired on the 6th of August 2013. Therefore, classification could be performed and crop map could be made available within August-September of the same vegetation year that is extremely important within operational context. For comparison, preliminary official statistics was only available in January 2014.

6. CONCLUSIONS

Knowledge on the area and distribution of crops is extremely important for many applications. To enable crop mapping at large scale, remote sensing images from space present the only source of reliable, continuous and human independent information. Optical images are contaminated by the presence of clouds and shadows that introduce missing values in the datasets. These missing values need to be properly processed to enable further classification of satellite imagery. This paper provides an integrated use of unsupervised and supervised neural networks in order to classify multi-temporal optical satellite images with the presence of missing data. First, SOMs are trained on available datasets with non-missing components, and are used to restore missing values. This restoration technique is universal, computationally effective and could be used for multiple scenes and satellite sensors. In the case of Landsat-8 multi-temporal images that are used in this study, it is possible to restore spectral bands with up to 19.7% relative RMSE error. Afterwards, a supervised classification is performed with the use of committee of MLP classifiers. This approach is applied for the JECAM test site in Ukraine for large area crop mapping (more than 28,000 km2). The committee of MLPs outperforms the best single MLP classifier and reaches a threshold of 85% of overall classification accuracy (OA=85.32% and Kappa 0.8235). For the following agriculture classes an 85% threshold of producer's and user's accuracies is achieved: winter wheat, winter rapeseed, maize, and sugar beet. Such crops as sunflower, soybeans, and spring crops show worse performance. The resulting crop map is used to derive crop area estimates that are compared to the official statistics. Results show good correspondence with 28% of relative error.

ACKNOWLEDGEMENTS

This work was supported by the EC under FP7 Grant "Stimulating Innovation for Global Monitoring of Agriculture and its Impact on the Environment in support of GEOGLAM" (SIGMA) [number 603719].

REFERENCES

Becker-Reshef, I., Vermote, E., Lindeman, M., Justice, C., 2010. A generalized regression-based model for forecasting winter wheat yields in Kansas and Ukraine using MODIS data. *Remote Sensing of Environment*, 114(6), pp. 1312–1323.

Bishop, C., 2006. *Pattern Recognition and Machine Learning*, Springer, New York, USA.

Boryan, C., Yang, Z., Mueller, R., Craig, M., 2011. Monitoring US agriculture: the US Department of Agriculture, National Agricultural Statistics Service, cropland datalayer program. *Geocarto International*, 26(5), pp. 341–358.

Boryan, C.G., Zhengwei Y., 2013. Deriving crop specific covariate data sets from multi-year NASS geospatial cropland data layers. In: *Proc. of 2013 IEEE International Geoscience and Remote Sensing Symposium (IGARSS)*, Melbourne, Australia, 21-26 July, pp. 4225–4228.

Chen, J., Zhu X., Vogelmann J.E., Gao F., Jin S. (2011) A simple and effective method for filling gaps in Landsat ETM+ SLC-off images. *Remote Sensing of Environment*, 115(4), pp. 1053–1064.

Conrad, C., Fritsch, S., Zeidler, J., Rücker, G., Dech, S., 2010. Per-field irrigated crop classification in arid Central Asia using SPOT and ASTER data. *Remote Sensing*, 2(4), pp. 1035–1056.

Foody, G.M., 2004. Thematic map comparison: evaluating the statistical significance of differences in classification accuracy. *Photogrammetric Engineering & Remote Sensing*, 70(5), pp. 627–633.

Fritz, S., See, L., You, L., et al., 2013. The need for improved maps of global cropland. *Eos, Transactions American Geophysical Union*, 94(3), pp. 31–32.

Gallego, F.J., Carfagna, E., Baruth, B., 2010. Accuracy objectivity and efficiency of remote sensing for agricultural statistics. In: Benedetti, R., Bee, M., Espa, G., Pier-simoni, F. (Eds.), *Agricultural Survey Methods*. John Wiley & Sons, pp. 193–211.

Gallego, J., Kravchenko, A.N., Kussul, N.N., Skakun, S.V., Shelestov, A.Y., Grypych Y.A., 2012. Efficiency assessment of different approaches to crop classification based on satellite and ground observations. *Journal of Automation and Information Sciences*, 44(5), pp. 67–80.

Gallego, F.J., Kussul, N., Skakun, S., Kravchenko, O., Shelestov, A., Kussul, O., 2014. Efficiency assessment of using satellite data for crop area estimation in Ukraine. *International Journal of Applied Earth Observation and Geoinformation*, 29, pp. 22–30.

Kogan, F., Kussul, N., Adamenko, T., Skakun, S., Kravchenko, O., Kryvobok, O., Shelestov, A., Kolotii, A., Kussul, O., Lavrenyuk, A., 2013a. Winter wheat yield forecasting in Ukraine based on Earth observation, meteorological data and biophysical models. *International Journal of Applied Earth Observation and Geoinformation*, 23, pp. 192–203.

Kogan F, Kussul N, Adamenko T, Skakun S, Kravchenko O, Krivobok O, Shelestov A, Kolotii A, Kussul O, Lavrenyuk A.

2013b. Winter wheat yield forecasting: a comparative analysis of results of regression and biophysical models. *Journal of Automation and Information Sciences*, 45(6), pp. 68–81.

Kohonen, T., 1995. *Self-organizing maps*. Series in information sciences 30, Springer, Heidelberg, Germany.

Kravchenko, A.N., Kussul, N.N., Lupian, E.A., Savorsky, V.P., Hluchy, L., Shelestov, A.Yu., 2008. Water resource quality monitoring using heterogeneous data and high-performance computations. *Cybernetics and System Analysis*, 44(4), pp. 616–624.

Kussul, N., Shelestov, A., Skakun, S., 2009. Grid and sensor web technologies for environmental monitoring. *Earth Science Informatics*, 2(1-2), pp. 37–51.

Kussul, N., Shelestov, A., Skakun, S., Kravchenko, O., Gripich, Y., Hluchy, L., Kopp, P., Lupian, E., 2010a. The Data Fusion Grid Infrastructure: Project Objectives and Achievements. *Computing and Informatics*, 29(2), pp. 319–334.

Kussul, N.N., Shelestov, A.Y., Skakun, S.V., Li, G., Kussul, O.M., 2012. The wide area grid testbed for flood monitoring using earth observation data. *IEEE Journal of Selected Topics in Applied Earth Observations and Remote Sensing*, 5(6), pp. 1746–1751.

Kussul, N.N., Sokolov, B.V., Zyelyk, Y.I., Zelentsov, V.A., Skakun, S.V., Shelestov, A.Y., 2010b. Disaster Risk Assessment Based on Heterogeneous Geospatial Information. *Journal of Automation and Information Sciences*, 42(12), pp. 32–45.

Latif, B.A., Lecerf, R., Mercier, G., Hubert-Moy, L., 2008. Preprocessing of low-resolution time series contaminated by clouds and shadows. *IEEE Transactions on Geoscience and Remote Sensing*, 46(7), pp. 2083–2096.

McNairn, H., Champagne, C., Shang, J., Holmstrom, D.A., Reichert, G., 2009. Integration of optical and Synthetic Aperture Radar (SAR) imagery for delivering operational annual crop inventories. *ISPRS Journal of Photogrammetry and Remote Sensing*, 64(5), pp. 434–449.

Meier, U., Ciresan, D.C., Gambardella, L.M., Schmidhuber, J., 2011. Better digit recognition with a committee of simple neural nets. In: *Proc. IEEE 2011 International Conference on Document Analysis and Recognition (ICDAR)*, Beijing, China, 18-21 September, pp. 1250–1254.

Moré, G., Pons, X., Serra, P., 2006. Improvements on Classification by Tolerating NoData Values - Application to a Hybrid Classifier to Discriminate Mediterranean Vegetation with a Detailed Legend Using Multitemporal Series of Images. In: *Proc. IEEE International Conference on Geoscience and Remote Sensing Symposium (IGARSS)*, Denver, CO, USA, 31 July - 4 August, pp. 192–195.

Peña-Barragán, J.M., Ngugi, M.K., Plant, R.E., Six, J., 2011. Object-based crop identification using multiple vegetation indices, textural features and crop phenology. *Remote Sensing of Environment*, 115(6), pp. 1301–1316.

Pittman, K., Hansen, M.C., Becker-Reshef, I., Potapov, P.V., Justice, C.O., 2010. Estimating Global Cropland Extent with Multi-year MODIS Data. *Remote Sensing*, 2(7), pp. 1844–1863.

Rahman, H., Dedieu, G., 1994. SMAC: a simplified method for the atmospheric correction of satellite measurements in the solar spectrum. *International Journal of Remote Sensing*, 15(1), pp. 123–143.

Roy, D.P., Wulder, M.A., Loveland, T.R., et al., 2014. Landsat-8: Science and product vision for terrestrial global change research. *Remote Sensing of Environment*, 145, pp. 154–172.

Shelestov, A.Y., Kravchenko, A.N., Skakun, S.V., Voloshin, S.V., Kussul, N.N., 2013. Geospatial information system for agricultural monitoring. *Cybernetics and System Analysis*, 49(1), pp. 124–132.

Shelestov, A.Y., Kussul, N.N., 2008. Using the fuzzy-ellipsoid method for robust estimation of the state of a grid system node. *Cybernetics and System Analysis*, 44(6), pp. 847–854.

Shelestov, A., Kussul, N., Skakun, S., 2006. Grid Technologies in Monitoring Systems Based on Satellite Data. *Journal of Automation and Information Sciences*, 38(3), pp. 69–80.

Simard, P.Y., Steinkraus, D., Platt, J.C., 2003. Best practices for convolutional neural networks applied to visual document analysis. In: *Proc. Seventh International Conference on Document Analysis and Recognition*, 3-6 Aug. 2003, pp. 958–963.

Skakun, S., Basarab, R., 2015. Reconstruction of Missing Data in Time-Series of Optical Satellite Images Using Self-Organizing Kohonen Maps. *Journal of Automation and Information Sciences*, 46(12), pp. 19–26.

Wardlow, B.D., Egbert, S.L., 2008. Large-area crop mapping using time-series MODIS 250 m NDVI data: An assessment for the US Central Great Plains. *Remote Sensing of Environment*, 112(3), pp. 1096–1116.

Yan, L., Roy, D.P., 2014. Automated crop field extraction from multi-temporal Web Enabled Landsat Data. *Remote Sensing of Environment*, 144, pp. 42–64.

Yang, C., Everitt, J.H., Murden, D., 2011. Evaluating high resolution SPOT 5 satellite imagery for crop identification. *Computers and Electronics in Agriculture*, 75(2), pp. 347–354.

Yu, L., Wang, J., Clinton, N., Xin, Q., Zhong, L., Chen, Y., Gong, P., 2013a. FROM-GC: 30 m global cropland extent derived through multisource data integration. *International Journal of Digital Earth*, 6(6), pp. 521–533.

Yu, L., Wang, J., Gong, P., 2013b. Improving 30 m global land-cover map FROM-GLC with time series MODIS and auxiliary data sets: a segmentation-based approach. *International Journal of Remote Sensing*, 34(16), pp. 5851-5867.

Zhang, C., Xie, Z., 2014. Data fusion and classifier ensemble techniques for vegetation mapping in the coastal Everglades. *Geocarto International*, 29(3), pp. 228–243.

Zhu, Z., Woodcock, C.E., 2012. Object-based cloud and cloud shadow detection in Landsat imagery. *Remote Sensing of Environment*, 118, pp. 83–94.

11

MULTI-YEAR GLOBAL LAND COVER MAPPING AT 300 M AND CHARACTERIZATION FOR CLIMATE MODELLING: ACHIEVEMENTS OF THE LAND COVER COMPONENT OF THE ESA CLIMATE CHANGE INITIATIVE

S. Bontemps [a,*], M. Boettcher [b], C. Brockmann [b], G. Kirches [b], C. Lamarche [a], J. Radoux [a], M. Santoro [c], E. Van Bogaert [a], U. Wegmüller [c], M. Herold [d], F. Achard [e], F. Ramoino [f], O. Arino [f], P. Defourny [a]

[a] Université catholique de Louvain, Earth and Life Institute, Belgium - (Sophie.Bontemps, Céline.Lamarche, Julien.Radoux, Eric.Vanbogaert, Pierre.Defourny)@uclouvain.be
[b] Brockmann Consult GmbH, Hamburg, Germany - (Martin.Boettcher, Carsten.Brockmann, Grit.Kirches)@brockmann-consult.de
[c] Gamma Remote Sensing, Switzerland – (santoro, wegmuller)@gamma-rs.ch
[d] Wageningen University, the Netherlands – martin.herold@wur.nl
[e] Joint Research Centre, Italy – frederic.achard@jrc.ec.europa.eu
[f] European Space Agency, European Space Research Institute, Italy – (Fabrizio.Ramoino, Olivier.Arino)@esa.int

KEY WORDS: Land Cover, Global, Time Series, Consistency, Essential Climate Variable

ABSTRACT:

Essential Climate Variables were listed by the Global Climate Observing System as critical information to further understand the climate system and support climate modelling. The European Space Agency launched its Climate Change Initiative in order to provide an adequate response to the set of requirements for long-term satellite-based products for climate. Within this program, the CCI Land Cover project aims at revisiting all algorithms required for the generation of global land cover products that are stable and consistent over time, while also reflecting the land surface seasonality. To this end, the land cover concept is revisited to deliver a set of three consistent global land cover products corresponding to the 1998-2002, 2003-2007 and 2008-2012 periods, along with climatological 7-day time series representing the average seasonal dynamics of the land surface over the 1998-2012 period. The full Envisat MERIS archive (2003-2012) is used as main Earth Observation dataset to derive the 300-m global land cover maps, complemented with SPOT-Vegetation time series between 1998 and 2012. Finally, a 300-m global map of open permanent water bodies is derived from the 2005-2010 archive of the Envisat Advanced SAR imagery mainly acquired in the 150m Wide Swath Mode.

1. INTRODUCTION

The demand for information on climate has never been greater than today (GCOS, 2010; IPCC, 2014). In 2004, the Global Climate Observing System (GCOS) established a first list of Essential Climate Variables (ECV), selected to be critical for a full understanding of the climate system and currently ready for global implementation on a systematic basis (CEOS, 2008).

In this context, the European Space Agency (ESA) initiated a new program of ECV global monitoring – known for convenience as the Climate Change Initiative (CCI) – which aims at providing a comprehensive and timely response to the need for long-term satellite-based products in the climate domain (ESA, 2009). The ESA-CCI program focuses, through individual projects, on 13 ECVs selected in the atmospheric, oceanic and terrestrial domains. The selection was driven by GCOS demands and the capabilities of ESA space mission. One project is dedicated to land cover. Land cover is indeed referred to as one of the most obvious and commonly used indicators for land surface and the associated human induced or naturally occurring processes, while also playing a significant role in climate forcing (Herold et al., 2009).

The overall objective of the CCI Land Cover (CCI-LC) project is to critically revisit all algorithms required for the generation of a global land cover product in the light of GCOS requirements, to be generated from data of various Earth Observation (EO) instruments and matching the needs of key users belonging to the climate modelling community.

2. CLIMATE MODELLING COMMUNITY CONSULTATION

During the six first months of the project, a user requirements analysis was conducted to derive the specifications for a new global Land Cover (LC) product to address the needs of key-users from the climate modelling community. The objective was twofold: (i) understand how LC data were used by the modellers and (ii) identify the future expectations for LC data in the context of climate and Earth system modelling. This user assessment was built upon the general guidance from the GCOS and its related panel activities and provided the next step to further derive more detailed characteristics and foundations to observe LC as an ECV.

The consultation was performed by surveys dedicated to specific user groups (Figure 1): i) a group of key users, ii) associated climate users who are involved in the CCI-LC project and are leading the development of relevant key climate models and applications and iii) the broad LC data user community represented by users of the ESA GlobCover product (Arino et al., 2008; Defourny et al., 2009). The GlobCover community was indeed identified as the key one as it counted around 8000 users at the time of making the survey and today,

---

* Corresponding author

it still counts more than 500 hits by week on the project webpage (http://due.esrin.esa.int/page_globcover.php). In addition, a detailed literature review was carried out with special attention to innovative concepts and approaches to better reflect land dynamics in the next generation climate models.

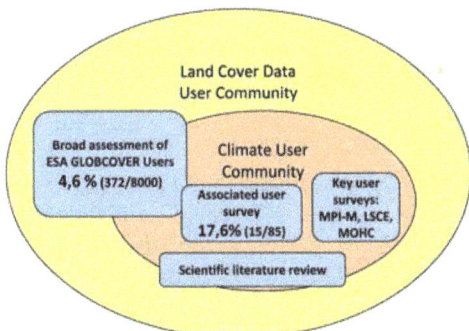

Figure 1. User consultation mechanism established to assess LC requirements. Percentages indicate the number of responses received from each user's group

This consultation showed that although the range of requirements coming from the climate modelling community is broad, there was a good match among the requirements coming from different user groups and the broader requirements derived from international organizations. Detailed user requirements assessment can be found in Herold et al. (2011). One finding of particular interest was the need for successive LC products stable over time (i.e. free from any temporary variability). In climate models, LC maps are indeed often used as a consistent basis for land surface parameterization and thus need to be stable to avoid – as much as possible – introducing inconsistencies in model inputs.

3. A NEW LAND COVER CONCEPT

The users' requirements analysis highlighted expectations for an improved LC product which would be more integrative than the current one. Indeed, there was a clear requirement for a land cover which includes both stable and dynamic components, while making the difference between LC change and natural variability.

The existing suite of global LC products doesn't meet this requirement: successive annual LC maps are contaminated by significant inter-annual variations, due to phenology and disturbances rather than to LC changes (Bontemps et al., 2009; Friedl et al., 2010). As a result, the whole LC concept was revisited with the aim of defining two distinct products to represent the stable and dynamic components of the land surface (Defourny and Bontemps, 2012).

On one hand, global LC maps should refer to the set of LC features remaining stable over time which define the LC independently of any sources of temporary or natural variability. On the other hand, a complementary product should be generated to describe the temporary or natural variability of LC features that can induce some variation in land surface over time without changing the LC in its essence. This second product could encompass different observable variables such as the green vegetation phenology, snow coverage, open water presence, burned areas occurrence, etc.

From the remote sensing point of view, the global LC maps should be derived from multi-year observation dataset to reduce the sensitivity of the classification methods to the date(s) of observation (Bontemps et al., 2012). Conversely, the natural variability of the land surface has to be considered within the perspective of a time cycle (typically a year) precisely in order to reflect the above-mentioned temporary conditions.

4. A NEW GENERATION OF GLOBAL LAND COVER PRODUCTS

Based on this innovative result, the CCI-LC project delivered in October 2014 global LC databases made of LC maps at 300m spatial resolution for three 5-year epochs – centred around 2000, 2005 and 2010 – and of land surface seasonality products. The surface reflectance (SR) time series which served as input for generating the global LC databases were also delivered as CCI-LC products. In addition, a 300-m global map of open permanent water bodies was derived from the 2005-2010 archive of the Envisat Advanced SAR (ASAR) imagery mainly acquired at 150m.

4.1 Surface reflectance time series

The main source of input EO data for the global LC maps is the full archive (2003-2012) of MERIS instrument.

The MERIS Full and Reduced resolution (FR and RR respectively - 300m and 1000m) time series are pre-processed in the framework of the project. The completed automated pre-processing chain performs the following operations (Figure 2): radiometric, geometric correction, identification of water/snow/cloud/cloud shadow/invalid pixels, atmospheric correction with aerosol retrieval as well as compositing and mosaicking. The output time series are made of temporal syntheses obtained over a 7-day compositing period (Figure 3).

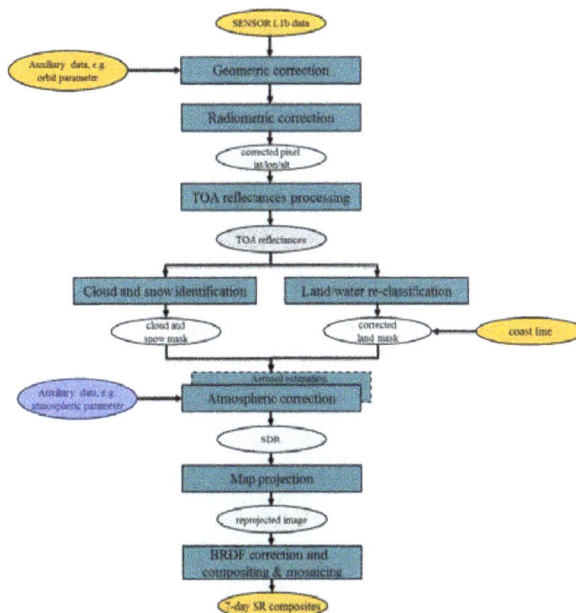

Figure 2. Schematic representation of the CCI-LC pre-processing chain (based on GlobAlbedo project, 2013)

MERIS FR and RR global time series from 2003 to 2012 are official outputs of the project. The spectral content encompasses

13 of 15 MERIS spectral channels, bands 11 and 15 being removed.

Figure 3. Example of SR 7-day composite, at 300m spatial resolution generated in the CCI-LC project

4.2   Global land cover maps

In order to meet the requirement to have successive global LC maps stable over time (see section 2), several years of EO dataset are used to generate each global LC map. This is the reason why the project delivers maps which are not related to single years but which are representative of 5-year epochs: 1998-2002, 2003-2007 and 2008-2012. Furthermore, these 3 maps are not derived independently but from a unique "baseline" LC map generated using the full MERIS archive (i.e. 10 years of data from 2003 to 2012). This "baseline" LC map is then back- and up-dated using a change detection approach based on SPOT-VGT time series. The whole workflow is presented in Figure 4.

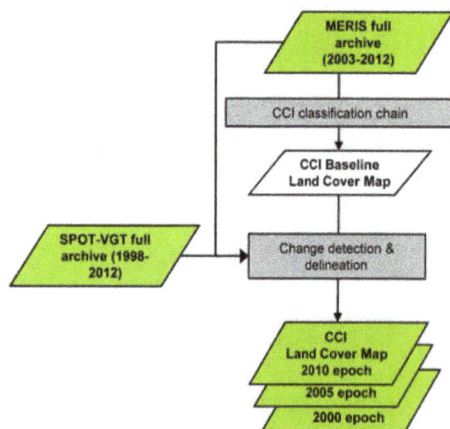

Figure 4. Classification workflow developed to generate global LC maps stable over time

The classification chain transforms the MERIS 7-day composites produced by the pre-processing module into meaningful global land cover products in 4 major processing steps. In the first and second steps, both machine learning and unsupervised classification algorithms are run using the spectral properties of MERIS composites. The 2 first steps result in two different classifications which are then merged in a third step based on objective rules. The fourth step finalizes the "baseline" LC maps through a set of post-classification editions.

The classification process is based on a-priori stratification of the world in equal-reasoning areas from an ecological and a remote sensing point of view. The classification process has been designed to run independently for each delineated equal-reasoning area.

The 3 epochs are then derived from the 10-year MERIS-based "baseline" LC map (Figure 5 and Figure 6). Up to now, only macroscopic forest cover changes have been targeted. Forest changes are identified through a specific analysis of successive annual classifications of SPOT-VGT time series (1998-2012). The changes identified at the SPOT-VGT 1km resolution are then re-mapped at the MERIS 300m resolution in the corresponding epoch (Figure 6).

The typology is made of 22 classes defined using the UN Land Cover Classification System (Di Gregorio, 2005) with the view to be as much as possible compatible with the GLC2000, GlobCover 2005 and 2009 products (Table 1). This system has been found quite compatible with the Plant Functional Types used by most climate modellers (Herold et al. 2011).

Label	Color
No Data	
Cropland, rainfed	
Cropland, irrigated or post-flooding	
Mosaic cropland (>50%) / natural vegetation (<50%)	
Mosaic natural vegetation (>50%) / cropland (<50%)	
Tree cover, broadleaved, evergreen, closed to open	
Tree cover, broadleaved, deciduous, closed to open	
Tree cover, needleleaved, evergreen, closed to open	
Tree cover, needleleaved, deciduous, closed to open	
Tree cover, mixed leaf type (broadleaved and needleleaved)	
Mosaic tree and shrub (>50%) / herbaceous (<50%)	
Mosaic herbaceous (>50%) / tree and shrub (<50%)	
Shrubland	
Grassland	
Lichens and mosses	
Sparse vegetation (tree, shrub, herbaceous) (<15%)	
Tree cover, flooded, fresh or brakish water	
Tree cover, flooded, saline water	
Shrub or herbaceous cover, flooded, fresh/saline/brakish water	
Urban areas	
Bare areas	
Water bodies	
Permanent snow and ice	

Table 1. Legend of the global LC maps, based on UN-LCCS

Among these LC classes, three are largely identified thanks to external dataset: the "tree cover, flooded, saline water" class which is based on the global mangrove atlas (Giri et al., 2011), the "water bodies" which have been inherited from the CCI-LC WB product (see section 4.4) and the "snow and ice" class which comes from the Randolf Glaciers Inventory (Arendt et al., 2014).

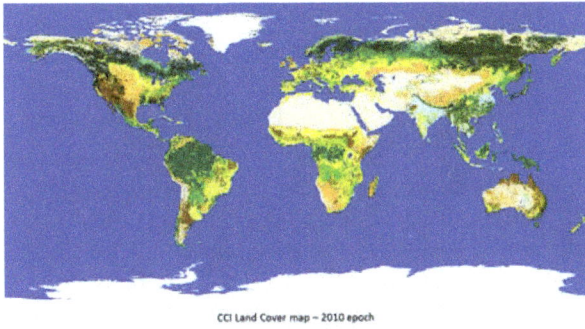

Figure 5. The global CCI-LC map from the 2010 epoch (2008-2012)

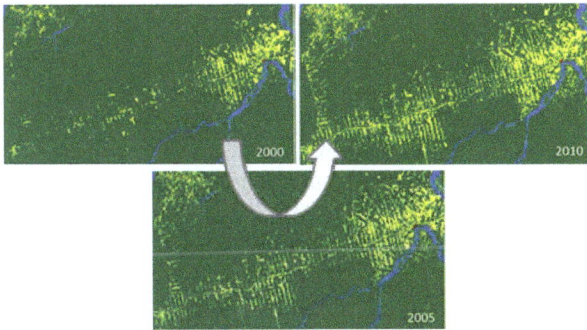

Figure 6. The CCI global land cover maps at 300m spatial resolution from the 2010, 2005 and 2000 epochs, over part of the Amazon basin

The accuracy of the 2008-2012 global LC map was assessed using the GlobCover 2009 validation dataset (Bontemps et al. 2009), resulting in a weighted-area overall accuracy of 74.1% which is slightly better than previous products. In addition, visual analysis revealed a much better delineation of landscape patterns. A more complete validation, based on a global dataset specifically collected within the CCI-LC project, is currently in progress; accuracy figures will be presented at the symposium.

4.3   Global land surface seasonality products

To characterize the typical seasonal dynamics of the land surface at the pixel level, three global climatology products are also generated: the vegetation greenness as described by the Normalized Vegetation Index (NDVI), the snow occurrence and the burned areas distribution over the 1998-2012 period (Figure 7). They are expressed as 7-day time series of the mean and standard deviation for continuous variables (NDVI) or as temporal series of occurrence probabilities for discrete variables (snow and burned areas). Particular emphasis is put on the consistency between these 3 products.

These are compiled from existing global datasets: SPOT-Vegetation daily top of canopy surface reflectance syntheses, the MODIS Direct Broadcast Monthly Burned Area Product being part of the Global Fire Emissions Database version 3 and the 8-day maximum snow extent product for the NDVI, burned areas and snow seasonality products respectively.

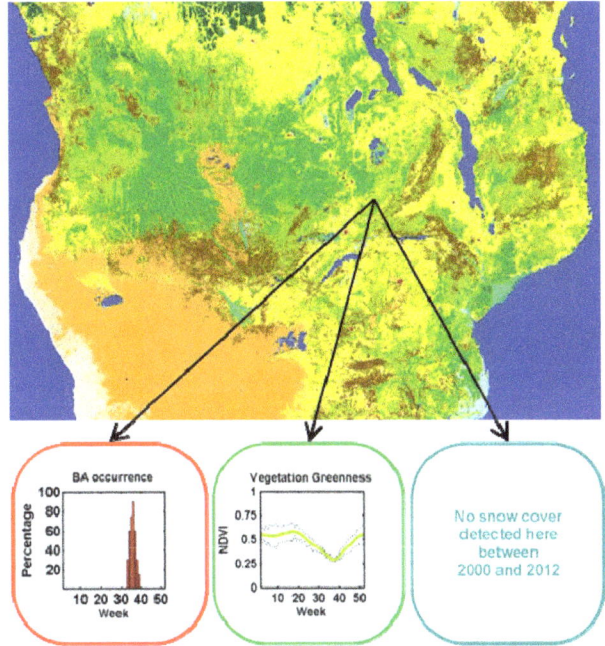

Figure 7. Climatological 7-day time series describing consistently and on a per-pixel basis the natural variability of the vegetation (NDVI), the snow cover and the burned areas

As an example, Figure 8 shows NDVI profiles extracted from 3 pixels belonging to 3 LC classes of the 2010 CCI-LC map. The variety of the dynamic of vegetation is clearly well-captured.

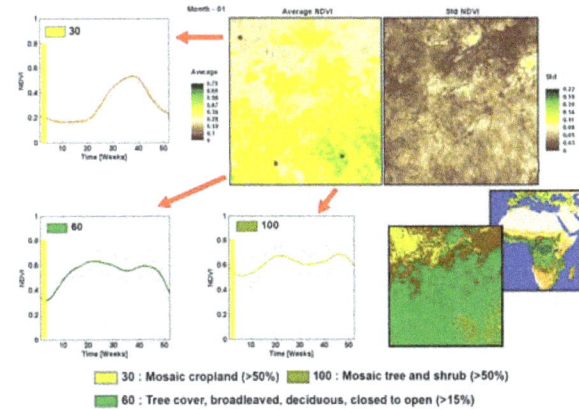

Figure 8. Detailed spatial example of NDVI climatological profiles - mean (plain line) and standard deviation (dotted line) - extracted in a region of Central Africa.

4.4   Global map of water bodies

Another key output of the project is a global SAR-based Water Bodies (WB) product, which gives the repartition of open and permanent water bodies (inland water and oceans) at 300m spatial resolution and global scale.

The product is generated using the Envisat ASAR Wide Swath Mode (WSM - 150m) dataset for the 2005-2010 period as the main source of imagery. As the coverage of WSM is insufficient in some places, imagery in the Image Mode (IMM - 75m) and Global Mode (GMM - 500m) are used in complement. The water/land classification scheme relies first on the temporal variability of the SAR backscatter and a measure of the minimum backscatter. As a result, a WB

Indicator is obtained. Refinements of the WB Indicator are then applied based on visual and inconsistency assessments. The CCI-LC WB product is finally obtained after resampling to the 300m spatial resolution of the global LC maps. Figure 9 gives an example of the thematic precision of the WB product when overlaid on optical imagery.

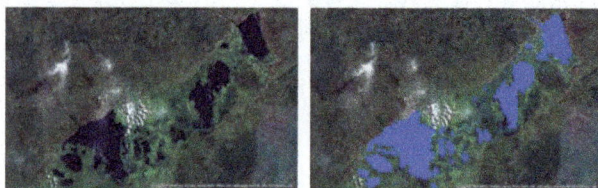

Figure 9. Detail of the global WB product, showing on left very high spatial resolution imagery from Google Earth and on right the 300m product

4.5   User tool

The project also delivers a user tool to allow users re-sampling, sub-setting and re-projecting the LC products. Indeed, the LC map and condition products are delivered at a given spatial resolution, all as global files, in a Plate-Carrée projection. However, climate models may need products associated with a coarser spatial resolution, over specific areas (e.g. for regional climate models) and/or in another projection. The developed tool allows them adjusting these three parameters in a way which is suitable to their models.

Furthermore, the tool offers the possibility to couple the aggregation with the conversion from the LC classes expressed using the UN-LCCS to user-specific Plant Functional Types.

5.   MODELLING ASSESSMENT

Three different Earth System models respectively from the Laboratoire des Sciences du Climat et l'Environnement (LSCE), the United Kingdom Met Office (MOHC) and the Max Planck Institute (MPI) for Meteorology are adjusted to use as input the new CCI-LC dataset in order to assess the possible simulation improvement against a set of benchmarks. Model experiments include offline as well as coupled carbon-climate simulations, and a dynamic vegetation simulation. For each simulation, the LC maps are converted into Plant Functional Types and merged with climate zones linking with biome specific parameters for structural and physiological traits.

The already obtained results are quite promising. For instance, an initial assessment of the global offline simulations, run by the LSCE with the WATCH-Forcing-Data-ERA-Interim climate forcing from 1979-2009, shows an improvement in modelled aboveground biomass stocks. Most recent forest inventory estimates of total woody biomass are around 363 ± 28 Pg C. With the new CCI-LC product, a 56 Pg C reduction in total biomass simulated by ORCHIDEE is found, from 688 to 632 PgC (Figure 10). Much of the reduction in biomass comes from improvements in mapping tropical land cover, where recent land-use transitions due to deforestation processes are included in the CCI-LC dataset but not in the original ORCHIDEE LC dataset.

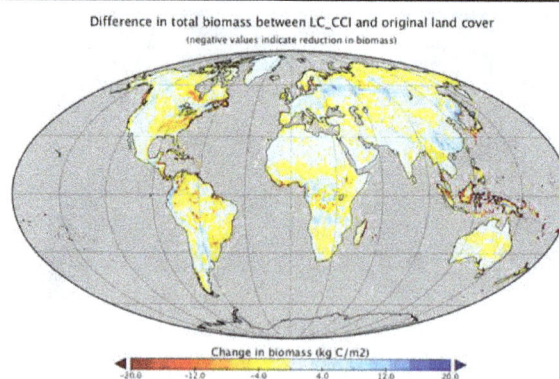

Figure 10. Difference in total biomass between CCI-LC and original LC products (negative values indicate reduction in biomass)

Yet, the results obtain by the different climate groups and simulations slightly differ according to the respective models sensitivity. Additional analysis and simulations are in progress to better understand the results and compare the models and datasets.

6.   CONCLUSIONS AND NEXT STEPS

In October 2014, the CCI-LC product made the first official release of its five key products: (i) 3 global LC maps at 300m spatial resolution corresponding to the 1998-2002, 2003-2007 and 2008-2012 epochs, (ii) 3 global land cover seasonality products describing the vegetation greenness, the snow and the burned areas occurrence along the year, (iii) a global map of open permanent WB at 300m spatial resolution, (iv) the full archive (2003-2012) of MERIS time series processed in 7-day composites and (v) a user tool for re-sampling, sub-setting, re-projecting and converting the products into climate model inputs. These products match the needs expressed by the climate research community, but are also of interest for the wider LC community. They can be freely and easily visualize and download online at: http://maps.elie.ucl.ac.be/CCI/viewer.

A second 3-year phase has started now for the project with key objectives: improving the methodologies developed for generating the CCI-LC products, extending in the past and in the future the CCI-LC database, moving to higher spatial resolution.

With respect to the second objective, it is planned to go back to the 1990s (and possibly 1980s) using AVHRR dataset, thus resulting for the first time in a 20 (or 30) year-long LC dataset. The challenge will be to remain consistent with the already existing 2000, 2005 and 2010 global LC maps. To this end, an independent classification is not foreseen.

It is also planned to generate a new global LC map over a 2015 epoch, using the successors of Envisat MERIS and SPOT-VGT, namely Sentinel-3 and PROBA-V.

The third objective will clearly be the most challenging one as it will require significant methodological adjustments and innovations. It will allow us addressing the requirement of higher spatial resolution expressed by the climate science community. Relying on the coming Sentinel-2 mission, the project will aim at demonstrating the feasibility and the performance of a continental LC map at 10-20 m spatial resolution, with the coming Sentinel-2 coverage of Africa. The contribution of Sentinel-1 to the optical mapping will finally be tested.

REFERENCES

Arendt, A., Bliss, A., Bolch, T., Cogley, J.G., Gardner, A.S., Hagen, J.-O., Hock, R., Huss, M., Kaser, G., Kienholz, G., Pfeffer, W.T., Moholdt, G., Paul, G., Radić, V. et al., 2014. Randolph Glacier Inventory – A Dataset of Global Glacier Outlines: Version 4.0. Global Land Ice Measurements from Space, Boulder Colorado, USA. Digital Media. Available at: http://www.glims.org/RGI/

Arino, O., Bicheron, P., Achard, F., Latham, J., Witt, R., and Weber, J. L., 2008. Globcover: the most detailed portrait of Earth, ESA Bulletin, 136, pp. 24-31. Available at: http://www.esa.int/esapub/bulletin/bulletin136/bul136d_arino.pdf

Bontemps, S., Defourny, P., Van Bogaert, E., Kalogirou, V. and Arino, O., 2010. GlobCover 2009 - Products Description and Validation Report. Available at: http://due.esrin.esa.int/files/GLOBCOVER2009_Validation_Report_2.2.pdf

Bontemps, S., Herold, M., Kooistra, L., van Groenestijn, A., Hartley, A., Arino, O., Moreau, I., and Defourny, P., 2012. Revisiting land cover observations to address the needs of the climate modelling community. Biogeosciences, 9, pp. 2145-2157

CEOS – Committee on Earth Observation Satellites, 2008. The Earth Observation Handbook, Climate Change Special Edition. Available at: http://www.eohandbook.com/eohb2008/

Defourny, P., Bicheron, P., Brockman, C., Bontemps, S., Van Bogaert, E., Vancutsem, C., Pekel, J.F., Huc, M., Henry, C.C., Ranera, F., Achard, F., Di Gregorio, A., Herold, M., Leroy, M. and Arino, O., 2009. The first 300 m global land cover map for 2005 using ENVISAT MERIS time series: A product of the GlobCover system. Proceedings of the 33rd International Symposium on Remote Sensing of Environment, pp. 205-208

Defourny, P. and Bontemps, S., 2012. Revisiting Land-Cover Mapping Concepts. In: Remote Sensing of Land Use and Land Cover : Principles and Applications. CRC Press - Taylor and Francis group, pp. 49-63.

Di Gregorio A., 2005. UN Land Cover Classification System (LCCS) – Classification concepts and user manual for Software version 2. Available at: http://www.glcn.org/sof_1_en.jsp

ESA – European Space Agency, 2009. ESA Climate Change Initiative description, EOP-SEP/TN/0030-09/SP, Technical Note. Available at: http://ionia1.esrin.esa.int/files/ESACCIDescription.pdf

Friedl, M. A., Sulla-Menashe, D., Tan, B., Schneider, A., Ramankutty, N., Sibley, A., and Huang, X., 2010. MODIS Collection 5 global land cover: Algorithm refinements and characterization of new datasets, Remote Sens. Environ., 114, pp. 168–182

GCOS, 2010. Implementation plan for the Global Observing System for Climate in Support of the UNFCCC, August 2010 (update), World Meteorological Organisation. Available at: http://www.wmo.int/pages/prog/gcos/Publications/gcos-138.pdf

Giri, C., Ochieng, L., Tieszen, L., Zhu, Z., Singh, A., Loveland, T., Masek, J. and Duke, N. 2011. Status and distribution of mangrove forests of the world using earth observation satellite data. Global Ecol. Biogeogr, 20, pp. 154-159

GlobAlbedo Project, 2013. GlobAlbedo Algorithm Theoretical Basis Document, Version 4.12. Available at http://www.globalbedo.org/docs/GlobAlbedo_Albedo_ATBD_V4.12.pdf

Herold, M., Woodcock, C., Wulder, M., Arino, O., Achard, F., Hansen, M., Olsson, H., Schmulllius, C., Brady, M., Di Gregorio, A., Latham, J. and Sessa, R., 2009. GTOS ECV T9: Land Cover - Assessment of the status of the development of standards for the Terrestrial Essential Climate Variables. Available at: http://www.fao.org/gtos/doc/ECVs/T09/T09.pdf

Herold, M., van Groenestijn, A., Kooistra, L., Kalogirou, V. and Arino, O., 2011. User Requirements Document, Report of the CCI Land Cover project, version 2.2 (23/02/2011)

IPCC, 2014. Climate Change 2014: Synthesis Report. Contribution of Working Groups I, II and III to the Fifth Assessment Report of the Intergovernmental Panel on Climate Change [Core Writing Team, R.K. Pachauri and L.A. Meyer (eds.)]. IPCC, Geneva, Switzerland, 151 pp. Available at: https://www.ipcc.ch/report/ar5/

12

MONITORING THE URBAN TREE COVER FOR URBAN ECOSYSTEM SERVICES – THE CASE OF LEIPZIG, GERMANY

E. Banzhaf, H. Kollai

Department Urban and Environmental Sociology, Working Group Geomatics
UFZ - Helmholtz – Centre for Environmental Research
Permoserstr. 15, D-04318 Leipzig
Ellen.banzhaf@ufz.de

Commission VIII, WG VIII/7 and WG VIII/8

KEY WORDS: Mapping Urban Tree Cover, Urban Ecosystem Functions, Urban Ecosystem Services (urban ESS), Digital Ortho Photos (DOP), Digital Surface Model (DSM), Object-based Image Analysis (OBIA), Inner Urban Differentiation

ABSTRACT:

Urban dynamics such as (extreme) growth and shrinkage bring about fundamental challenges for urban land use and related changes. In order to achieve a sustainable urban development, it is crucial to monitor urban green infrastructure at microscale level as it provides various urban ecosystem services in neighbourhoods, supporting quality of life and environmental health. We monitor urban trees by means of a multiple data set to get a detailed knowledge on its distribution and change over a decade for the entire city. We have digital orthophotos, a digital elevation model and a digital surface model. The refined knowledge on the absolute height above ground helps to differentiate tree tops. Grounded on an object-based image analysis scheme a detailed mapping of trees in an urbanized environment is processed. Results show high accuracy of tree detection and avoidance of misclassification due to shadows. The study area is the City of Leipzig, Germany. One of the leading German cities, it is home to contiguous community allotments that characterize the configuration of the city. Leipzig has one of the most well-preserved floodplain forests in Europe.

1. INTRODUCTION

Simultaneous urban shrinkage and (re)growth have consequences for changes in land use, ecosystem services and related societal impacts. Synergies and trade-offs between land use changes and the provision of urban ecosystem services (UES) as well as the consequences of these interactions for different forms of urban land uses (housing areas, public green spaces, tree coverage etc.) and socio-demographic information need detailed investigations. In order to achieve sustainable urban land use and an appropriate provision of ecosystem services, the monitoring of urban vegetation must be reflected against the background of inner urban differentiation.

Scientific knowledge needs to be produced on land use changes that also include urban tree cover. Urban trees do not only serve as woodland for recreational purpose and nature conservation, they also provide shade in parks and on other green spaces, and thus mitigate urban heat island. Beyond, trees along streets facilitate as a carbon sequestration pool and improve air quality. To estimate the quality of life in different neighbourhoods, tree cover densities help to explain urban areas and their configuration. Beyond, different kinds of vegetation help to explain how the urban fabric is formally organized, how this formal spatial organization characterizes urban neighborhoods in terms of socio-spatial differentiation, and how and which vegetation can contribute to the city in terms of biodiversity. Climate change and urban induced developments from urbanization force science and planners to continuously update

their monitoring of the natural environment and to evaluate natural environment.

In Germany, a new kind of administration has been launched under the term of "Doppik" that assigns monetary value to each public property which then provides a nature-based economic mapping of communal assets. As a conclusion, the awareness of trees, bushes and other natural communal assets has been rising, and ecosystem services are being implemented in urban planning.

At an early stage, ecosystem services were explained as "… the conditions and processes through which natural∗ecosystems and the species that make them up sustain and fulfil human life" (Daily, 1997: p. 3), followed by the Millennium Ecosystem Assessment (2005) that characterized their four different scopes: provisioning services, regulating services, supporting services and cultural services. Bolund and Hunhammar (1999) emphazise the merit of ecosystem services for urban areas, especially when facing urban environmental challenges evoked by urban growth and climate change (Kabisch, 2015).

2. STATE OF THE ART

The role of land use / land cover (LULC) is important to understand the urban ecosystem and to set ecosystems in the context to ecosystem services for urban residents. In the research context of urban greenhouse gas emissions, Baur et al (2015) set the specific focus on spatial structures, but in their study they rather concentrate on urban built-up density regarding the urban fabrics than on the vegetation structure and on urban trees as a sequestration source, although urban trees are a spatial determinant for ecosystem services.

Of course, it is important to set the urban fabric in relationship to the urban vegetation pattern to understand urban densification processes and their impact on greenhouse gas emissions. But, in addition, the tree cover, its distribution and development would serve as an essential indicator for this kind of ecosystem service. At a regional scale, Maimaitijiang et al. (2015) discuss the drivers of urban land cover and land use changes by subdividing vegetation cover urban into deciduous and mixed forest, various types of agricultural land, and thus regarding the built-up area in this environmental context to understand urban spatial heterogeneity over time.

In Germany, there is a long-standing research on land-use dynamics at a regional scale, developing new indicators and functionalities (Meinel et al. 2014), and thus providing data of the IÖR-Monitor with an INSPIRE metadata set at Geoportal.de. A newly integrated category is the relief, i.e. the absolute height of a land use above ground, described as indicator relief energy and relief diversity. This indicator pays tribute to the third dimension of land use and land cover for sensible areas such as ecosystems and their services. Furthermore, Walz et al. (2014) incorporates 3D- structural measurements into a raster-based landscape analysis to differentiate landscape structures more appropriately, according to their real-world conditions.

Our approach also makes use of the data set derived from surface models from Airborne Laserscanning (ALS) to get a better picture of vegetation levels, and especially distinguish between young trees and bushes and trees defined as such by their minimum height of 5 meters (EEA convention). As a sophisticated mapping tool, remotely sensed data and techniques serve to differentiate trees from other vegetation structure, as well as from buildings and further anthropogenic elements. Most recently, ALS and LiDAR data are used as ancillary information to identify above-ground LULC elements and distinguish spectrally similar land-use categories by their height information (O'Neil-Dunne et al. 2014). By applying these data sets, the absolute height of single elements is integrated into segmentation processes, following the principles of object-based image analysis, to calculate the delineated segments at a higher differentiation level, and to classify the elements of interest (Rutzinger et al. 2007). Especially for ecosystem management it is understood, that a three-dimensional model with fused data from very high resolution imageries and LiDAR data sets are important to reconstruct forest canopies (Chen et al. 2012; Secord and Zakhor, 2007). In this study, we monitor urban trees over a decade to understand their inner urban differentiation and their local contribution to ecosystem services.

3. STUDY AREA

Germany does not possess a mega city, but is rather composed of four major cities with more than a million inhabitants (Berlin, Hamburg, Munich, Cologne), and about 15 urban agglomerations with more than 500,000 inhabitants, amongst them is the City of Leipzig (Fig. 1). It is located in East Germany, 180 km south of Berlin. This city underwent a severe shrinkage phase after reunification in 1990, and is now one of the fastest re-growing cities in Germany. It is composed of large residential areas with Wilhelmenian multi-storey buildings

constructed between 1850 and 1915 (more than 22,000 buildings of that type). As one of the East German leading cities it is home to contiguous community allotment as neighbourhood open spaces that characterize the configuration of the city (Table 1).

Population in 2013	531,562 inh.
Area in 2013	297.38 km²
Population density	1,787 inh./km²
Unemployment rate in 2014	9.4 %
Latitude	51° 19′ 44″ N
Climate	Transitional Continental
Altitude	113 m
Rainfall	595 mm/a
Mean temperature in 2014	11.0 °C
Forest area in 2012	20.81 km²
Area of public green spaces	121 m² per capita
Area of community allotments	8.43 km²

Table 1. Socio-demographic and environmental indicators for the City of Leipzig (Source: City Council of Leipzig)

Leipzig has one of the most well-preserved floodplain forests in Europe that crosses the urbanized area from south to north, and northwest bound, that acts as the green lung of the city. The City Council subdivided the urban area into 10 urban districts which are central planning spaces for urban development. The historical centre is in the middle of the city as central district, surrounded by nine other administrative urban districts (Fig. 1 (c)).

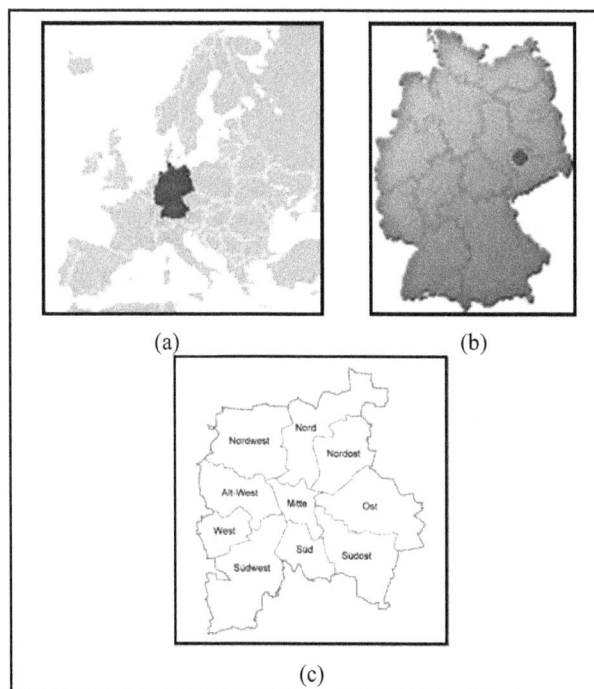

(a)   (b)

(c)

Figure 1. Location of case study, (a) in Europe, (b) in Germany, (c) City of Leipzig with its 10 urban districts

4. METHODOLOGY

We present the update of a monitoring system in which single standing trees and woodland serve as environmental indicator for ecosystem service in the spatial context of urban districts in the City of Leipzig. Therefore urban trees are monitored by means of multiple data sets to get a detailed knowledge on its distribution and change over the entire city and its 10 urban districts. The presented study is based on digital orthophotos (DOP) at the spectral resolution of Colorinfrared Imageries (CIR in 2002; RGBI for 2012) that originally possess a ground resolution of 20 cm for the years 2002 and 2012.

For the first tree canopy layer the pixel size needed to be comprised to 80 cm due to data processing limitations. This spatial resolution still served well to distinguish treetops, but additional height information is lacking for this point in time. To some extent, the resampling of DOPs also filtered from heterogeneities (e.g. roof windows) inside objects that where defined previously. In this preprocessing step all DOPs have been mosaicked to cover the whole extent of Leipzig. As the most appropriate processing methodology, we applied the object-based image analysis (OBIA) in eCognition. The segmentation process is explained in detail for the classification in 2012.

For the analysis of 2012 data set, data fusion of very high-resolution DOP and LiDAR derivatives (2m DEM, DSM) was feasible in OBIA. Hence, trees can be differentiated from other vegetated areas such as bushes and lawn. Young trees are still hard to be differentiated from bushes due to their similar height. Non-vegetated surfaces can be separated into buildings and other anthropogenic surfaces.

Thresholds of indices and height measures are used to classify the initial objects that are created within the segmentation process. Not part of the presented study is the differentiation of buildings, and only mentioned briefly: the ranges of height serve well to differentiate types of housing as height of buildings is typically similar within a certain era and often distinct between different periods.

The first preprocessing step is alike the one of 2002, i.e. mosaicking and resampling of the DOPs. The normalized difference vegetation index (NDVI) is then calculated from DOP (Fig. 2 (b)). To match pixel size DEM and DSM have been resampled to 80cm as well. To receive true objects heights DEM and DSM have been subtracted (DSM-DEM) to create a normalized Digital Surface Model (nDSM). A multi direction Lee filter is applied to reduce local noise while saving edges. Then, NDVI and nDSM image layers are normalized to a value range of 0-255 (8bit). This step seems to be important for the segmentation process to ensure that none of the input data sets is weighted differently because of a different value range. All input data sets were matched to the same georeference system: ETRS89/UTM Zone 33N (EPSG:25833).

In the following, a Multi-Resolution Segmentation (MRS) is generated with the equally weighted bands Red, Green, Blue, Infrared, normalized NDVI, normalized DSM (scale 15, shape 0.2, compactness 0.3). Stepwise, non-vegetated and vegetated classes are differentiated, followed by further subdivision of the vegetated surface. Here, the normalized bands DSM and NDVI are used to segment the vegetated areas by height and vitality. Because of the rather compact shape of trees and small sized elements, the following parameters were set to scale 5, shape 0.1, and compactness 0.8.

The subsequent classification is based on different measures as the mean height of objects and other statistical measures of height as the 25 or 50% quantile of pixels in the objects. Quantile parameters are introduced to reduce the effects of

mismatching object borders between DOP and nDSM that are result of the central camera perspective and different acquisition technologies. Not only the possibly biased mean values are used, but also thresholds of a minimum amount of pixels above a certain height value are set. In the classification process true value NDVI and nDSM layers are utilized.

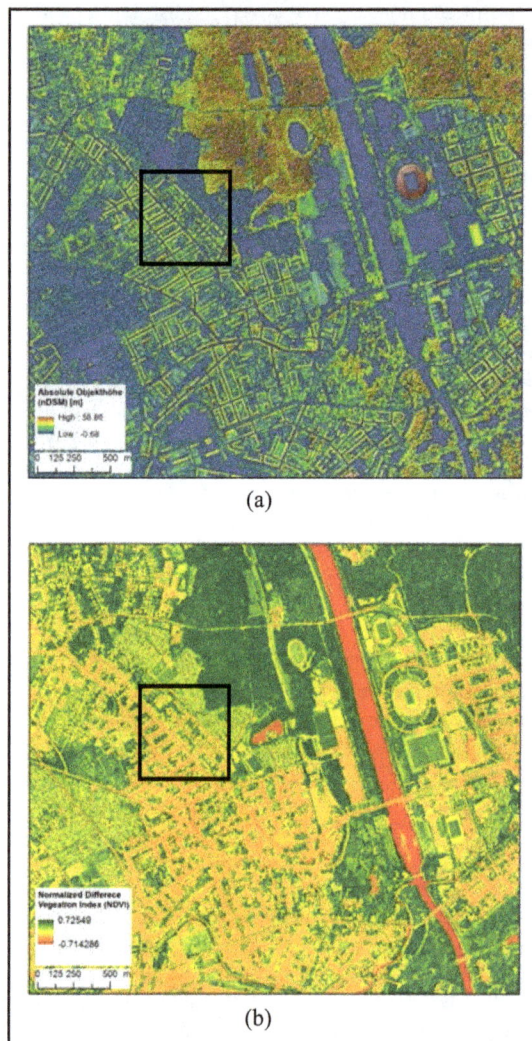

(a)

(b)

Figure 2. Display of preprocessed input data (a) the absolute height above ground, (b) the NDVI, both depicting an extraction of the study area: Wilhelmenian style residential buildings (1870-1920) in the west and east, the river "Weiße Elster" from south to north next to the stadium, and the floodplain forest in the northern part

5. RESULTS

As the refined knowledge on the absolute height above ground only exists for the more recent time slot in 2012, it was only then possible to differentiate tree tops from young trees and bushes. So just the general monitoring of the class trees is possible for 2002 and 2012. According to different data sets there is some uncertainty left for the comparison between the two time steps. This is a matter of fact that is inherent in monitoring over time, as methodologies and data quality get continuously enhanced. Visually, both results show high accuracy of tree detection, but only for the year 2012 misclassification due to shadows could be avoided entirely. Fig.

3 not only presents the distinguished classes of trees, young trees and bushes, as well as lawn and meadows, but also the delineation of residential buildings in a central residential area.

Figure 3. Classified image for a residential area in the City of Leipzig; see square in Fig. 2 for the zoomed-in location

To distinguish the tree cover within the City of Leipzig, the two derived data sets undergo a GIS analysis in which the amount of tree cover is calculated for each urban district. In addition, the differentiation between trees and young trees / bushes is depicted for the year 2012 (Tab. 2).

Urban district	Trees 2002 [ha] Total	Trees 2012 [ha]		
		Total	Young trees and bushes	Mature trees >5m
Alt-West	1001,0	1130,9	393,9	737,0
Mitte	424,0	480,9	120,8	360,0
Nord	664,7	801,3	459,9	341,3
Nordost	687,9	825,7	458,1	367,6
Nordwest	1038,9	1199,9	473,9	726,0
Ost	895,7	1025,9	610,8	415,1
Süd	856,7	977,8	241,3	736,5
Südost	908,5	1011,6	472,6	539,1
Südwest	883,0	1106,6	526,8	579,8
West	421,2	534,1	299,4	234,7

Table 2. Area covered by trees within the urban districts of the City of Leipzig for the years 2002 and 2012. In 2012 a further differentiation is undertaken for mature trees > 5m, and young trees and bushes

Soon after reunification in 1990, urban environment has got a significant push in urban planning due to shrinkage processes until after the turn of the millennium, and the aim to perform attractively in local neighbourhoods, produce greening on brownfields as an interim use, raise the vegetation connectivity in the city, and rouse the awareness of residents for a green city. In this context, the City Council has promoted a strong campaign for a "Baumstarke Stadt" (in English: trees for a stronger city suggesting environmentally, socially and health-wise) which obviously supports the success of gaining more trees in the different districts. In 2002 the total amount of urban trees was set as 7,781.6 ha, while it rose to 9,094.7 ha in 2012

(summing up the total figures of trees in the respective years, Tab. 2). Hence, in each of the 10 urban districts an increase is observed, some stating a stronger increase than others.

A strikingly higher differentiation can be undertaken for the assignment of young trees and bushes and the mature trees. Young trees and bushes cover a large area and represent planting activities during the last two decades. These figures witness the effort of the City council to develop and strengthen a prosperous environment for Leipzig, enhance the ecosystem services and increase the life quality in neighbourhoods through tree planting. But inherent in the figures of young trees and bushes is also the inaccuracy to which share the green infrastructure is covered by younger trees and by bushes respectively.

6. CONCLUSIONS

The urban tree cover is an important piece of information to differentiate types of residential areas, to characterize urban forest, and delineate infrastructural development. It serves as an environmental indicator to understand the impact of land use changes due to the highly dynamic urban re-growth on ecosystem function and the special challenges of urban ecosystem services for human well-being. Critical methodological aspects are that first, in each monitoring sequence over many years the quality and quantity of available input data improves and evokes the ambition to enhance the result. This is a very prosperous procedure for one step in time, but inherently to long-term monitoring, not all distinguished details of the classified results will then remain comparable. Second, the presented methodology defines one single class for young trees and bushes. This is a weakness in the quantitative analysis. It still needs to enhance the differentiation between young trees and bushes, to make a clearer statement on the actual area covered by younger trees only. In this context, LiDAR point clouds could support the subdivision of young trees and bushes and will be tested in our ongoing research.

ACKNOWLEDGEMENTS
We want to thank the Ordnance Survey of the State of Saxony, Germany for the kind appropriation of the above-mentioned data sets DOP, DEM and DSM (© Staatsbetrieb Geobasisinformation und Vermessung Sachsen).

REFERENCES

Baur, A.H., Förster, M., Kleinschmit, B. 2015. The spatial dimension of urban greenhouse gas emissions: analyzing the influence of spatial structures and LULC patterns in European cities. *Landscape Ecology*, DOI 10.1007/s10980-015-0169-5.

Bolund, P., Hunhammar, S., 1999. Ecosystem services in urban areas. *Ecol. Econ.* 29, 293–301.

Chen, L.C., Huang, C.Y., and Teo, T.A., 2012. Multi-type change detection of building models by integrating spatial and spectral information, *International Journal of Remote Sensing*, Vol. 33, No. 6, pp. 1655-1681. (SCI/EI).

Daily, G. (ed.), 1997. Nature's Services: Societal Dependence of Natural Ecosystems.Island Press, Washington, DC, p. 392.

European Environmental Agency (EEA). EUNIS categories. Habitat Types Key Navigation. Category: (G) Woodland, forest and other wooded land. http://eunis.eea.europa.eu/habitats-key.jsp?level=2&idQuestionLink=--%3E&pageCode=G; accessed 2015/03/24.

Kabisch, N. 2015. Ecosystem service implementation and governance challenges in urban green space planning – The case of Berlin, Germany. *Land Use Policy*, 42, 557-567.

Maimaitijiang, M., Ghulam, A., Onésimo Sandoval, J.S. 2015. Drivers of land cover and land use changes in St. Lous metropolitan area over the past 40 years characterized by remote sensing and census population data. *Int. Journal of Applied Earth Observation and Geoinformation*, 35, 161-174.

Meinel, G., Krüger, T., Schumacher, U., Hennersdorf, J., Förster, J., Köhler, C., Walz, U., Stein, C. 2014. Aktuelle Trends der Flächennutzungsentwicklung, neue Indikatoren und Funktionalitäten des IÖR-Monitors. *IÖR-Schriften*, 35-43.

Millennium Ecosystem Assessment, 2005. *Ecosystems and Human Well-being: Synthesis.*

O'Neil-Dunne, MacFaden, S., Royar, A. 2014. A versatile, production-oriented approach to high-resolution tree-canopy mapping in urban and suburban landscapes using GEOBIA and data fusion. *Remote Sensing,* 6, 12837-12865, doi:10.3390/rs61212837.

Rutzinger, M., Höfle, B. & Pfeifer, N. (2007): Detection of high urban vegetation with airborne laser scanning data. In: *Proceedings forestsat 2007.* Montpellier, France, November, 2007., pp. digital media.

Secord, J., Zakhor, A. 2007. Tree detection in urban regions using aerial LiDAR and image data. *IEEE Geoscience and Remote Sensing Letters,* Vol. 4, No. 2, 196-200.

13

CATCHMENT PROPERTIES IN THE KRUGER NATIONAL PARK DERIVED FROM THE NEW TANDEM-X INTERMEDIATE DIGITAL ELEVATION MODEL (IDEM)

J. Baade [a, *], C. Schmullius [b]

[a] Department of Geography, Physical Geography, Friedrich-Schiller-University Jena, 07737 Jena, Germany – cub@uni-jena.de
[b] Department of Geography, Earth Observation, Friedrich-Schiller-University Jena, 07737 Jena, Germany – c.schmullius@uni-jena.de

KEY WORDS: TanDEM-X, ASTER GDEM2, SRTM, RTK-GNSS, validation, watershed delineation, Savanna, South Africa,

ABSTRACT:

Digital Elevation Models (DEM) represent fundamental data for a wide range of Earth surface process studies. Over the past years the German TanDEM-X mission acquired data for a new, truly global Digital Elevation Model with unprecedied geometric resolution, precision and accuracy. First processed data sets (i. e. IDEM) with a geometric resolution of 0.4 to 3 arcsec have been made available for scientific purposes. This includes four 1° x 1° tiles covering the Kruger National Park in South Africa. Here we document the results of a local scale IDEM validation exercise utilizing RTK-GNSS-based ground survey points from a dried out reservoir basin and its vicinity characterized by pristine open Savanna vegetation. Selected precursor data sets (SRTM1, SRTM90, ASTER-GDEM2) were included in the analysis and highlight the immense progress in satellite-based Earth surface surveying over the past two decades. Surprisingly, the high precision and accuracy of the IDEM data sets have only little impact on the delineation of watersheds and the calculation of catchment size. But, when it comes to the derivation of topographic catchment properties (e.g. mean slope, etc.) the high resolution of the IDEM04 is of crucial importance, if - from a geomorphologist's view - it was not for the disturbing vegetation.

1. INTRODUCTION

Digital Elevation Models (DEM) represent fundamental data for a range of applications including Earth surface process studies in the field of ecology, geology, geomorphology and hydrology, among others. For some countries high resolution Digital Terrain Models (DTM) representing the solid Earth surface derived from topographic maps or aerial surveys (photo-grammetry, laser) are available. But, for vast regions of the Earth this fundamental data is missing. Starting with the Shuttle Radar Topographic Mission (SRTM) in February 2000, the past two decades have witnessed a continuous growth in the use of satellite-based data for the production of DEMs. Being based on either Synthetic Aperture Radar (SAR) interferometry like the suite of SRTM SIR-C products (Farr et al., 2007), or optical images like the ASTER-GDEM (ASTER GDEM Validation Team, 2011) these elevation models represent the surface of the Earth including the height of the land cover (vegetation, buildings and other objects). Thus, these DEMs are often considered Digital Surface Models (DSM) as compared to DTMs.

Since December 2010 the German radar satellite mission TanDEM-X acquired data for a new and truly global Digital Elevation Model (DEM) with unprecedied geometric resolution, precision and accuracy (Krieger et al., 2013; Zink and Moreira, 2014). According to Bräutigam et al. (2014) data acquisition was expected to conclude by August 2014 and the finalization of the global DEM by the end of 2015. Since November 2014 processed data sets from the first year's acquisition (i. e. the Intermediate DEM, IDEM) with a geometric resolution of up to 0.4 arcsec (~12 m)(IDEM04) at the equator are available for selected regions of the World for scientific purposes. This includes four 1° x 1° tiles covering almost the entire Kruger National Park (KNP) in South Africa. In addition, IDEM tiles with 1 arcsec (IDEM10, ~30 m) and 3 arcsec (IDEM30, ~90 m) resolution were made available. Due to the fact that the TanDEM-X derived height measurements include land cover, the IDEM data sets have to be considered DSMs.

This paper reports on the application of the three IDEM data sets for the delineation of hydrological catchments and the derivation of catchment properties in a pristine Savanna environment, i.e. the Kruger National Park (KNP) in the North-east of the Republic of South Africa (RSA). This includes an accuracy assessment utilizing RTK-GNSS-based ground survey points from a dried out reservoir basin and its vicinity. In order to test the connotation of an unprecedied precision and accuracy, the pertinent, global, open access precursor DSMs, i.e. SRTM1, SRTM90 and ASTER-GDEM2, were included in the analysis.

2. MATERIAL AND METHODS

2.1 Study site

The study is conducted in the southern part of Kruger National Park (KNP) located in the northeastern part of the Republic of South Africa (KNP). In total, the KNP occupies about 19,500 km² of the undulating Lowveld Savanna between the foot slopes of the Drakensberg Escarpment to the west and the coastal plains of Mozambique to the east (Figure 1). Its N–S extension is about 350 km and its W–E extension ranges from 35 to 70 km. The landscape pattern within KNP follows in general the NNW–SSE strike of the major geological units characterized by granitic rocks and the basement complex in the west and the Karoo sedimentary and volcanic rocks in the east (Venter et al., 2003). Most parts of what became KNP in 1926

---

* Corresponding author.

have not been attractive to white farmers because of mosquito and tsetse fly infestation and were set aside for the recovery of wildlife in the first decade of the 20th century (Carruthers, 1995). Thus, most parts of the KNP have never experienced an enhanced European style agricultural development and can be considered to represent a pristine Savanna environment. In particular, this study focusses on the catchment areas of 21 fresh water reservoirs (Figure 1) constructed in the second half of the last century in the framework of a water provisioning programme. At the time of the IDEM data acquisition some of the reservoirs were breached and dried out while others were still intact and contained water.

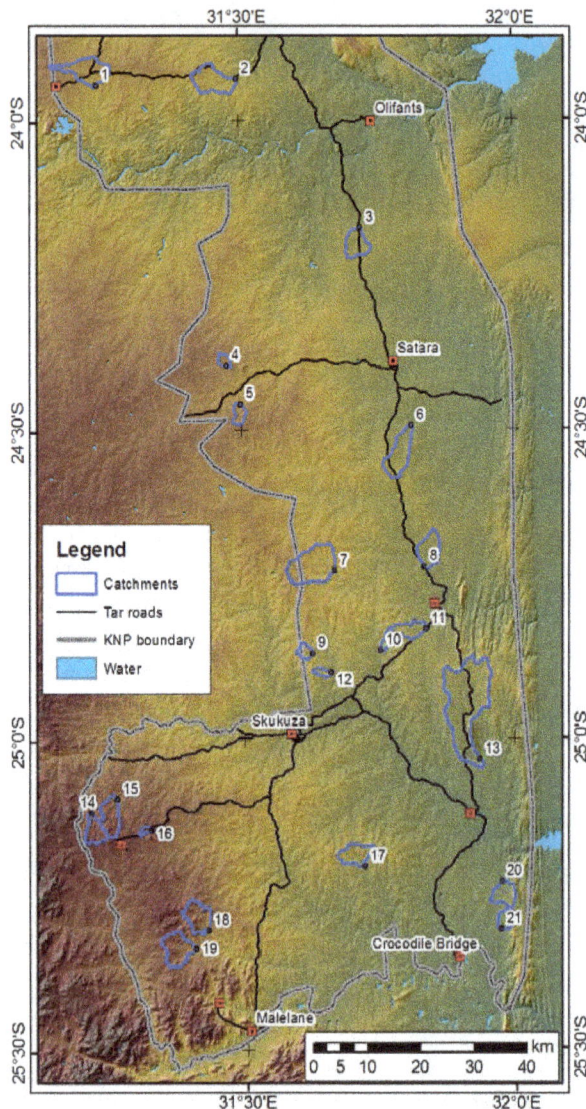

Figure 1. Location of studied catchments within the southern part of Kruger National Park, South Africa (IDEM30 source: ©DLR 2014).

2.2 Properties of the DSM data sets

For this study three high resolution IDEM04, three medium resolution IDEM10 as well as six low resolution IDEM30 tiles as well as the corresponding medium resolution ASTER GDEM2 and SRTM1 and the low resolution SRTM90 data sets (Table 2) were mosaicked. The general properties of the IDEM data sets were analysed for the wider study area, i.e. the

southern part of the Kruger National Park, covering 11,500 km². The general properties of the IDEM data sets are documented in a number of auxiliary files, e. g. layover and shadow mask (LSM), coverage map (COV), height error map (HEM), and a water indication mask (WAM) (Wessel et al., 2013). Due to the rather flat terrain of the Lowveld (Figure 1), layover and shadow is not an issue of concern. Less than 0.01 % of all pixel are affected. The coverage map indicates, that 60 %, 36 % and 4 % of all pixel represent data from one, two and three acquisitions, respectively.

The HEM, representing the uncertainty induced by the interferometric coherence and the geometry, provides values between < 0.01 m and 115 m with a mean value of 0.56±0.25 m for the IDEM04 mosaic covering the southern part of KNP. About 1 % of all pixel exhibit a height error ≥ 1.0 m, but less than 0.01 % have an error ≥ 10.0 m (Figure 3). Analysing the HEM for the IDEM10 and IDEM30 provides very similar mean uncertainties, but higher minimum and lower maximum errors. This provides first evidence for the overall high precision of the TanDEM-X instrument and the data processing chain.

Pixels affected by water are identified in the WAM based on several criteria. However, one need to bear in mind that *islands with an area smaller 1 hectar and water bodies with an area smaller 2 hectar are not considered* (Wessel et al., 2013, p. 18). This threshold does not pose a problem when dealing with compact water bodies like reservoirs and lakes. But it clearly makes it more difficult to identify height pixels affected by water along the course of rivers, often inducing disrupted sinks when pixels affected by water are masked (Figure 1).

	Resolution		
	Geometric (arcsec)	data format	Acquisition
IDEM04	0.4	float	X-SAR
IDEM10	1	float	X-SAR
IDEM30	3	float	X-SAR
ASTER GDEM2	1	integer	Optical
SRTM1	1	integer	SIR-C
SRTM90	3	integer	SIR-C

Table 2. Properties of the DEM data sets used in this study

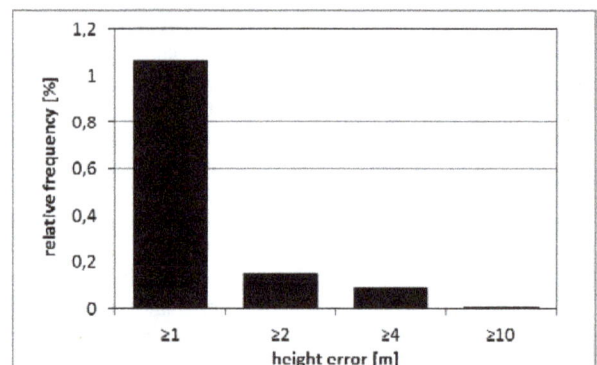

Figure 3. Relative frequency [%] distribution of IDEM04 height error classes (≥ 1 m) within the southern part of KNP. Note: over 98 % of all pixel are associated with a height error < 1 m.

Unfortunately, none of the other DSM data sets include detailed, i.e. pixel specific, quality assessment files. Therefore, a study site specific a priory introduction to the properties of these data sets is inhibited. Nonetheless, pertinent accuracy assessment reports are available and will be used in the discussion of the results.

2.3 Accuracy assessment using ground measurements

Recently, Wessel et al. (2014) used ICESat validation points, kinematic GPS (KGPS) tracks with an accuracy of < 1 m and SRTM data to assess the accuracy of IDEM04 tiles on the global scale. Here, we utilize Real Time Kinematic (RTK) GNSS based survey points from a dried out reservoir basin (Silolweni, no. 11 in Figure 1) devoid of woody vegetation and its vicinity characterized by open, woody Savanna vegetation (Figure 4) for a local scale assessment of the accuracy of the IDEM data sets. The survey points (N = 1088 points) were measured in February and September 2014 using LEICA GS10 and GS15 GNSS receivers equipped with a radio connection. The base station position was determined with reference to the South African network of permanent GNSS stations (Trignet). The accuracy of the point heights is < 0.05 m for height above ellipsoid (HAE) and ~ 0.10 m for height measures referenced to the geoid (ALT). Thus, the absolute accuracy of the survey points is much better than the precision of the height measures from the IDEM data sets as well as the other DSM data sets.

Figure 4. View of the dried out Silolweni reservoir basin and its vicinity used for IDEM accuracy assessment. The distance to the far end is about 650 m. The bigger trees at the far end reach heights of about 10 m (photo: J. Baade, Sep. 2014).

In order to weight each DSM pixel equally, multiple survey point samples within a DSM pixel were averaged (Rodrigues et al., 2006) taking into account the detailed geometry of the raster data sets to compare with. Thus, the ~ 1100 individual survey points were reduced to 765 pixel with 0.4 arc sec resolution and 370 and 75 pixel with 1 arc sec and 3 arc sec resolution, respectively. The difference between ground survey heights (RTK heights) and DSM heights was calculated by subtracting RTK heights from DSM heights. In accordance with other validation efforts (e.g. ASTER GDEM Validation Team, 2011) this yields positive differences for all DSM pixel characterized by a vegetation cover dense and high enough to bias the height measure in the DSM.

2.4 Watershed delineation and catchment properties

In order to derive a hydrological correct elevation model the IDEM data sets representing heights above the WGS84 ellipsoid (HAE) (Wessel et al., 2013) were transformed to heights above the geoid (ALT) using the most recent hybrid

geoid model for South Africa, the SAGEOID10 (Chandler and Merry, 2010). This geoid model is provided with a 1 arcsec geometric resolution and an accuracy of about 7 cm. The geoid model was accordingly resampled and finally the transformed DSM was projected to UTM36S. No attempts were made to remove or manipulate pixel, especially from the high resolution IDEM04 clearly representing canopy height or single trees. However, all sinks were filled prior to the performance of the hydrological analysis.

The calculation of the flow direction and flow accumulation grids as well as the consecutive batch watershed delineation for the reservoirs was conducted using standard routines implemented in Arc Hydro Tools for ArcGIS 10.2 (Esri Water Resources Team, 2014). Due to the fact that the routing of runoff over the digital elevation model changes with the resolution of the grid, some manual user interaction was needed to adjust the catchment outflow point and ensure a correct delineation of the watersheds.

In addition to catchment area representing the primary catchment property, we report here on the variation of catchment wide slope estimates.

3. RESULTS AND DISCUSSION

3.1 Accuracy assessment based on ground survey points

The RTK-GNSS based bare ground height measurements averaged for the IDEM04 conform 0.4 arc sec pixel (N = 767) in the dried out Silolweni reservoir basin and its vicinity range from 282.80 to 294.60 m HAE (Figure 4) over an area of about 0.5 km². Compared to this, the IDEM04 height readings range

	N	Height measure	RMSE
IDEM04	767	HAE	1.55
IDEM10	368	HAE	1.46
IDEM30	76	HAE	1.65
GDEM2	368	ALT	5.9
SRTM1	368	ALT	5.7
SRTM90	72	ALT	5.9

Table 5. Summary statistics of the height difference between pixel based averaged ground survey point measures and DEM pixel values [in m], HAE refers to the WGS84 ellipsoid, ALT refers to the EGM96 geoid.

	Min	Mean	1 σ	90 %	Max
RTK - IDEM04	-3.65	0.70	1.40	< 2.51	7.85
IDEM04 (HEM)	0.26	0.4	0.1	< 0.52	1.30
RTK – IDEM10	-2.05	0.83	1.20	< 2.45	7.88
IDEM10 (HEM)	0.29	0.41	0.06	< 0.49	0.75
RTK – IDEM30	-0.78	1.25	1.07	< 2.75	4.28
IDEM30 (HEM)	0.32	0.41	0.04	< 0.46	0.52
RTK – GDEM2	-14.2	1.9	5.6	< 9.60	18.7
RTK - SRTM1	0.6	5.4	1.9	< 7.60	12.3
RTK - SRTM90	1.6	5.6	1.7	< 7.55	10.6

Table 6. Detailed statistics of the height difference between pixel based averaged ground survey point measures and DEM pixel values [in m]

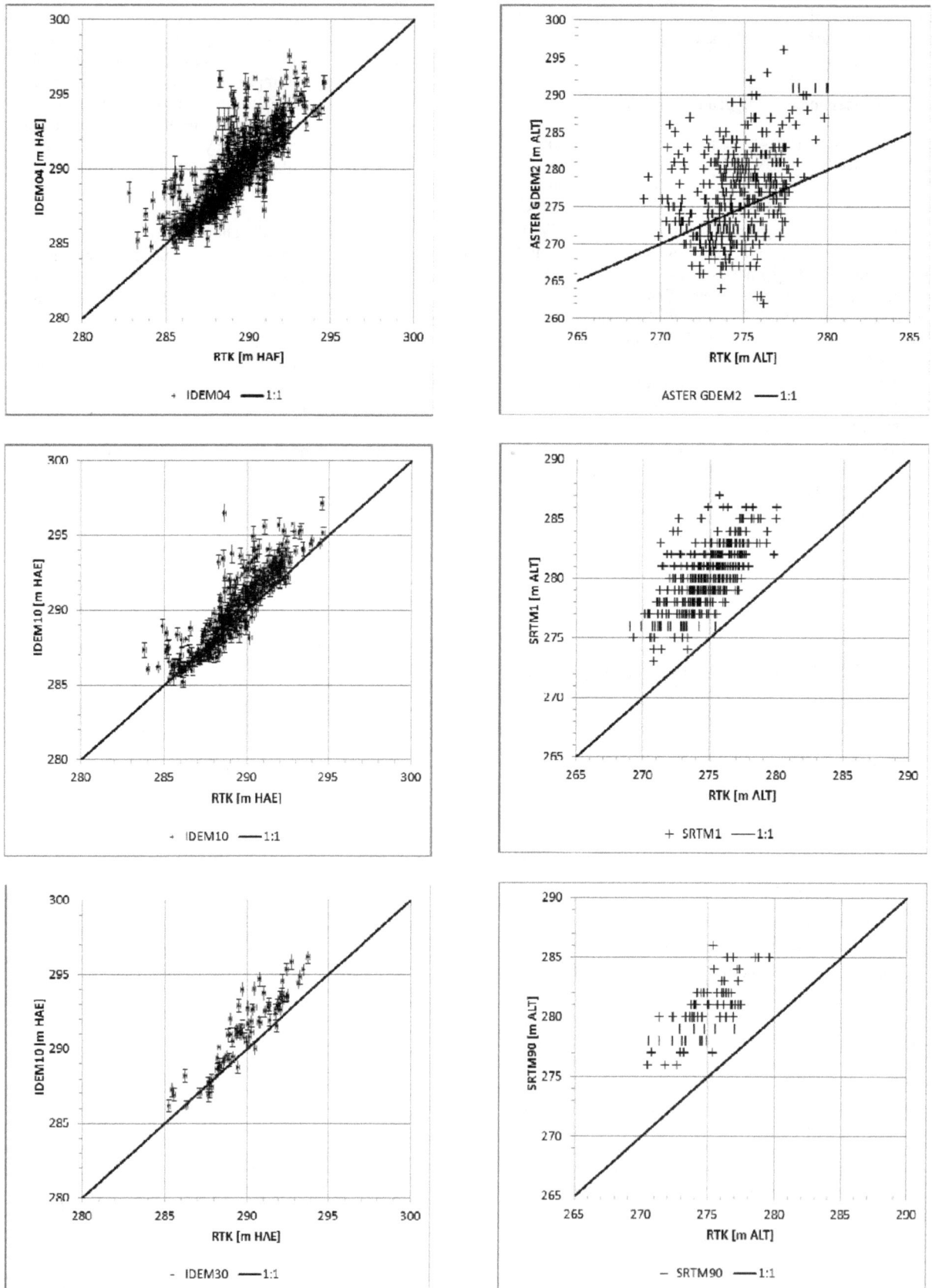

Figure 7. Absolute height difference scatter plots derived from the comparison of RTK-GNSS based ground survey points and DEM pixel heights. For the IDEM data sets the height is given in m HAE and the precision of the measurement is indicated with whiskers. For the other data sets height is given in m ALT (data sources: IDEM: ©DLR 2014, ASTER GDEM2 is a product of METI and NASA, SRTM1: USGS 2014, SRTM90: Jarvis et al. 2008).

from 284.70 to 297.60 m HAE. The RMSE yields a value of 1.55 m (Table 5). The mean difference is 0.7 m and thus slightly larger than the mean precision of the IDEM04 data (HEM) for all pixel analysed here (0.4±0.1 m)(Table 6). With a maximum height difference of ~ 7.9 m and 90 % of all absolute height errors < 2.5 m, the IDEM04 data set clearly fulfils the accuracy benchmark for the TanDEM-X mission (Krieger et al., 2013). This holds as well for the lower resolution IDEM products showing very similar RMSE values (Table 5) as well as detailed statistical characteristics (Table 6).

Inspection of the absolute height difference scatter plots (Figure 7) provides evidence for the IDEM products of a distribution in accordance with the expectation. There is a clear general trend of increasing IDEM height readings with increasing ground survey point heights. This indicates that the IDEM products represent the changes in relief in this rather flat terrain quite well. Basically, the height differences are distributed symmetrically around the one-to-one line. However, there is an increase of the height difference with increasing ground survey point height. This is in accordance with the landscape characteristics within the dried out Silolweni reservoir basin and its vicinity. The reservoir is virtually bare of woody cover while the elevated surroundings of the reservoir basin are characterized by an open woody Savanna vegetation reaching heights of about 10 m (Figure 3). The high scatter at the lower end of the point cloud can be explained by pixel representing thalwegs leading in and out of the reservoir basin covered by rather dense riparian forests. Thus, positive height differences can be easily explained by the fact that the X-band radar backscatter originates from the vegetation canopy and not the bare ground (Wessel et al., 2013; Baade & Schmullius, 2014).

However, the scatter plot shows as well a considerable number of observations with rather large negative deviations, i.e. well beyond the maximum height error (1.3 m acc. to HEM), especially for the IDEM04 data set. An in depth analysis of these phenomenon is beyond the scope of this paper. But, it seems that these pixels are often associated with surface features like the earthen dam of Silolweni reservoir or gullies orientated perpendicular to the line of sight of the TanDEM-X instrument and being smaller than the resolution cell.

The statistics (Table 5, Table 6) and the scatter plots (Figure 7) for the other three data sets show clearly higher RMSE values as well as much stronger scatter of the height differences. This holds especially for the ASTER GDEM2 data set. Here, the scatter plot basically suggests strong random errors. Fitting a linear model to the point cloud reveals that only 15 % of the variation in ASTER GDEM2 height readings is attributable to changes in bare ground heights and suggests that the ASTER GDEM2 does not represent the changes in relief in this rather flat terrain adequately.

Despite of the higher RMSE values for the SRTM1 and SRTM90 data set, the corresponding scatter plots resemble the IDEM scatter plots in appearance. Both SRTM scatter plots show a general trend of increasing DSM height readings with increasing bare ground height measurements. This, again, indicates that both SRTM products reproduce the changes in relief. In contrast to the IDEM products, height differences for the SRTM products are all positive. The mean height difference is about 5.5 m. About 10 % of the observed height difference might be attributed to the different geoids applied. The SRTM products are referenced to the EGM96 geoid, while we used the SAGEOID10 to transfer RTK-GNSS height measurements from

m HAE to m ALT. Apart from this, the mean height difference corresponds well to the 5.6 m absolute height error identified for the African continent during the SRTM performance analysis (Farr et al., 2007). The fact that the maximum height differences are larger for the SRTM product than the ones observed for the IDEM products is of a surprise. Due to the longer radar wavelength used in the SRTM mission (C-band with 5.66 cm)(Farr et al., 2007) compared to the TanDEM-X mission (X-band with 3.1 cm)(Krieger et al. 2013) one would expect a better penetration of the canopy cover during the SRTM mission. However, the timing of the SRTM mission corresponds to the leaf-on season in this part of South Africa and might have offset the effects of the different wave lengths.

3.2 Watershed delineation and catchment area

The application of the three IDEM products and the other DEMs for the delineation of the watershed using Arc Hydro Tools for ArcGIS 10.2 (Esri Water Resources Team, 2014) was tested using 21 catchments of reservoirs within the KNP (Figure 1) ranging in size from a few km² to about 100 km². Minor manual user interaction was used to adjust the catchment outflow points to the automatically derived thalweg pattern.

With one exception, i.e. the ASTER GDEM2, all analysed DSM provided reasonable and consistent results (Figure 8). The analysis based on the ASTER GDEM2 failed in two cases to provide any catchment area extending beyond the immediate vicinity of the outflow points. This was due to the development of an erroneous thalweg pattern (Figure 9). In addition, the ASTER GDEM2 provided in some other cases rather large deviations from the SAR-based DSM estimations (Figure 8).

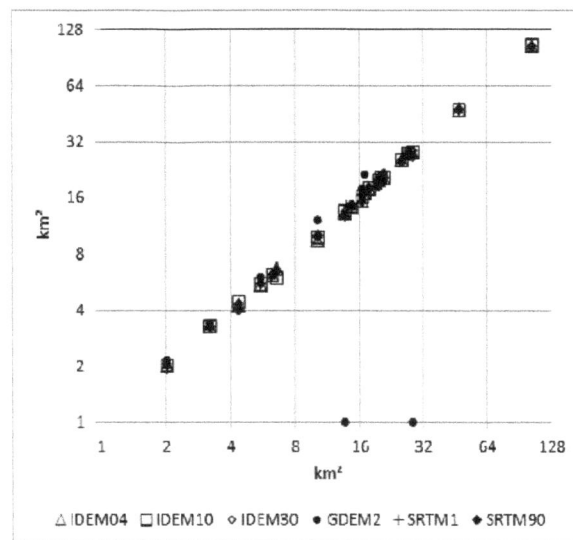

Figure 8. Comparison of catchment size derived from the six different DSM analysed in this study. Catchment delineation based on the ASTER GDEM2 failed in two cases to produce a reasonable result. In order to visualize these two cases, the corresponding data points were assigned a value of 1.

Consecutively, the statistical analysis of the different catchment area estimates was restricted to the three IDEM products (IDEM04, IDEM10, IDEM30) and the two SRTM products (SRTM1, SRTM90). Table 10 summarizes the main statistical outcome. Unfortunately, two cases (i.e. Nhlanganzwani and

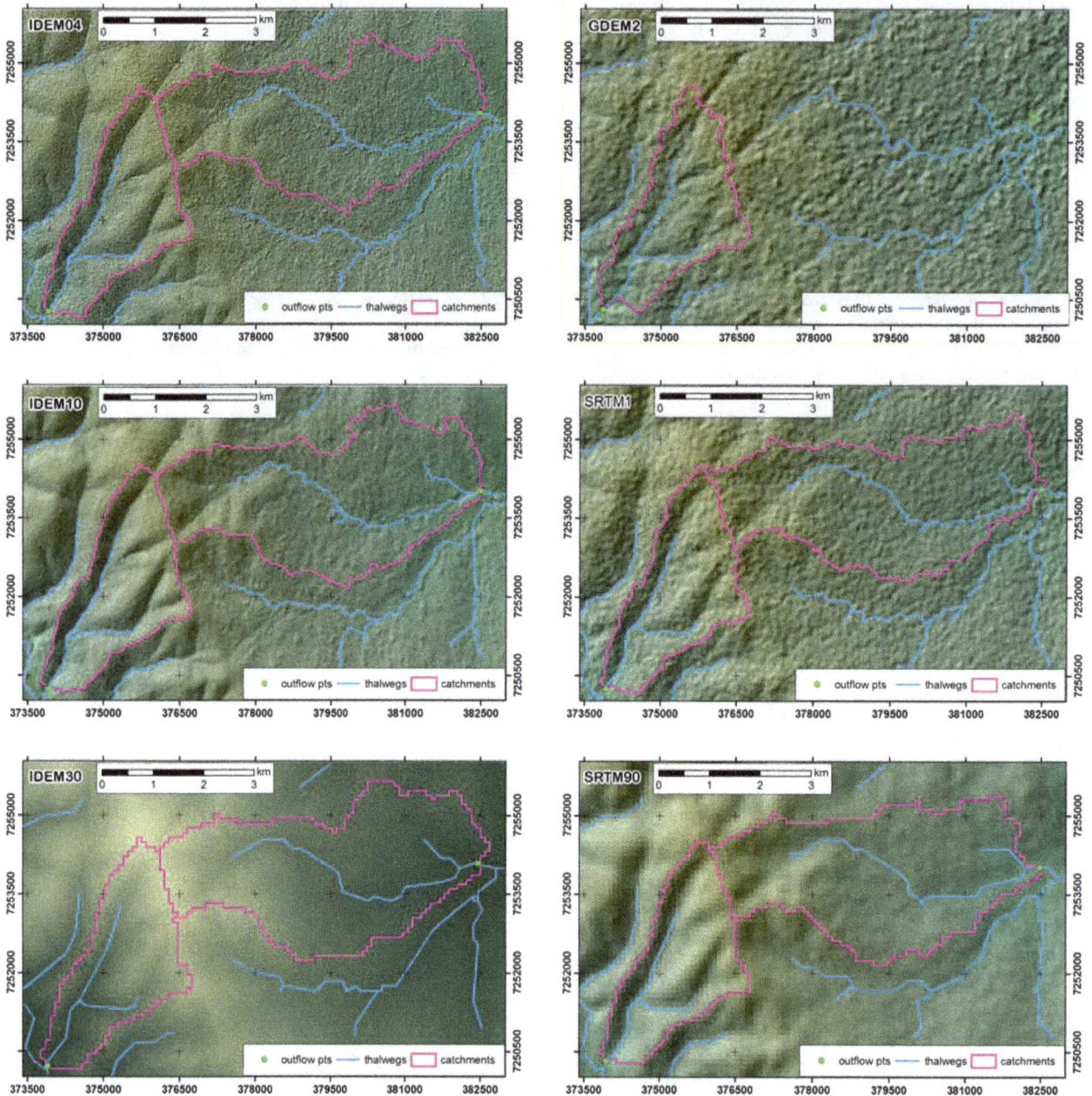

Figure 9. Detailed comparison of DEMs (combined height and hill shade representation) used in this study for runoff routing (thalwegs with 1 km² upstream contributing area) and catchment area delineation. The catchments shown in this figure are the Jones-se catchment (ID No. 10 in Figure 1 and Table 10) to the left and Silolweni (ID No. 11) to the right (data sources: IDEM: ©DLR 2014, SRTM1: USGS 2014, ASTER GDEM2 is a product of METI and NASA, SRTM90: Jarvis et al. 2008).

Mpanamana reservoir) are biased by the fact that the Eastern boundary of the catchments extends just beyond the Eastern boundary of the high and medium resolution IDEM tiles, i.e. IDEM04 and IDEM10, respectively.

Excluding these two catchments from the analysis of the variation of catchment area, provides evidence for overall consistent estimates. In many cases the coefficient of variation (CV) is less than 1 %. In the worst case (i.e. Rabelais reservoir) the CV reaches a value of 4 %. Analysis of the individual results did not provide any clear evidence for systematic deviations due to the data sets used. However, analysis of the minimum and maximum estimates of catchment size showed that ~ 50 % of the maximum values originated from the use of

the IDEM30 data set and another 30 % from the IDEM10 data set. At the same time, 50 % of the lowest estimates were attributable to the SRTM1 data set. If we assume that the mean catchment area calculated is a good estimate of the true catchment area, then this indicates a tendency of the low and medium IDEM products to overestimate and the SRTM1 product to underestimate catchment area in this rather flat terrain.

Taking into account the number of observations (N = 5) the coefficient of variation transforms into a rather small relative uncertainty (Taylor, 1997) of the catchment area estimation between 1 and 5 % on a 95 % confidence level. This indicates a rather high precision of the catchment area estimates and conforms to the findings of Oksanen and Sarjakoski, 2005. They investigated error propagation in digital elevation models of small catchments (<5 km²) derived from high resolution topographic maps and determined the error of catchment size estimations to be 10 % at a maximum on the 95 % confidence level. Unfortunately, no high resolution DTMs are available for extensive areas of the KNP to test the accuracy of the catchment area estimates.

The comparison of the detailed maps presenting the results of the catchment delineation for Jones-se (West) and Silolweni (East) reservoirs based on all discussed DSMs (Figure 9) and reproduced at a scale of ~ 1:120.000, provide visual evidence for the new quality of the digital representation of the Earth implemented by the IDEM products. Scrutinizing the IDEM04 map, it becomes clear that the short length roughness elements on the surface, exhibiting clear differences between Jones-se and Silolweni catchment, indeed represent canopy cover structures due to single trees and groups of trees. In a more generalized way, these structures are still discernible in the IDEM10 map, but are basically smoothed out in the IDEM30. This smoothing actually might be of advantage for the routing of discharge. Although it might be difficult to retrace in the medium scale printed maps, there is evidence that the drainage pattern in the higher resolution IDEM maps is quite often laterally shifted due to high and dense riparian vegetation following the thalwegs. This phenomenon is clearly more common in the flatter Silolweni catchment as compared to the steeper Jones-se catchment (Table 11). Obviously, the rather flat terrain in Silolweni catchment is as well the explanation for the poor performance of the ASTER GDEM2 in the catchment delineation exercise. For the steeper Jones-se catchment the DSM derived from optical images provides a result comparable in catchment form and size (Figure 9).

Finally, the comparison of the medium and low resolution SRTM and IDEM products clearly provide evidence for the engineering progress in the field of satellite-based SAR technology. The IDEM products are obviously largely free of speckle artefacts. This clearly provides new opportunities to characterize the canopy cover of the Earth and geomorphological features in arid areas devoid of vegetation.

3.3 Catchment relief

The high precision and accuracy of the IDEM data sets resulting in comparable catchment delineations and catchment rea estimates (Figure 9, Table 10) provides the opportunity to further investigate the variation of catchment property estimates due to changes in the geometric resolution of the fundamental DSM. Generally, it can be anticipated, that estimates of local relief, terrain roughness as well as local and mean slope are reduced with increasing pixel size (Hengl and Evans, 2009).

Name	ID	Mean [km²]	STD [km²]	CV [%]
Sable	1	27,688	0,894	3,2
Nhlanganini	2	27,501	0,167	0,6
Ngotso_B	3	16,930	0,068	0,4
Hartbeesfontein	4	4,270	0,131	3,1
Rabelais	5	6,429	0,257	4,0
Marheya	6	27,473	0,135	0,5
Lugmag	7	47,405	0,114	0,2
Mazithi	8	19,909	0,097	0,5
Ntswiri	9	5,533	0,030	0,5
Jones-se	10	6,205	0,051	0,8
Silolweni	11	13,193	0,353	2,7
N'wanetsana	12	3,248	0,026	0,8
Mlondozi	13	104,191	0,384	0,4
Mestel	14	14,462	0,156	1,1
Mtshawu	15	20,338	0,045	0,2
Shitlhave	16	2,019	0,033	1,6
Mpondo	17	17,996	0,088	0,5
Newu	18	20,752	0,273	1,3
Stolsnek	19	25,570	0,073	0,3
Nhlanganzwani*	20*	15,952	0,539	3,4
Mpanamana*	21*	9,853	0,242	2,5

ID	IDEM04	IDEM10	IDEM30
1	3.7±2.2	2.1±1.0	1.8±0.7
2	4.9±2,7	3.6±1.7	2.8±1.2
3	3.1±1.7	2.0±0.9	1.6±0.6
4	4.7±3.1	2.9±1.7	2.4±1.1
5	5.4±2.7	4.0±1.3	3.3±1.1
6	3.6±2.1	2.0±1.1	1.8±0.8
7	4.4±2.3	3.3±1.3	2.8±1.1
8	2.5±1.8	1.2±0.7	1.1±0.4
9	5.6±2.5	4.5±1.6	3.8±1.3
10	6.1±3.0	4.8±1.6	4.0±1.3
11	4.6±2.6	2.4±1.2	1.9±1.0
12	5.6±2.8	4.4±1.4	3.6±1.2
13	4.1±6.8	3.2±6.5	2.9±5.6
14	9.1±4.9	7.2±3.0	6.2±2.7
15	8.3±5.4	6.8±4.2	5.8±3.2
16	6.1±3.8	5.2±2.2	4.3±1.6
17	6.0±2.7	5.0±1.7	4.2±1.4
18	12.5±8.1	11.1±7.0	9.0±5.2
19	12.1±9.0	10.5±7.7	8.5±5.5
20*	4.7±5.7	3.8±5.3	3.9±4.7
21*	6.2±5.6	5.2±5.0	5.3±4.7

Table 10. Statistical analysis of the catchment size [km²] variation derived from the IDEM and SRTM products (N = 5) using Arc Hydro Tools. * denotes catchment areas were the Eastern catchment boundary is affected by the Eastern boundary of the IDEM04 and IDEM10 tiles.

Table 11. Statistical analysis of the mean slope ± STD [%] estimates based on the three IDEM products.
* denotes catchment areas were the Eastern catchment boundary is affected by the Eastern boundary of the IDEM04 and IDEM10 tiles.

Table 11 compiles the mean slope measures, reported in percent rise, derived from the three IDEM data sets for 21 catchments located in the southern part of the KNP. Often, the mean slope derived from the low resolution IDEM30 is about 50 % of the value derived from the high resolution IDEM04 data set. However, due to the fact that the open canopy induces in some locations, i.e. at the edge of trees and group of trees, quite steep 'slopes', it remains unclear, to which extend catchment wide steeper mean slopes derived from the high resolution IDEM04 are representative of the bare ground gradient. Further work is needed to elucidate this question in order to fully exploit the potential of the IDEM products to provide consistent, high resolution catchment properties applicable to geomorphological and hydrological studies in the vast parts of the World were high resolution DTMs from other sources are missing.

4. CONCLUSIONS

The local scale validation of TanDEM-X derived IDEM products with a geometric resolution from 0.4 to 3 arc sec (IDEM04, IDEM10, and IDEM30) using moderate terrain RTK-GNSS based ground survey points with an absolute accuracy of < 0.1 m from a pristine Lowveld Savanna environment provides evidence for the high accuracy of this new and truly global digital elevation model. Including selected, global, open access precursor DEMs, i.e. SRTM1, SRTM90 and ASTER-GDEM2 in the analysis highlights the engineering progress in the field of satellite-based surveying of the Earth.

However, and quite surprisingly, the high precision and accuracy of the IDEM data sets has only little impact when it comes to the delineation of watersheds and the calculation of surface water catchment area. Using 21 catchments of reservoirs within the Kruger National Park ranging in size from a few km² to about 100 km² resulted in differences of often only 1 % when comparing the results derived from the IDEM and the SRTM data sets.

ACKNOWLEDGEMENTS

The IDEM data was provided by the German Aerospace Center (Deutsches Zentrum für Luft- und Raumfahrt (DLR)) under the grant IDEM_Other0118, Kruger National Park Erosion Research DEM (KNPErosIDEM). ASTER GDEM2 is a product of METI and NASA. The SRTM1 data was released by the USGS in September 2014 and the SRTM90 data used here is courtesy of CGIAR-CSI. Field work was conducted in the framework of the KNP Erosion Research Project funded by the German Research Foundation (Deutsche Forschungs-gemeinschaft, DFG, grant: BA 1377/12-1). Ample support by SANParks Scientific Services in Skukuza, South Africa, is sincerely acknowledged.

REFERENCES

ASTER GDEM Validation Team, 2011. ASTER Global Digital Elevation Model Version 2 – Summary of Validation Results. http://www.jspacesystems.or.jp/ersdac/GDEM/ver2Validation/Summary_GDEM2_validation_report_final.pdf (25 Mar 2015).

Baade, J., and C. Schmullius, 2014. Uncertainties of a TanDEM-X derived Digital Surface Model. A Case Study from the Roda Catchment, Germany. *Proceedings IGARSS*, 2014: 4327-4330.

Bräutigam, B., M. Bachmann, D. Schulze, D. Borla Tridon, P. Rizzoli, M. Martone, C. Gonzales, M. Zink, and G. Krieger, 2014. TanDEM-X global DEM Quality status and aquisition completion. *Proceedings IGARSS*, 2014, pp. 3390-3393.

Chandler, G., and C. Merry, 2010. The South African Geoid 2010: SAGEOID10. *PositionIT*, 22 (June 2010), pp. 29-33.

Carruthers J., 1995. *The Kruger National Park. A social and political history.* Univ. of Natal Press, Pietermaritzburg, South Africa.

Esri Water Rescources Team, 2014. Arc Hydro Overview Document #1. Environmental Systems Research Institute, Inc. (Esri), Redlands, CA, USA.

Farr, T.G., P.A. Rosen, E. Caro, R. Crippen, R. Duren, S. Hensley, M. Kobrick, M. Paller, E. Rodriguez, L. Roth, D. Seal, S. Shaffer, J. Shimada, J. Umland, M. Werner, M. Oskin, D. Burbank and D. Alsdorf, 2007. The Shuttle Radar Topography Mission. *Reviews of Geophysics* 45(2), RG2004, pp. 1-33.

Hengl, T. and I.S. Evans, 2009. Mathematical and digital models of the land surface. In: Hengl, T. and H.I. Reuter, Eds., Geomorphometry. Concepts, Software, Applications. Elsevier, Amsterdam, NL, pp. 31-63.

Jarvis A., H.I. Reuter, A. Nelson, and E. Guevara, 2008. *Hole-filled seamless SRTM data V4*, International Centre for Tropical Agriculture (CIAT), available from http://srtm.csi.cgiar.org.

Krieger, G., M. Zink, M. Bachmann, B. Bräutigam, D. Schulze, M. Martone, P. Rizzoli, U. Steinbrecher, J.W. Antony, F. De Zan, I. Hajnsek, K. Papathanassiou, F. Kugler, M.R. Cassola, M. Younis, S. Baumgartner, P. Lopez-Dekker, P. Prats, and A. Moreira, 2013. TanDEM-X: A radar interferometer with two formation-flying satellites. *Acta Astronautica* 89(2), pp. 83-98.

Oksanen, J. and T. Sarjakoski, 2005. Error propagation of DEM-based surface derivatives. *Computers & Geosciences* 31, pp. 1015–1027.

Rodriguez, E., C.S. Morris, and J.E. Belz, 2006. A global assessment of the SRTM performance. *Photogrammetric Engineering & Remote Sensing*, 72(3), pp. 249-260.

Venter, F.J., R.J. Scholes and H.C. Eckhardt, 2003. The abiotic template and its associated vegetation pattern. In: Du Toit, J.T., K.H. Rogers and H.C. Biggs, Eds., *The Kruger Experience. Ecology and Management of Savanna Heterogeneity.* Island Press, Washington, USA, pp. 83-129.

Wessel, B., T. Fritz, T. Busche, B. Bräutigam, G. Krieger, B. Schättler, and M. Zink, 2013. TanDEM-X Ground Segment DEM Products Specification Document. Report TD-GS-PS-0021. Issue 2.0. Deutsches Zentrum für Luft- und Raumfahrt, Oberpfaffenhofen, Germany.

Wessel, B., A. Gruber, M. Huber, M. Breunig, S. Wagenbrenner, A. Wendleder, and A. Roth, 2014. Validation of the absolute height accuracy of TanDEM-X DEM for moderate terrain. *Proceedings IGARSS*, 2014, pp. 3394-3397.

Zink, M., and A. Moreira, 2014. TanDEM-X Mission status: The new topography of the Earth takes shape. *Proceedings IGARSS*, 2014, pp. 3386-3389.

14

ASTER AND WORLDVIEW-2 SATELLITE DATA COMPARISON FOR IDENTIFICATION OF GROUNDWATER SALINIZATION EFFECTS ON THE CLASSE PINE FOREST VEGETATION (RAVENNA, ITALY)

M. Barbarella[a], M. De Giglio[a, *], N. Greggio[b], L. Panciroli[a]

[a] Civil, Chemical, Environmental and Materials Engineering Department (DICAM), University of Bologna, Viale Risorgimento 2, 40136 Bologna, Italy - maurizio.barbarella@unibo.it, michaela.degiglio@unibo.it, lorenzo.panciroli@studio.unibo.it
[b] Interdepartmental Centre for Environmental Science Research (CIRSA), Lab. IGRG, University of Bologna, Via S. Alberto 163, 48100 Ravenna, Italy - nicolas.greggio2@unibo.it

BIOD-8

KEY WORDS: Aster, WorldView – 2, NDVI, stressed vegetation, groundwater salinity

ABSTRACT:

The availability of a large number of data acquired by satellite sensors with different spatial and spectral resolutions has always required an evaluation of their synergistic use. The integration of dataset of images coming from different sources can be an optimal solution for the study of various environmental problems which need a continuous monitoring (coastal development, forest evolution, land use changes etc.). The Classe pinewood, an important safeguarded biodiversity hot spot near Ravenna city (Italy), is historically affected by the groundwater salinization. Since changes in the water concentration are able to induce variations of the leaf properties and vegetation cover, recognizable by surveys carried out with different spectral bands, the comparison between ASTER and Worldview-2 data was performed using the (Normalized Difference Vegetation Index) NDVI. For each satellite data, the same Areas of Interest (AOIs) were selected within the most widespread cover, Thermophilic Deciduous Forest (TDF). The NDVI was calculated, statistically evaluated and the AOI rankings were built. In order to evaluate the difference between the results provided by the two images, statistical tests were applied on the average NDVI values. Finally the calculated NDVI were compared with groundwater salinity data collected during a contemporary field monitoring campaign. Based on groundwater salinity the same AOIs ranking was reached for both satellite sensors. This study suggests the opportunity to employ the medium resolution Aster images in continuity with high resolution WarldView-2 dataset.

1.1 INTRODUCTION

The availability of a constantly increasing number of satellites observing the Earth (Bailey et al., 2001) require tools and procedures to integrate the results obtained from data with different spectral and spatial resolutions (Wald et al., 1997; Chander et al., 2008). The main purpose is to achieve the continuity of data by temporal infill for the monitoring and modelling of natural resources (Yin et al., 2012). In fact, the use of data from different sensors can maximize the chances of obtaining a cloud-free image and to meet time requirements for information. In particular, several studies reported the inter-comparison of the Normalized Difference Vegetation Index (NDVI) data (Thenkabail, 2004; Abuzar et al., 2014). Many researchers used the NDVI as a biophysical indicator to analyze indirect effects of environmental changes (Aguilar et al. 2012; Barton, 2012), including those due to processes of salinization (Naumann et al. 2008; Zhang et al. 2011). The delineation of type and status of vegetation could provide a spatial overview of salinity distribution (Dehaan and Taylor, 2002; Tilley et al. 2007) and support land planners to reduce the risk resulting from salinization (Wiegand et al. 1994). Increased water salinity induces changes in chlorophyll concentration and therefore a photosynthesis slowdown (DeLaune et al. 1987). By measuring the relative difference between responses of chlorophyll and cellular structure in the red and near-infrared bands (Peñuelas, 1998), the NDVI analyses the greenness and al. productivity (Reed et 1994) of the plants.

The roman-time Classe pinewood (Ravenna, Italy), selected as study area, has been affected by groundwater salinization for several years (Antonellini et al. 2008). Included in the Po Delta Park, with other natural features of this region (wetlands, dunes, river mouths), it is classified as a protected area (EU site of importance and special area of conservation), in conformity with the Council Directive 92/43/EEC (http://ec.europa.eu/environment/nature/legislation/habitatsdirective/). Here, natural and anthropogenic land subsidence, low topography and the artificial drainage system led to a widespread saltwater intrusion (Antonellini et al. 2008). Antonellini and Mollema (2010) found that groundwater salinity and the water table level are the main causes of a progressive degradation of this ancient pine forest and of the coastal zones. Furthermore, according to the results of Giorgi and Lionello (2008), climate change will have a large influence on the water budget of Mediterranean countries, leading to an increase in the dry periods and to a subsequent increase of sea water intrusion.

Therefore, in order to monitor the temporal evolution of the pine forest health status and groundwater quality an upgradable dataset is necessary. Furthermore, the availability of comparable data from different sensors would provide comprehensive and continuous information over time. In this work, a comparison between two multispectral satellite data has been conducted with the aim to assess their potential integrated use. The presented analysis is based on ASTER and Worldview-2 (below WV-2) images, acquired in May 2011. In order to study the possible advantages of high spatial resolution in face of the

---

* Corresponding author. Michaela De Giglio, DICAM – University of Bologna, Viale Risorgimento 2, 40136 Bologna (Italy) michaela.degiglio@unibo.it.

medium resolution, the same procedure to identify portions of pinewood affected by groundwater salinization (Barbarella et al., 2015) was applied to each satellite data and the results were statistically compared (Pu and Landry, 2012). Given that, within the Classe pinewood, the same stressed areas were recognized, this study can suggest the use of the medium resolution in continuity with high resolution dataset.

2. STUDY AREA

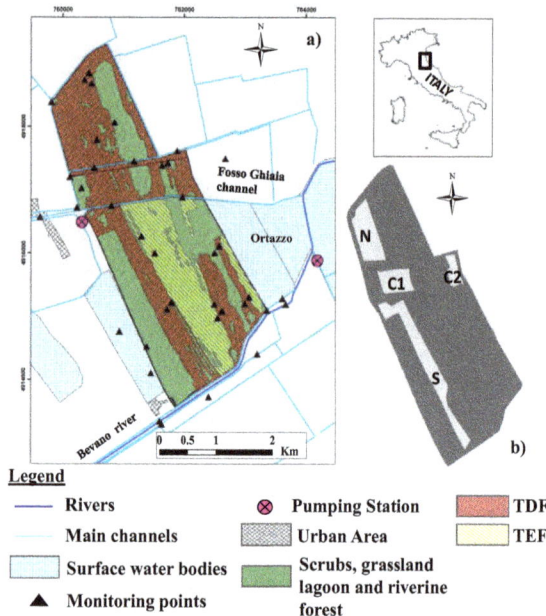

Figure 1. Classe pinewood (a) and the Areas of Interest (b)

The historical Classe pinewood (Fig. 1 (a)), included in the Po Delta Regional Park (Ravenna, Italy) because of its considerable plant and animal biodiversity, is 5 km long and 2 km large (900 ha). The pine forest is mainly surrounded by agricultural land except for the South-East part where it borders with the Ortazzo freshwater lagoon. The thickness of the underlying aquifer varies from a minimum of 6 to a maximum of 22 m (Amorosi et al. 1999).

Because of the low topography (about 2 m amsl) the area is strongly drained and a pumping station is present in the western part of the pinewood. Furthermore, the Bevano River and Fosso Ghiaia channel flow respectively in the southern boundary and in the center of the pinewood. Those two channels are directly open to the sea and especially during drought period, seawater can encroach the riverbed and reach the pineforest (Antonellini et al. 2008).

Based on the official vegetation map from Regione Emilia Romagna (1999), two main vegetation types are present in the Classe pineforest: "Thermophilic Deciduous Forest" (below TDF) and "Thermophilic Evergreen Forest" (below TEF). The main species found in both classes are *Quercus robur, Quercus pubescens, Fraxinus ornus, Populus alba, Ulmus minor, Salix cinerea*. The only difference between TDF and TEF is the presence of *Quercus ilex* (evergreen species) which indicates a drier and more elevated habitat. Because of its wider extension the TDF was selected as the target vegetation in this study. The *Pinus pinea* species is not classified in the vegetation map because it is not able to reproduce itself inside these natural area. Moreover this species, planted by monks in the 13th AD

(Ginanni, 1774), is stressed because it grows outside of its original climax (Piccoli et al. 1991).

3. METHODS

Before comparing the ASTER (VNIR sensor only, spatial resolution of 15m) and the WV-2 data (spatial resolution of 2m), the procedure developed on the nearby San Vitale Pinewood (Barbarella et al., 2015) was applied to each satellite data. Since the two images were acquired respectively on 05/18/2011 and on 05/29/2011, the atmospheric correction to retrieve surface reflectance was required (Yuan and Niu 2008, Abuzar et al., 2014). This pre-processing step was performed by the MODTRAN4 module as implemented into ENVI FLAASH (FLAASH Module, 2009).

For both ASTER and WV-2 images, the NDVI was calculated with ENVI software (below NDVI_A11 and NDVI_W11 respectively) for the same four Areas of Interest (AOIs, Fig. 1(b)), selected inside the TDF class (N, C1, C2, S). For each AOI, the NDVI values were statistically evaluated (mean, standard deviation). In order to understand if the difference between results provided by the two satellite data was significant, a statistical test was applied on the average NDVI values. In this case, considering that the two unknown population variances are not assumed to be equal, the Student's t-test (Eq. 1) was used to verify whether the population means were different, based on the statistic:

$$t = \frac{(\bar{x}_1 - \bar{x}_2)}{\sqrt{\frac{s_1^2}{n_1} + \frac{s_2^2}{n_2}}} \tag{1}$$

where $\bar{x}_1$, $\bar{x}_2$, s_1^2, s_2^2 = population means and variances,

n_1, n_2 = number of pixel

and for the degree of freedom (v) computation the Welch formula was used

$$v \approx \frac{(\frac{s_1^2}{n_1} + \frac{s_2^2}{n_2})^2}{(\frac{s_1^2}{n_1})^2 / (n_1 - 1) + (\frac{s_2^2}{n_2})^2 / (n_2 - 1)} \tag{2}$$

Subsequently, the frequency histograms of AOI NDVI values were plotted to explain the data distributions and higher order moments, Skewness and Kurtosis shape factors, were obtained to evaluate a possible deviation from the Guassian trend. In every image, using the AOI with the highest average NDVI, the NDVI value corresponding to the 5% of the pixels was used as threshold to compare the health vegetation status. Later, the percentages of pixels that fall below this limit were classified as stressed vegetation in all the other AOIs. Finally, the AOI rankings for each satellite image was carried out (Steps shown in the right column of flow chart., Fig 2).

In order to compare the results obtained from the two different sensors, further phases have been added to the original procedure (Steps shown in the left column of flow chart, Fig 2). The first analysis consisted in the statistical comparison of the average NDVI values for each AOI.

Afterwards, to compare the two satellite data at the same spatial resolution, the WV-2 pixels (2m) were resampled and re-projected to the ASTER projection and pixel size (15m). Based on the same AOIs the statistical NDVI analysis were repeated. The availability of NDVI values comparable for all pixels allowed to assess the correlation pixel by pixel between the two images.

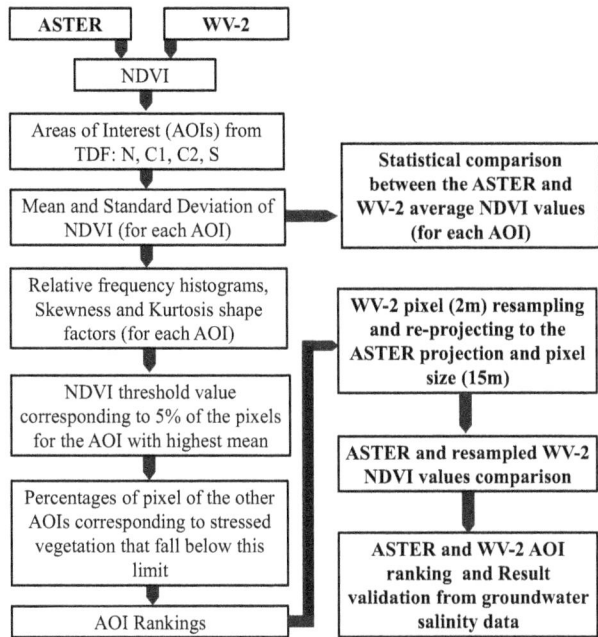

Figure 2. Flow chart of the whole procedure used to compare ASTER and WV-2 data

Finally, to validate the ASTER and WV-2 ranking salinity data collected from contemporary groundwater monitoring campaign were used.

During the Spring 2011, electrical conductivity and water table depth measures were collected within and surrounding the Classe pine forest from shallow piezometers (10 locations), surface water bodies (16 locations) and drainage channels (13 locations), (Fig. 1(a)). Electrical conductivity was converted into salinity using UNESCO methodology (1983) while water table depth was referred to the mean sea level. In order to produce the salinity maps only the top aquifer salinity were used. Firstly because of its contact with the trees roots and secondly because of its role in the supply of water during the evaporation processes (Buscaroli and Zannoni, 2010).

An interpolated grid was created for the month corresponding to the image acquisition dates (May 2011). Starting from sparse points, the Kriging algorithm with linear variogram was applied to obtain a continuous pattern (Akkala et al. 2010) using the software SURFER 11. The spacing of nodes is identical to the geometric resolutions of the satellite data. The saline contour line maps were overlaid with the AOIs location to highlight zones of pinewood affected by saltwater intrusion. Later from the original grid file, salinity values were extracted for each node within the AOIs boundaries. After this, means, standard deviations, minimums and maximums were computed for every area. Thus, the relation between NDVI and groundwater salinity data was analysed for each image. To study the relationship between the average NDVI (Y) and salinity (X) values, the errors that affect both variables must be considered. Therefore, assessing the linear fit, the use of the traditional formulas that consider the independent variable (X) as error-free was not correct. Accordingly, in this study, the following regression model was applied. The linear function:

$$Y = \alpha + \beta\, X \tag{3}$$

has been combined with a more complex statistical model.

For a generic point it was possible to consider that the value of the coordinate measured was affected by a normally distributed error, i.e. $\tilde{x}_k = X_k + \varepsilon_{Xk}$ and $\tilde{y}_k = Y_k + \varepsilon_{Yk}$, where $\varepsilon_k \in N(0, \sigma_{\varepsilon k}^2)$, $\eta_k \in N(0, \sigma_{\eta k}^2)$, $cov(\varepsilon_i, \eta_j) = 0$ with $i, j, k = 1, 2, ... n$.

Substituting each unknown regression parameter with an approximate value and an unknown correction, $\alpha = a_0 + \delta a$, $\beta = b_0 + \delta b$, and, eliminating the infinitesimal of higher order $\delta b\, v_x$, the linear function became

$$\begin{bmatrix} 1 & x_k \end{bmatrix} \begin{bmatrix} \delta a \\ \delta b \end{bmatrix} + \begin{bmatrix} b_o & -1 \end{bmatrix} \begin{bmatrix} v_{x_k} \\ v_{y_k} \end{bmatrix} + a_o + b_o\, x_k - y_k = 0 \tag{4}$$

with $k = 1, 2, .. n$.

For all points the functional model was:

$$\begin{bmatrix} 1 & x_1 \\ . & . \\ . & . \\ 1 & x_n \end{bmatrix} \begin{bmatrix} \delta a \\ \delta b \end{bmatrix} + \begin{bmatrix} b_o & -1 & 0 & 0 & . & . & 0 & 0 \\ 0 & 0 & b_o & -1 & . & . & 0 & 0 \\ . & . & . & . & . & . & 0 & 0 \\ . & . & . & . & . & . & . & . \\ 0 & 0 & 0 & 0 & b_o & -1 & 0 & 0 \\ 0 & 0 & 0 & 0 & 0 & 0 & b_o & -1 \end{bmatrix} \begin{bmatrix} v_{x_1} \\ v_{y_1} \\ . \\ . \\ . \\ . \\ v_{x_n} \\ v_{y_n} \end{bmatrix} + \begin{bmatrix} a_o + b_o\, x_1 - y_1 \\ . \\ . \\ a_o + b_o\, x_n - y_n \end{bmatrix} = \begin{bmatrix} 0 \\ . \\ . \\ 0 \end{bmatrix} \tag{5}$$

That is:

$$\underset{n*2}{B}\, \underset{2*1}{\delta u} + \underset{n*2n}{A}\, \underset{2n*1}{v} + \underset{n*1}{f} = \underset{n*1}{0} \tag{6}$$

with the associated stochastic model

$$\Sigma_{ff} = diag(\sigma_{x_1}^2, \sigma_{y_1}^2, ..., \sigma_{x_n}^2, \sigma_{y_n}^2) \equiv \sigma_o^2\, Q_{ff} \tag{7}$$

The model solution obtained by applying the method of weighed least squares was:

$$\delta \underline{u} = -B^t (A Q_l\, A^t)^{-1} B]^{-1} B^t (A Q_l\, A^t)^{-1}\, \underline{f} \tag{8}$$

with the cofactor matrix of the unknowns given by

$$Q_{\delta u} = [B^t (A Q_l\, A^t)^{-1} B]^{-1} \tag{9}$$

To initialize the interactive process, the fit according to traditional formulas was considered while the end of the iteration criterion was based on the negligible increase of the estimated parameters.

4. RESULTS AND DISCUSSION

The result section is divided into three subparagraphs: ASTER and WV-2 comparison; ASTER and resampled WV-2 comparison and the finding validation by groundwater salinity data.

4.1 ASTER and WV-2 comparison

The NDVI results relative to ASTER and WV-2 images are reported as a map in figure 3. It is possible to see, according to the chromatic scale, an apparent improvement in the NDVI_W11. This last aspect could be related to the mediation of the reflectance value assigned to the Aster pixel corresponding to a larger area on the ground compared to the pixel of the WV-2.

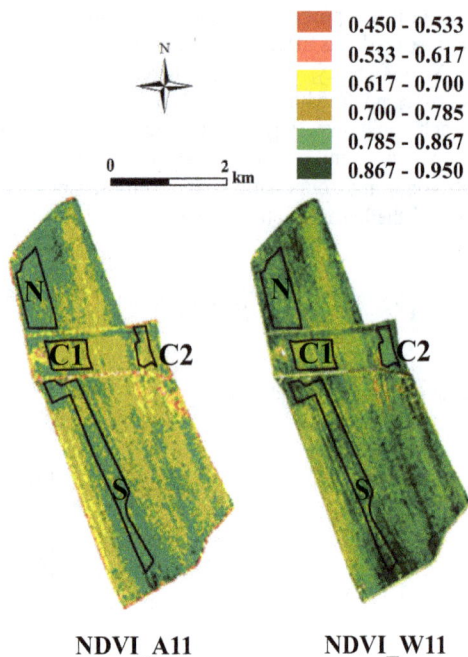

Figure 3. Classe NDVI maps

For each AOI, the mean and the standard deviation of vegetation index values are shown in table 1. Within the same scene the four AOIs are statistically separated. In each satellite data, N and S AOIs are identified as less stressed areas while C1 and C2 AOIs show the lower average NDVI values.

AOI	NDVI_A11		NDVI_W11	
	N° pixel	Mean/St. Dev.	N° pixel	Mean/St. Dev.
N	2029	0.811 / 0.025	114826	0.831 / 0.048
C1	1239	0.763 / 0.030	70271	0.788 / 0.046
C2	551	0.773 / 0.024	22908	0.819 / 0.029
S	3104	0.809 / 0.033	176248	0.832 / 0.055

Table 1. Basic statistics of NDVI values

To evaluate the difference between results shown in Table 1 the Student t-test was applied.

Given N, C1, C2, S AOIs, the sample consisted of the NDVI values for the two data. Due to the different resolution, the largeness of samples differed by one or two orders of magnitude. Moreover, the element number of the samples was elevated. The results reported in Table 2 show that the NDVI averages are significantly different for each AOI. In fact, the NDVI differences exceeded the limit $t_\alpha = 3.3$ corresponding to the significance level of $\alpha = 0.1$ % conservatively assumed for the test.

AOI	t	v
N	34.9	2300.6
S	37.9	3413.9
C2	66.6	1302.5
C1	19.0	613.9

Table 2. Student t-test results relative to the average NDVI differences

The real distribution of NDVI_A11 and NDVIW11 values were analyzed by relative frequency histograms shown in Fig.4. Regarding the NDVI_A11 graphs, it is possible to distinguish three different histogram shapes related to several AOI behaviors. The most stressed area has a right tail, while the healthiest shows a left tail. The intermediate area does not show any tail. Instead, from the NDVI_W11 graphs is less evident the presence of tail, but S and C1 AOIs show a bimodal trend and, probably due to the different vegetation species present in TDF. In the histograms, the grey line represents the NDVI value corresponding to the 5th percentile computed in the AOI with the highest average NDVI value, i.e. the N area for ASTER image (0.767) and the S area for in the WV-2 data (0.742). For each data, the percentage of pixels below this limit was calculated for the remaining AOIs (Tab. 3). For the NDVI_A11, the percentages of pixels below the limit increases moving from the areas with higher average NDVI values (N and S) towards the lower NDVI values (C1 and C2).

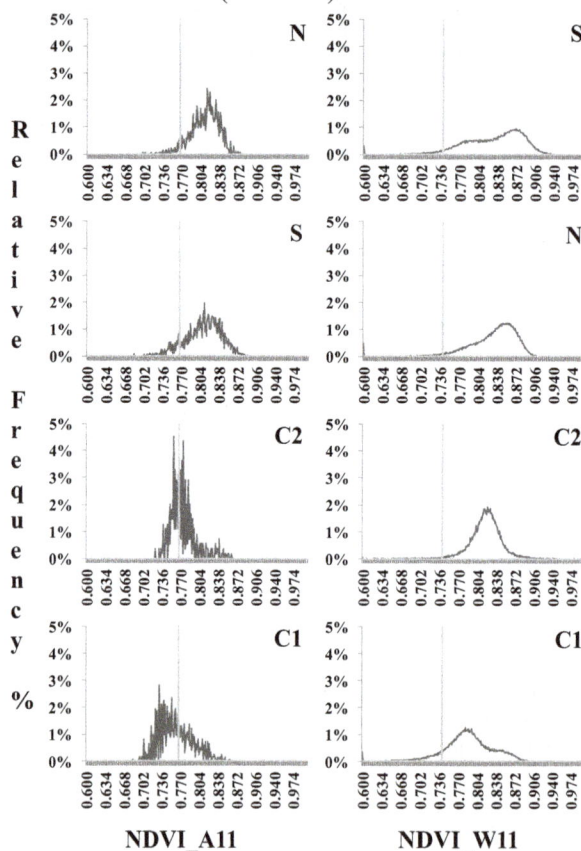

Figure 4. Relative frequency histograms of NDVI values. Based on average NDVI the AOIs are reported in descending order

This behaviour is less evident in the NDVI_W11 histograms where the worst area (C2) has only 12% of pixel below the threshold.

		% pixels of stressed vegetation			
	Threshold	N	S	C1	C2
NDVI_A11	N: 0.767	5.00	11.60	59.00	47.60
NDVI_W11	S: 0.742	3.40	5.00	12.00	1.20

Table 3. Percentage of pixels below the threshold determined on the AOI with higher average NDVI value

Additional information can be obtained by studying the Skewness and Kurtosis shape factors (Tab. 4).The NDVI_A11 Skewness results confirm that the N and S areas have a left tail, whereas C2 have a right tail.

AOI	Shape Factors	NDVI_A11	NDVI_W11
N	Skewness	-0.594	-2.640
	Kurtosis	0.528	15.474
C1	Skewness	0.432	-0.847
	Kurtosis	-0.153	3.784
C2	Skewness	1.194	-0.641
	Kurtosis	1.752	5.090
S	Skewness	-0.466	-1.253
	Kurtosis	0.079	4.547

Table 4. Higher order moment results

With regard to the NDVI_A11 Kurtosis results, they are close to zero, with the C1 factor negative, demonstrating that their value distribution get closer the Gaussian distribution. Instead, the NDVI_W11 Skewness results are negative for all AOIs and S and N have the longest tail. Finally, all the Kurtosis values are largely positive showing that their NDVI value distributions moves away from Gaussian due to more acute peaks around the mean.

The following rankings were derived from the previous results (Tab.5). Based on the average NDVI the AOIs are reported in descending order.

Order	NDVI_A11	NDVI_W11
1	N	S
2	S	N
3	C2	C2
4	C1	C1

Table 5. ASTER and WV-2 AOI ranking based on average NDVI values.

Table 5 shows that both satellite data recognize as the more suffering areas the AOIs located between two drainage channel crossing the pine forest, i.e. C1 and C2. However, areas N and S are inverted in the two rankings. Therefore, in order to assess if the difference between N and S average NVDI values could be considered negligible, the previous Student t-test was separately applied on each satellite data. From Table 1, it is possible to deduce that the difference between N and S average NDVI values in absolute terms is of 0.002 for NDVI_A11 and of 0.001 for NDVI_W11.

	t
N-S NDVI_A11	2.5
N-S NDVI_W11	-5.2

Table 6. Student t-test results relative to the N and S average NDVI differences for each image.

For the Aster data, the test result is lower than the limit t_α= 3.3 corresponding to the significance level of α= 0.1 % assumed for the test. Instead, for the WV-2 image the equality hypothesis between the two average NDVI values is not acceptable (Tab. 6). The problem is probably related to the large amount of data that, reducing the standard deviation in the denominator, raises the relationship value.

4.2 ASTER and resampled WV-2 comparison

After the WV-2 pixels resampling and re-projection to the ASTER projection and pixel size (15m), the new AOI average NDVI values were equal to the average NDVI values of no-resampled WV-2 AOIs (Tab. 7).

Resampled WV-2			
AOI	N° pixel	Mean	Dev. St
N	2029	0.8323	0.0494
S	3104	0.8316	0.0541
C1	1239	0.7879	0.0299
C2	551	0.8199	0.0161

Table 7.Basic statistics of NDVI values for the resampled WV-2

In this case, the result of the Student t-test (t=0.5) applied on the new N and S average NDVI difference demonstrated that the new NDVI averages of the two WV-2 AOIs could be considered equal. Therefore the previous AOI rankings obtained from the two sensors can be considered coincident.

The availability of NDVI values comparable between ASTER and WV-2 allowed the pixel to pixel correlation between the all AOI pixels of NDVI_A11 and all AOI pixels of resampled NDVI_W11 (Fig. 5). The graph shows that the limited NDVI_A11 range of variability (0.683 - 0.890) corresponds to a very large range of values assumed by NDVI_W11(0.283-0.992). The likely cause is the original WV-2 high spatial resolution that makes the satellite data more sensitive to the vegetation variability within the TDF class. However, the global regression line coefficient is high, almost 0.84 demonstrating a good correlation between the two satellite images.

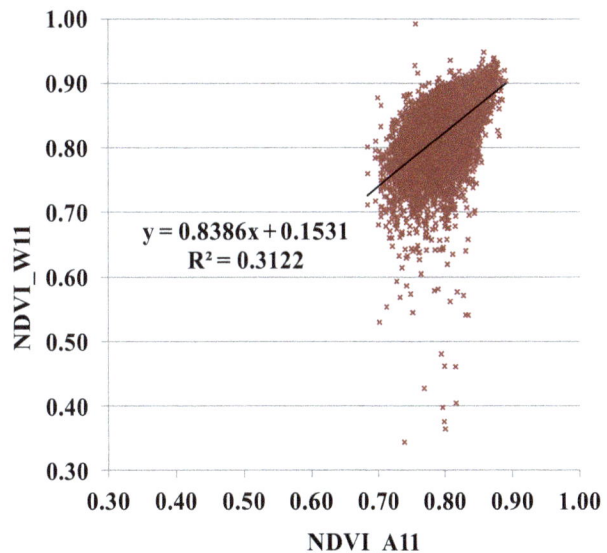

Figure 5. Pixel to pixel correlation between the all AOI pixels of NDVI_A11 and the all AOI pixels of resampled NDVI_W11.

However, the single AOI behavior is different. While for the pixel to pixel linear correlations of N, S and C1 areas the angular coefficient ranges from 0.71 to 0.99, the C2 coefficient (0.2) proves the almost total invariance of the NDVI_W11 compared to NDVI_A11. This anomalous behavior is evident also in Fig.6. The graph shows the relationship between the average NDVI_W11 AOI values and the average NDVI_A11 AOI values. The error bars are a graphical representation of the standard deviations of each average value while the red line is the regression line relative to N, S and C1 values. The linear correlation between the average NDVI values of these areas is remarkable, instead the C2 behavior, excluded from the calculation, differs greatly. For comparison, the bisector indicated by the dashed line was reported.

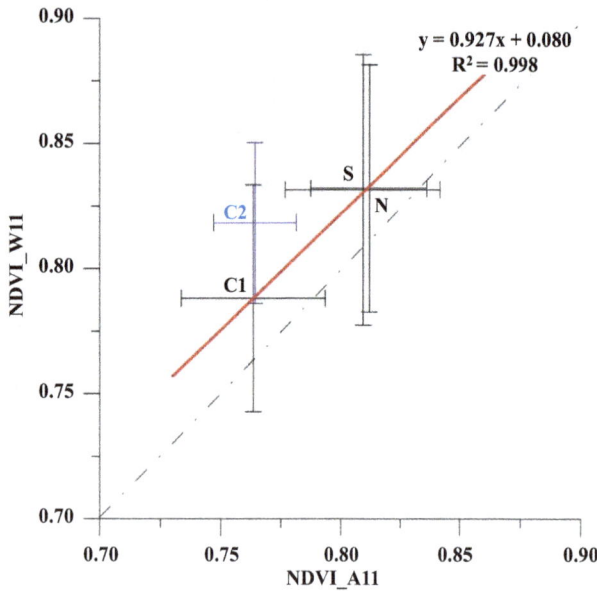

Figure 6. Relationship between the average NDVI_W11 AOI values and the average NDVI_A11 AOIs values

4.3 Result validation by groundwater salinity data

Despite the different spatial resolution that implies a different response at individual pixel level, both satellite data provided the same AOI ranking, recognizing N and S areas as the areas covered by healthier vegetation while C1 and C2 those covered by more suffering vegetation.

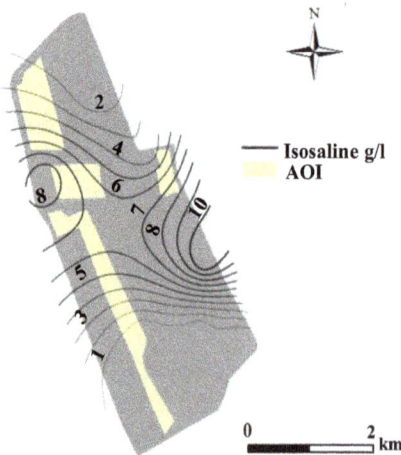

Figure 7. Overlay between the Classe AOI locations and the groundwater salinity isosalines (g/l) of May 2011

These findings were validated by the contemporaneous groundwater salinity data. Figure 7 shows the spatial salinity distributions relative to May 2011 through the isosaline.

The AOIs characterized by higher salinity (from 5 to 8 g/l) were C1 and C2 while a wider salinity variability range involved the remaining AOIs. Everywhere in the pine forest the surface salinity is below 12 g/l. The mean and the standard deviation salinity values extracted for each AOI are reported in Tab. 9. The maximum average salinity value for May 2011 was recorded within the C1 area (6.94 g/l) whereas S and N had the lower average salinity values, respectively, 3.46 and 3.94 g/l.

Salinity(g/l)			
AOI	N° pixel	Mean (g/l)	Dev. St
N	2029	3.94	1.27
C1	1239	6.94	0.77
C2	3104	5.99	0.95
S	551	3.46	2.84

Table 9. Basic statistics of salinity values of the Classe AOIs

For every satellite data, the relationship between the average salinity and NDVI values of each AOI were verified (Fig. 8 and Fig. 9). Taking into account that the salinity standard deviations were highly variable between the AOIs while the NDVI standard deviations were comparable from sample to sample, within each satellite data, the Equation 6 was applied.

The slope of the regression line is negative for both considered images, confirming the relation between lower NDVI and higher salinity values and vice versa.

However, the variability range of the NDVI_W11 regression line slope is wider than that of NDVI_A11.

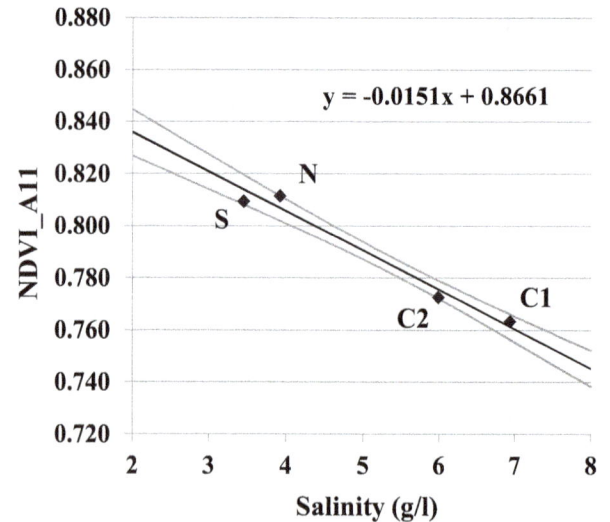

Figure 8. Correlation between the average NDVI_A11 and groundwater salinity values for Classe AOIs

Finally, after all the assessments made, a comparison of the AOI rankings in terms of NDVI_A11, NDVI_W11 and groundwater salinity data can be done. Table 10 shows a perfect agreement between the results obtained.

In particular, the most stressed areas (C1 andC2) are located in the portion of pinewood define by the two drainage channels. This consideration agree with the findings of Antonellini et al., (2008) who identify drainage channels as accountable of vertical seepage of saline water from the bottom part of the aquifer.

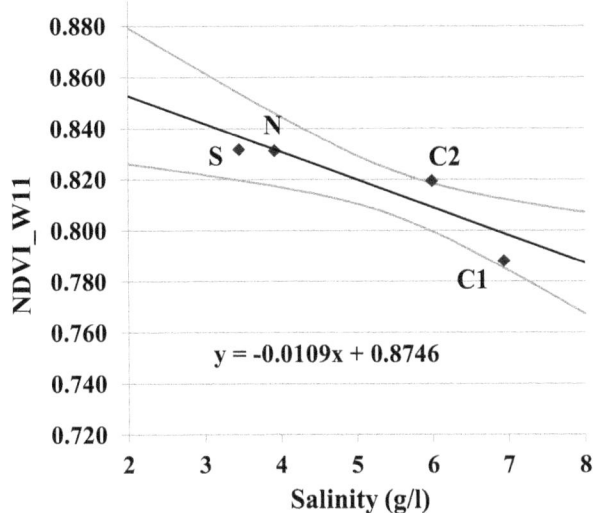

Figure 9. Correlation between the average NDVI_W11 and groundwater salinity values for Classe AOIs

Order	NDVI_A11/W11	Salinity
1	S	S
2	N	N
3	C2	C2
4	C1	C1

Table 10. Comparison between the rankings of AOIs in terms of NDVIs and salinity. The AOIs are reported in descending order.

CONCLUSIONS

The same NDVI ranking has been obtained considering both satellite datasets keeping the original spatial resolution, unless the areas N and S which are inverted. However the Student t-test reveals that the average AOIs NDVI calculated from ASTER data is statistically different from the average AOIs NDVI calculated from WV-2 data. The same test repeated to assess if the difference between the average NDVI values of N and S shows a significant difference for WV-2 image. For ASTER image the histogram analysis identifies the left tails for the less stressed AOIs and right tails for the most stressed; this different shape is less evident for WV-2 data. In the same way the percentage of pixels below the 5% threshold can be used as measure of stress conditions only for ASTER image; in fact for WV image the percentage of pixels are not related with the average AOIs NDVI.

After the WV re-sampling and re-projection to the ASTER resolution, the ranking was the same and the statistical evaluation of N and S average NDVI revealed that can be considered equal. The correlation between ASTER and resampled WV-2 shows high values for the regression line coefficients which considerably improve using the average AOIs NDVI, excluding C2. From the validation with the groundwater salinity emerges that the previous AOIs ranking based on average NDVI perfectly agree with the ranking of salinity, i.e. N and S AOIs have the lower average salinity and the higher NDVI values while the contrary for C1and C2. Finally, the use of the NDVI analysis allowed to identify the AOIs more affected by groundwater salinization based on the groundwater monitoring data. The same NDVI ranking has been obtained considering either by keeping the original spatial resolution of the satellite data that with the WV-2 re-sampling and re-projection. After these findings is possible to conclude that ASTER and WV-2 sources can be integrated with the aim to study environmental problems which require a long dataset even back in time.

ACKNOWLEDGEMENTS

A special thanks to Prof. Giovanni Gabbianelli for having made available the WorldView-2 data and for the constant supervision to the work.

REFERENCES

Abuzar, M., Sheffield, K., Whitfield, D., O'Connell, M., McAllister, A., 2014. Comparing inter-sensor NDVI for the analysis of horticulture crops in south-eastern Australia. *American Journal of Remote Sensing*, 2(1), pp.1-9.

Aguilar, C., Zinnertb, J.C., Poloa, M.R., Young, D.R., 2012. NDVI as an indicator for changes in water availability to woody vegetation. *Ecological Indicators*, 23, pp. 290–300.

Akkala, A., Devabhaktuni, V., & Kumar, A., 2010. Interpolation techniques and associated software for environmental data. *Environmental progress & sustainable energy*, 29 (2), pp. 134-141.

Amorosi, A., Colalongo, M.L., Pasini, G., Preti, D., 1999. Sedimentary response to Late Quaternary sea-level changes in the Romagna coastal plain (northern Italy). *Sedimentology*, 46(1), pp. 99-121.

Antonellini, M., Mollena, P., Giambastiani, B.M.S., Bishop, K., Caruso, L., Minchio, A., Pellegrini, L., Sabia, M., Ulazzi, E., Gabbianelli, G., 2008. Salt water intrusion in the coastal aquifer of southern Po Plain, Italy. *Hydrogeology journal*, 16, pp.1541–1556.

Antonellini, M., Mollema, P., 2010. Impact Of Groundwater Salinity On Vegetation Species Richness In The Coastal Pine Forests And Wetlands Of Ravenna, Italy. *Ecological Engineering*, 236(9), pp. 1201-1211.

Bailey, G. B., Lauer, D. T., Carneggie, D. M., 2001. International collaboration: the cornerstone of satellite land remote sensing in the 21st century. *Space Policy*, 17(3), pp. 161-169.

Barbarella, M., De Giglio, M., Greggio, N., 2015. Effects of salt water intrusion on pinewood vegetation using satellite ASTER data: the case study of Ravenna (Italy). *Environmental Monitoring and Assessment*. DOI: 10.1007/s10661-015-4375-z.

Barton, C.WM., 2012. Advances in remote sensing of plant stress. *Plant and Soil*, 354, pp. 41–44.

Buscaroli, A., Zannoni, D., 2010. Influence of ground water on soil salinity in the San Vitale Pinewood (Ravenna-Italy). *Agrochimica*, 54(5), pp. 303-320

Chander, G., Coan, M. J., Scaramuzza, P. L., 2008. Evaluation and comparison of the IRS-P6 and the Landsat sensors. *Geoscience and Remote Sensing*, IEEE Transactions on, 46(1), pp. 209-221.

Dehaan, R. L., Taylor, G. R., 2002. Field-derived spectra of salinized soils and vegetation as indicators of irrigation-induced soil salinization. *Remote Sensing of Environment*, 80(3), pp. 406-417.

DeLaune, R.D., Pezeshki, S.R., Patrick Jr., W.A., 1987. Response of coastal plants to increase in submergence and salinity. *Journal of Coastal Research*, 3(4), pp. 535–546.

FLAASH Module, 2009. Atmospheric Correction Module: QUAC and FLAASH User's Guide, ENVI Version 5.0. ITT Visual Information Solutions, Boulder, CO.

Ginanni, F., 1774. Istoria civile e naturale delle Pinete Ravennati nella quale si tratta della loro origine, situazione, fabriche antiche, e moderne, terre moltiplici, acqua, aria, fossili, vegetabili, &c: Opera postuma. Generoso Salomoni.

Giorgi, F., Lionello, P. (2008). Climate Change Projections For The Mediterranean Region. *Global Planet Change*, 63, pp. 90–104.

Naumann, J.C., Anderson, J.E., Young, D.R., 2008. Linking physiological responses, chlorophyll fluorescence and hyperspectral imagery to detect salinity stress using the physiological reflectance index in the coastal shrub, Myrica cerifera. *Remote sensing of environment*, 112(10), pp. 3865-3875.

Peñuelas, J., 1998. Visible and near–infrared reflectance techniques for diagnosing plant physiological status. *Trends in Plant Science,* 3, pp. 151–6.

Piccoli, F., Gerdol, R., Ferrari, C., 1991. Vegetation Map of St. Vitale pinewood (Northern Adriatic coast, Italy). *Phytocoenosis*, pp. 337-342.

Pu, R.; Landry, S., 2012. A comparative analysis of high spatial resolution IKONOS and WorldView-2 imagery for mapping urban tree species. *Remote Sens. Environ.*, 124: pp. 516–533

Reed, B.C., Brown, J.F., Vander Zee, D., Loveland, T.R., Merchant, J.W., Ohlen, D.O., 1994. Measuring phenological variability from satellite imagery. *Journal of Vegetation Science*, 5, pp. 703–714.

Regione Emilia Romagna - Servizio Cartografico e Geologico, 1999. Carta della Vegetazione del Parco Regionale del Delta del Po - Stazione "Pineta di Classe e Saline di Cervia". http://geoportale.regione.emilia-romagna.it/it/catalogo/dati-cartografici/ambiente/carta-della-vegetazione/carta-della-vegetazione-parco-regionale-del-delta-del-po-stazione-pineta-di-classe-e-saline-di-cervia-digitale-edizione-1999. Accessed 02/22/2015

Thenkabail, P. S. 2004. Inter-sensor relationships between IKONOS and Landsat-7 ETM+ NDVI data in three ecoregions of Africa. *International Journal of Remote Sensing*, 25(2), 389-408.

Tilley, D.R., Ahmed, M., Son, J.H., Badrinarayanan, H., 2007. Hyperspectral reflectance response of freshwater macrophytes to salinity in a brackish subtropical marsh. *Journal of Environmental Quality*, 36, pp. 780–789.

UNESCO, 1983. Algorithms for computation of fundamental properties of seawater. Unesco technical papers in marine science 44, Unesco/SCOR/ICES/IAPSO Joint Panel on Oceanographic Tables and Standards and SCOR Working Group 51.

Wald, L., Ranchin, T., Mangolini, M., 1997. Fusion of satellite images of different spatial resolutions: assessing the quality of resulting images. *Photogrammetric Engineering and Remote Sensing*, 63(6), pp. 691-699.

Wiegand, C.L., Rhoades, J.D., Escobar, D.E., Everitt, J.H., 1994. Photographic and videographic observations for determining and mapping the response of cotton to soil-salinity. *Remote Sensing of Environment*, 49, pp. 212–223.

Yin, H., Udelhoven, T., Fensholt, R., Pflugmacher, D., Hostert, P., 2012. How normalized difference vegetation index (ndvi) trendsfrom advanced very high resolution radiometer (AVHRR) and système probatoire d'observation de la terre vegetation (spot vgt) time series differ in agricultural areas: An inner mongolian case study. *Remote Sensing*, 4(11), pp. 3364-3389.

Yuan, J., Niu, Z., 2008. Evaluation of atmospheric correction using FLAASH. In: International Workshop on Earth Observation and Remote Sensing Applications, pp. 1–6.

Zhang, T., Zeng, S.L., Gao, Y., Ouyang, Z.T., Li, B., Fang, C.M., Zhao, B., 2011. Using hyperspectral vegetation indices as a proxy to monitor soil salinity. *Ecological Indicators*, 11, 1552–1562.

15

SATELLITE IMAGE SIMULATIONS FOR MODEL-SUPERVISED, DYNAMIC RETRIEVAL OF CROP TYPE AND LAND USE INTENSITY

H. Bach [a], *P. Klug* [a], *T. Ruf* [a], *S. Migdall* [a], *F. Schlenz* [b], *T. Hank* [b], *W. Mauser* [b]

[a] Vista Geowissenschaftliche Fernerkundung GmbH, Gabelsbergerstr. 51, 80333 München, Germany – bach@vista-geo.de, ruf@vista-geo.de, klug@vista-geo.de, migdall@vista-geo.de
[b] Ludwig-Maximilians-Universität München, Luisenstraße 37, 80333 München, Germany – f.schlenz@iggf.geo.uni-muenchen.de, tobias.hank@lmu.de, w.mauser@lmu.de

KEY WORDS: M4Land, land management classification system, Sentinel, canopy reflectance model SLC, crop growth model PROMET

ABSTRACT:

To support food security, information products about the actual cropping area per crop type, the current status of agricultural production and estimated yields, as well as the sustainability of the agricultural management are necessary. Based on this information, well-targeted land management decisions can be made. Remote sensing is in a unique position to contribute to this task as it is globally available and provides a plethora of information about current crop status.

M[4]Land is a comprehensive system in which a crop growth model (PROMET) and a reflectance model (SLC) are coupled in order to provide these information products by analyzing multi-temporal satellite images. SLC uses modelled surface state parameters from PROMET, such as leaf area index or phenology of different crops to simulate spatially distributed surface reflectance spectra. This is the basis for generating artificial satellite images considering sensor specific configurations (spectral bands, solar and observation geometries). Ensembles of model runs are used to represent different crop types, fertilization status, soil colour and soil moisture. By multi-temporal comparisons of simulated and real satellite images, the land cover/crop type can be classified in a dynamically, model-supervised way and without in-situ training data. The method is demonstrated in an agricultural test-site in Bavaria. Its transferability is studied by analysing PROMET model results for the rest of Germany. Especially the simulated phenological development can be verified on this scale in order to understand whether PROMET is able to adequately simulate spatial, as well as temporal (intra- and inter-season) crop growth conditions, a prerequisite for the model-supervised approach.

This sophisticated new technology allows monitoring of management decisions on the field-level using high resolution optical data (presently RapidEye and Landsat). The M[4]Land analysis system is designed to integrate multi-mission data and is well suited for the use of Sentinel-2's continuous and manifold data stream.

1. INTRODUCTION

To support food security, information products about the actual cropping area per crop type, the current status of agricultural production and estimated yields, as well as the sustainability of the agricultural management are necessary. Based on this information, well-targeted land management decisions can be made. Remote sensing is in a unique position to contribute to this task as it is globally available and provides a plethora of information about current crop status.

With the SENTINEL sensor family, a fleet of Earth Observation (EO) satellites is starting to become available, which will continuously monitor the land surface at different spatial scales (10 – 300 m) and with different systems (optical, microwave) (Berger, 2011). For an optimal translation of this data stream of different resolutions and wavelength ranges into land management information, an integrated analysis of the complete image data stream is required. This can be achieved through embedding the analysis in a continuous spatial modeling of land surface processes covering also the intervals between acquisitions.

In the frame of the M[4]Land project (**M**odel based, **M**ulti-temporal, **M**ulti scale and **M**ulti sensorial retrieval of continuous land management information), a method to derive products for a sustainable management of the land surface is being developed. The method combines the full bandwidth of the spatial information provided by the future SENTINEL series within a land surface process model to generate spatially explicit and temporally continuous land surface management information products, such as dynamic land use, degree of ecological intensification, irrigation status, calamities etc. The system uses a dynamic classification of land cover, which is physically based and without training by a combination of the reflectance model SLC (Soil-Leaf-Canopy) (Verhoef, 2003) and (Verhoef, 2007) and the land surface process model PROMET (Processes of Radiation, Mass and Energy Transfer) (Mauser, 2009).

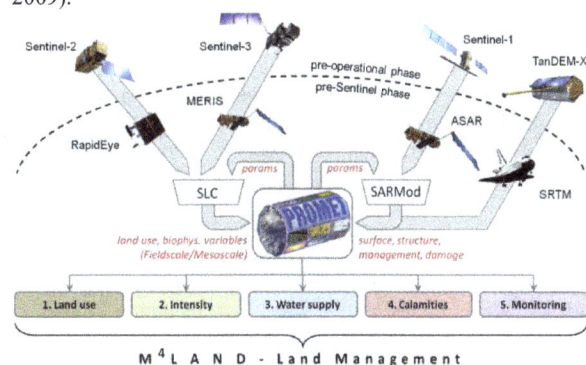

Figure 1. The M[4]Land concept, showing the sensors employed during the development phase as well as in the pre-operational phase after the SENTINEL launch (Klug, 2014)

This paper explains the M⁴Land concept and demonstrates it using time series of high resolution, optical data (RapidEye). Focus is laid on the principle of the new methodology as well as on its geographical transferability, for which the model-based approach is essential. This leads us to calling the methodology a "model-supervised" classification in contrast to common supervised or unsupervised classifications. Our assumption is that when we understand the land surface processes, which cause crop growth and phenological development as well as the radiative transfer of the canopy and soils (absorption, scattering and reflectance) adequately, we are able to simulate satellite images that are similar to real observations. Using this technique in an inverse mode we are then able to derive management information (like decisions on land use or seeding dates) that cannot be perfectly simulated and therefore rely on e.g. satellite image information. This synergistic concept shall be demonstrated in this paper.

2. METHODS

Two types of physically-based models are used in M⁴Land in an integrative way, a crop growth agro-hydrological model and a radiative transfer model for simulating satellite measurements of reflectances. They are introduced below.

2.1 Crop growth modeling with PROMET

PROMET allows simulating all relevant water and energy fluxes related to radiation balance, vegetation, soil, snow and aerodynamic exchange processes on the land surface in a spatially distributed way. A detailed description of the model physics and components is given in (Mauser, 2009). The model results have been validated in different test sites on different scales (from 5 m to 1 km) with good results (Hank, 2015), (Migdall, 2009), (Mauser, 2009).

PROMET uses spatial data like soil maps and a digital terrain model as well as meteorological forcing data as input for hourly simulations. The meteorological data consists of hourly information on temperature, precipitation, relative humidity, wind speed and cloud cover, as offered by national weather services.

The development of crops is simulated in PROMET dynamically depending on the environmental conditions (mainly temperature, radiation and moisture conditions) while standard farming practices (e.g. seeding and harvest dates) are taken into account. The growth and accumulation of biomass is the result of an explicit simulation of photosynthetic processes based on the Farquhar concept (Farquhar, 1980). The assimilates are distributed within the canopy depending on the phenological progress of the different crop types.

The necessary parameterization of the crop types (from which 23 are implemented in PROMET) are kept generic and not optimized for a specific site, in order to allow for the geographical transferability of the M⁴Land approach.

2.2 Radiative transfer modeling with SLC

The used surface reflectance model SLC (Soil-Leaf-Canopy) is an integrated radiative transfer model for the simulation of top-of-canopy spectral reflectance. The model consists of a modified Hapke soil BRDF model, a robust version of the PROSPECT leaf optical properties model, and the canopy radiative transfer model 4SAIL2, a two-layer robust version of SAILH (Verhoef, 2003).

In the M⁴Land system, SLC is configured to use spectral configurations and acquisition parameters from the used satellite sensors (in this case RapidEye), soil spectral properties (single scattering albedo values for various soil types), as well as leaf parameters like chlorophyll content, leaf water, leaf dry matter and mesophyll structure, which can be predefined for every simulated crop type. SLC also allows to use PROMET outputs as input, like canopy parameters such as leaf area index (LAI), leaf angle distribution (connected to phenological development) and degree of maturity (fraction of brown leaves).

2.3 Satellite data and test site

As test site for a first demo application an agricultural area near Neusling in Bavaria, Germany, is selected. Land use and crop type were mapped during the growing season of 2010 for an area of approx. 4 km by 3 km. Winter wheat, winter barley, silage maize, potato and sugar beet are the relevant crops in this region.

A total of 10 almost cloud free RapidEye scenes were available for the growing season of 2010 (Table 1). With exception of September 2010, at least one RapidEye image is available for every month, guaranteeing a good and evenly distributed coverage of the entire growing season. The satellite images were resampled to a 20 m grid and an atmospheric correction was carried out using a MODTRAN Interrogation Technique (Verhoef, 2003) to retrieve bottom of atmosphere reflectance values.

March 26th	July 11th
April 8th	July 31st
May 11th	August 21st
June 6th	October 12th
June 25th	October 22nd

Table 1. List of cloud-free RapidEye images used in the test site Neusling during the growing season 2010.

2.4 Classification approach

Figure 2 gives an overview on the methodology of the classification.

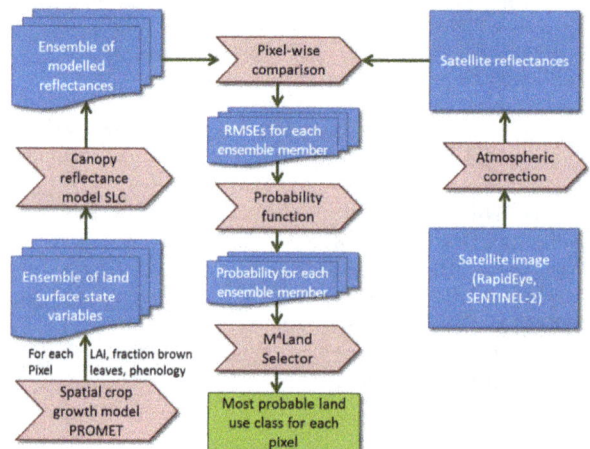

Figure 2. Flowchart illustrating the methodology of the model-supervised classification (Klug, 2014)

Surface state variables as modelled by PROMET (green leaf area index, phenology and degree of maturity) are used as input to the spectral reflectance simulations with SLC. This is the basis for allowing a pixel-wise comparison of the simulated to the measured reflectances (after atmospheric correction of satellite data). The RMSE criterion is used to compare these two sets of spectral reflectances and is converted into probability via an arithmetic function (exponential form with a RMSE of less than 1 assigned a probability of 1, and a RMSE higher than 5 a probability of 0). The probability thus determines how likely it is that the respective pixel belongs to a specific land use class. A multi temporal application of this procedure provides the most probable land use class for each pixel by averaging the probabilities in the M^4Land selector.

PROMET is used to model the temporal dynamics of state variables for various crop types in a spatially distributed way. Figure 3 illustrates such simulation results for the leaf area index and the crop types of one pixel in the test area and the investigation period 2010. LAI is selected in this figure, since this state variable is the most significant factor for the temporal variation of spectral reflectances on the land surface (in the absence of snow and flooding). Each crop type in Figure 3 shows a distinctively different temporal pattern of LAI development that is connected to their phenological development. These different temporal courses form the baseline that allows for a model-based multi temporal classification. In Figure 3 an idealized crop development is assumed, without nutrient stress and assuming normal phenological development. In reality, crops are very likely confronted with nutrient stress at some points during their life cycle. This can be caused by different fertilization intensities, but can also be a consequence of poor soil water holding capacities that lead to insufficient soil moisture. The phenological development also varies with seeding date, crop variety, or occurrence of water stress. Accordingly there is a variability of LAI development in reality that is not yet covered in Figure 3.

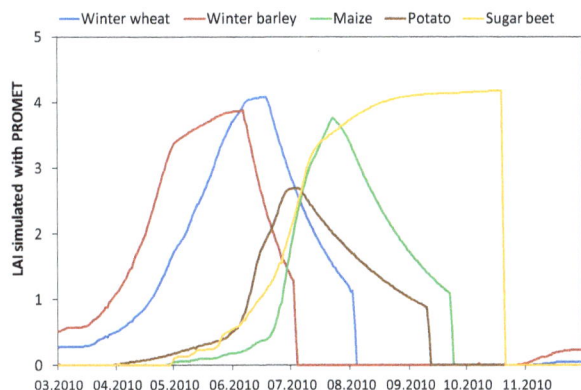

Figure 3. Modelled leaf area index development for the growing season of 2010 for the crop types in the test site, corresponding to scenarios with optimal plant development.

In order to consider this variability, the modeling for each crop type is carried out not only for optimal conditions but also for a variety of ensemble members (scenarios), in which the nutrition situation and the pace at which phenological development of the plants takes place can vary. The results are shown in Figure 4 for maize. Instead of a single curve for the LAI development of maize a set of possible courses is now provided. Reducing the nutrition supply of the maize plants, results in a decrease of biomass accumulation over the growing season and therefore in a decreased maximum leaf area index. A modified phenological development pace of the plants shifts the temporal course and with this also the date of maximum LAI. It can also have an

effect on the harvest date, which is however not the case for silage maize, since it is harvested before phenological maturity.

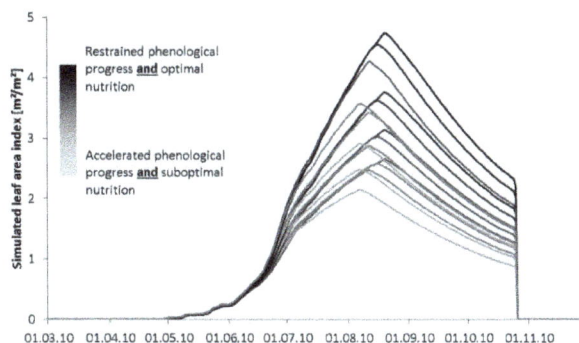

Figure 4. PROMET modeled leaf area index development ensemble for maize for the growing season 2010, with varying nutrition supply and phenological progress.

The use of scenarios thus allows for the representation of environmental conditions and management decisions of the farmer (e.g. fertilization level, crop variety or seeding date) providing a realistic range of possible land surface developments for each crop type.

These ensembles of crop developments are further depending on local meteorological conditions and thus are geographically variable. They also vary from year to year. This spatial and inter-annual heterogeneity is again simulated with PROMET, since the ensemble runs are performed for each individual pixel and variable meteorological conditions are thus considered.

In PROMET the phenological progress of agricultural crops is modelled using consecutive growth stages corresponding to the BBCH phenological classification system (Meier, 2001), a number system varying from 0 (seeding) to 100 (harvest). How PROMET is able to simulate geographical variations of phenological development is illustrated for model results for Germany in Figure 5. A point in time is selected (5th August 2014) when phenology of maize can range in Germany from leaf development to maturation. Accordingly also the temporal LAI development courses will strongly vary throughout Germany.

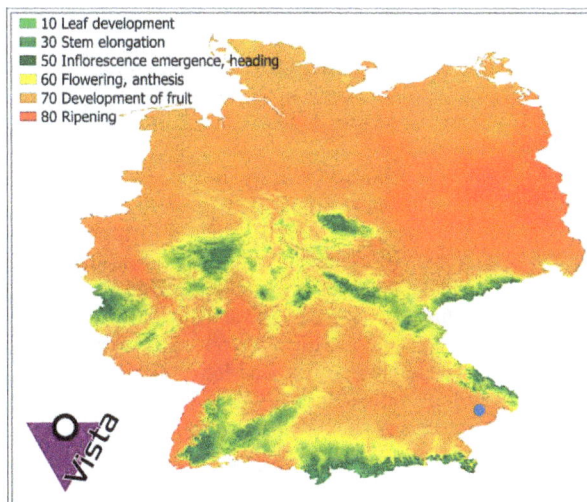

Figure 5. Simulated phenological stages for maize in Germany for the 5th August 2014 illustrating the heterogeneity of crop development (blue spot indicates location of Neusling)

3. RESULTS

3.1 Analyses of the transferability of the model-supervised approach

First the quality of the PROMET simulations shall be validated and their capability to adequately simulate the spatial and temporal patterns of crop growth. For this demonstration the phenological development for maize is selected. The first question is how well the intra-seasonal and inter-seasonal trends can be modelled.

For this we use in-situ measurements of phenology from 108 fields distributed over Germany that are provided by the German Weather Service DWD. To study the inter-annual variation, 4 years (2011 to 2014) were chosen. These in-situ observations are then compared with PROMET model results of the respective region. First averages of all fields were calculated for each year and related to the multi-annual average to understand how years can vary. These analyses based on the DWD observations are illustrated in Figure 6 as solid lines and compared to PROMET results (dashed lines).

Obviously 2011 showed Germany-wide a retarded phenological development of up to 7 days delay at stem elongation that is slowly caught up until ripening. This course is similarly simulated in PROMET. The accelerated phenological development of 6 days in 2013 is also simulated in PROMET but to a lesser degree. 2012 and 2014 are similar to the average, which is also depicted in the simulations. On average, measurements and simulated only differ by one day.

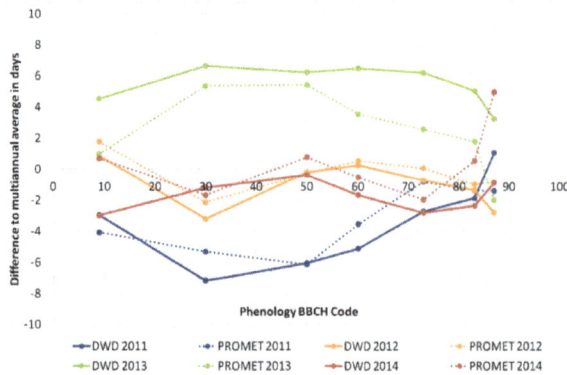

Figure 6. Validation of simulated phenological development of maize using in-situ measurements of the German Weather service DWD (averages over all 108 sites in Germany)

This validation on Germany-wide averages helps to study seasonal trends and inter-annual variations. Obviously both are well captured in the simulations. Another option is to compare the date of occurrence of a certain phenological state in measurement and simulation. This is illustrated in Figure 7 for all measurements of the 4 considered years and all 108 fields. It is evident that there is a very high concurrence. The points scatter very close around the 1:1 line. The Root Mean Square Error (RMSE) amounts to 10.9 days. This RMSE can be interpreted as the variability of the phenological development that is connected to management decisions of the farmer (seeding dates and crop variety for example) or local soil conditions (water stress leads to accelerated ripening).

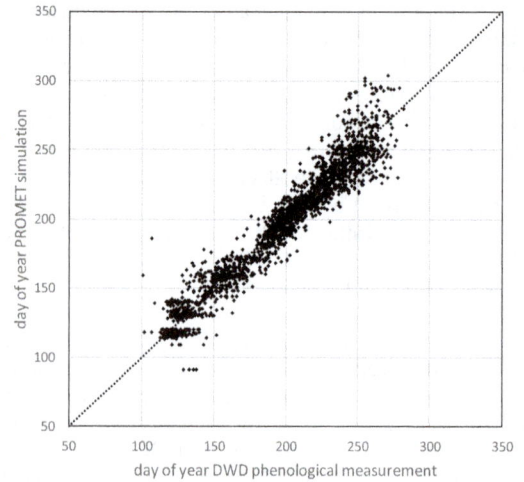

Figure 7. Simulated and measured dates (day of year DOY) of reaching a certain phenological stage for maize during the years 2011 to 2014.

3.2 Model-supervised classification of the Neusling test site

Results of the M[4]Land concept are presented for the Neusling test site in Bavaria. For the crop type classification the 10 RapidEye images of Table 1 were used. For each date of satellite acquisition, for each possible crop type and for each ensemble member spectral reflectances were simulated using SLC and the land surface state variables as provided by PROMET. These simulated spectral signatures can now be compared to the satellite measurement.

In a first step within one land use class the one ensemble member with the closest match with satellite derived spectrum is selected. An example for this step is illustrated in Figure 8. It shows, for a representative RapidEye acquisition date, the spectrum of each land use class for the most probable scenario in comparison to the RapidEye spectrum of one pixel. In this example the simulated spectrum of a maize pixel shows the closest congruence to the measured RapidEye spectrum. In order to quantify the match the RMSE criterion is used and the RMSE is transferred into a probability that ranges from 0 to 1.

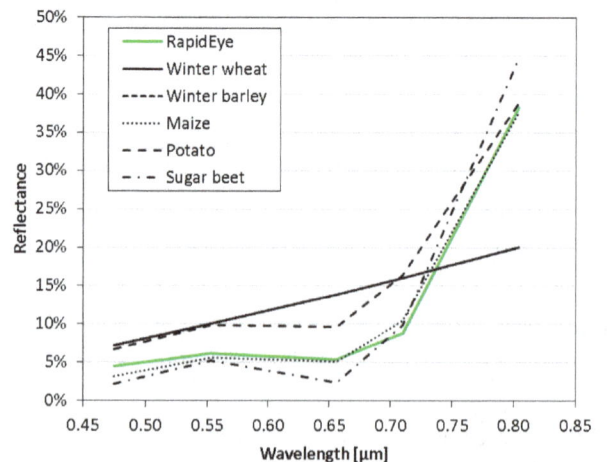

Figure 8. Comparison of a RapidEye observed spectrum (green) of one pixel on 21th of August 2010 with modelled spectra (black) for different land cover classes

The probabilities for each land use class and each acquisition date are aggregated by averaging the probabilities over all available image dates during the growing season. Figure 9 shows the aggregated mean probabilities of all crop types for the same pixel used in Figure 8. The probability for each acquisition date is calculated as the mean of all probabilities of the earlier acquisition dates including the current acquisition date. At the end of the growing season, the pixel is finally classified as the crop type with the highest aggregated probability. In our case it is maize. This is identical to the crop type that was mapped in the field.

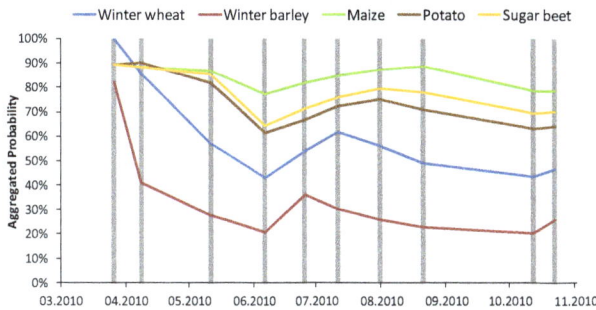

Figure 9. Aggregated probabilities for all modelled crop types of one maize pixel over the whole growing season of 2010. Grey vertical lines indicate satellite acquisitions.

The methodology demonstrated above for one pixel is repeated for each pixel in the satellite image. One of the intermediate outputs are simulated artificial satellite images that show the spectral reflectances with the best match to the EO data. For three selected dates these artificial images are compared with the measured satellite images in Figure 10. A false colour presentation was chosen with the green band in blue, the red band in green and the near infrared presented in red colour.

The images almost cover the whole crop cycle and illustrate well the high temporal dynamics of the spectral reflectance of agricultural areas. In the SLC simulation the soil reflectance is not assumed to be fix, however a soil background reflectance is selected pixel-wise from a set of typical soil spectra. Also the surface moisture of the soils that change the soil brightness is allowed to vary. This concept makes it possible to adequately simulate the soil reflectance and its effect on the canopy reflectance. Thus negative impacts on the classification accuracy can be avoided. This is also evident in the high

correspondence between simulated and measured images in Figure 10.

Figure 11 shows the resulting crop type map from the multi-temporal analyses. All pixels are classified with the most probable land cover class by the M^4Land framework. If the highest average probability is below a threshold of 70 %, pixels are left unclassified. They mostly occur in built-up areas and few fields that obviously are not sufficiently represented in the model setup.

The classification results were not filtered in order to be able to allow a fair evaluation of the model-supervised approach. The map reveals that most fields are uniformly classified. Only few fields share several different land cover classes, hardly any more than two.

Quantitative validation of the resulting land cover / crop type map was performed by pixel-wise comparing the classification with the mapped land use. A confusion matrix was created that allows the analysis of the product accuracy (see Table 2). The User's accuracy indicates how many pixels of a classified land cover class have actually been classified correctly, while the Producer's accuracy indicates how many pixels of the mapped land cover class have been classified correctly. User's and Producer's accuracies are both high. Mis-classifications occur to a larger extent for potato fields that were misinterpreted as maize or sugar beet. Winter wheat classification was almost 100 % correct, however some of the winter wheat fields were assigned winter barley which reduced the Producers'accuracy for winter wheat to 88 %. The overall accuracy of the achieved land cover map of the whole area is 85 %, which can be judged very high for an unsupervised autonomous methodology.

		GROUNDTRUTH					
		Winter wheat	Winter barley	Maize	Potato	Sugar beet	Total	Users's accuracy
CLASSIFIED	Winter wheat	3 249	8	3	2	0	3 262	100%
	Winter barley	269	305	0	0	0	574	53%
	Maize	8	1	1 354	293	113	1 769	77%
	Potato	169	4	42	1 082	288	1 585	68%
	Sugar beet	0	0	17	272	2 481	2 770	90%
	Total	3 695	318	1 416	1 649	2 882	9 960	
	Producer's accuracy	88%	96%	96%	66%	86%	Overall	85%

Table 2. Confusion matrix based on the comparison of in-situ-mapped and modelled land cover maps.

Measured by Satellite   Simulated in M4Land

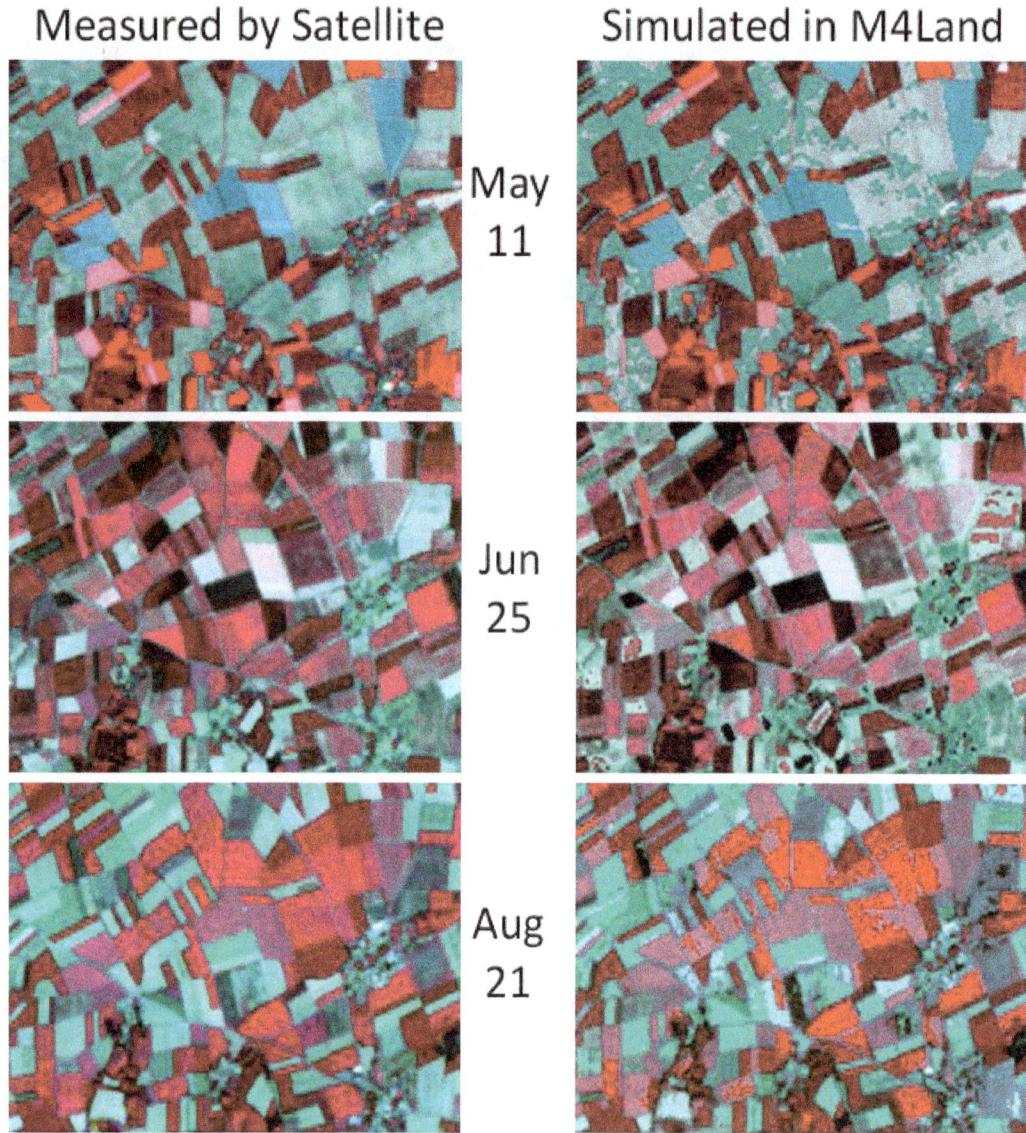

Figure 10. Real (left) compared with simulated (right) satellite images for selected dates during the growing season.

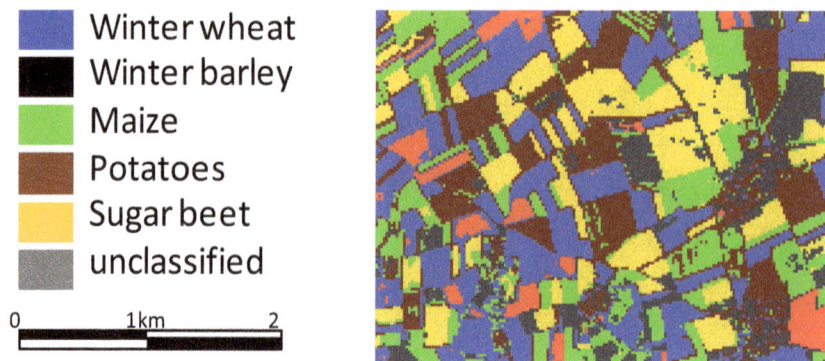

Figure 11. Land cover map derived by the model-supervised classification approach of assigning crop types to the highest average probability at the end of the growing season in 2010.

4. CONCLUSIONS AND OUTLOOK

The feasibility of the M⁴Land concept has been successfully demonstrated. The model-supervised approach is able to dynamically classify multi-temporal RapidEye images based on physical and physiological principles and without training. It uses a combination of the reflectance model SLC and the land surface process model PROMET. The classification results have a high overall accuracy of 85%.

The fact that most fields are uniformly classified even though the M⁴Land approach works pixel based and no post processing of the land cover product is performed, suggests that the model-supervised land cover classification is quite robust.

In a next step, the classification performance will be checked for other regions in Germany. Also several years shall be classified in order to allow the monitoring of the cropping cycle. It is further targeted to derive additional land surface management information products such as intensity of agricultural production, irrigation status or calamities using the ensemble information.

The M⁴Land system shall also be extended to natural environments in a mesoscale setup. Demonstrations in climatologically different areas are currently performed. The generic character of the M⁴Land approach will also allow for the extension towards the use of other satellite data apart from high resolution optical data (e.g. lower resolution optical or SAR data).

The required preprocessing chains for the inclusion of current and near-future optical Earth Observation Systems are already available within the M⁴Land system, so that for example the SENTINEL data sets will be integrated as soon as they become available.

The M⁴Land framework is designed to allow for an efficient handling of the rich data-stream of SENTINEL data that will soon be available. It therefore enables a continuous monitoring of non-linear processes at the land surface.

ACKNOWLEDGEMENTS

This work was funded by the German Federal Ministry of Economics and Technology through the Space Agency of the German Aerospace Center (DLR) (Grant code: 50 EE 1210). RapidEye data was kindly provided by ESA as Third Party Mission. Meteorological data was provided by DWD.

REFERENCES

Berger, M., Moreno, J., Johannessen, J. A., Levelt, P. F. & Hanssen, R. F., 2012. ESA's sentinel missions in support of Earth system science. *Remote Sensing of Environment*. **120**, pp. 84-90.

Farquhar, G. D., von Caemmerer, S. & Berry, J. A., 1980. A biochemical model of photosynthetic CO_2 assimilation in leaves of C3 species. *Planta*, **149**(1), pp. 78-90.

Hank, T., Bach, H., Mauser, W., 2014. Using a remote sensing supported hydro-agroecological model for field-scale simulation of heterogeneous crop growth and yield: application for wheat in Central Europe. *Remote Sens.* 2015, 7, doi:10.3390/rs60.

Klug, P., Schlenz, F., Hank, T. B., Migdall, S., Bach, H., Mauser, W., 2014. Generation of continuous agricultural information products using multi-temporal high resolution optical data in a model framework – The M4Land project. *ESA Special Publication SP-726, Frascati (Italy), Proceeding*, published.

Mauser, W., Bach, H., 2009. PROMET – Large scale distributed hydrological modeling to study the impact of climate change on the water flows of mountain watersheds. *J. of Hydrology*, 376, pp. 362-377.

Meier, U. 2001. Growth Stages of Mono- and Dicotyledonous Plants; Federal Biological Research Centre for Agriculture and Forestry: Braunschweig, Germany

Migdall, S., Bach, H., Bobert, J., Wehrhan, M., Mauser, W. 2009. Inversion of a canopy reflectance model using hyperspectral imagery for monitoring wheat growth and estimating yield. *Precision Agriculture*, DOI 10.1007/s11119-009-9104-6, 2009.

Verhoef, W., Bach, H., 2003. Simulation of hyperspectral and directional radiance images using coupled biophysical and atmospheric radiative transfer models. *Remote Sensing of Environment*, 87, pp. 23-41.

Verhoef, W., Bach, H., 2007. Coupled soil–leaf-canopy and atmosphere radiative transfer modeling to simulate hyperspectral multi-angular surface reflectance and TOA radiance data. *Remote Sensing of Environment*, 109, pp. 166-182.

16

VEGETATION HEIGHT ESTIMATION NEAR POWER TRANSMISSION POLES VIA SATELLITE STEREO IMAGES USING 3D DEPTH ESTIMATION ALGORITHMS

A. Qayyum [a, *], A. S. Malik [a], M. N. M. Saad [a], M. Iqbal [a], F. Abdullah [a], W. Rahseed [a], T. A. R. B. T. Abdullah [b], A. Q. Ramli[b]

[a] Centre for Intelligent Signal and Imaging Research (CISIR), Department of Electrical and Electronic Engineering, Universiti Teknologi PETRONAS 31750 Tronoh, Perak, Malaysia, aamir_saeed@petronas.com.my
[b] Universiti Tenaga Nasional ,43000 Kajang ,Selangor, Malaysia.

THEME: Airborne and innovative remote sensing platforms and techniques.

KEYWORDS: QuickBird Satellite sensor, Pleiades satellite Sensor, Stereo matching techniques, Dynamic Programming, Graph-Cut, 3D depth approach

ABSTRACT:

Monitoring vegetation encroachment under overhead high voltage power line is a challenging problem for electricity distribution companies. Absence of proper monitoring could result in damage to the power lines and consequently cause blackout. This will affect electric power supply to industries, businesses, and daily life. Therefore, to avoid the blackouts, it is mandatory to monitor the vegetation/trees near power transmission lines. Unfortunately, the existing approaches are more time consuming and expensive. In this paper, we have proposed a novel approach to monitor the vegetation/trees near or under the power transmission poles using satellite stereo images, which were acquired using Pleiades satellites. The 3D depth of vegetation has been measured near power transmission lines using stereo algorithms. The area of interest scanned by Pleiades satellite sensors is 100 square kilometer. Our dataset covers power transmission poles in a state called Sabah in East Malaysia, encompassing a total of 52 poles in the area of 100 km. We have compared the results of Pleiades satellite stereo images using dynamic programming and Graph-Cut algorithms, consequently comparing satellites' imaging sensors and Depth-estimation Algorithms. Our results show that Graph-Cut Algorithm performs better than dynamic programming (DP) in terms of accuracy and speed.

1. INTRODUCTION

Vegetation or trees may pose a major risk to the reliability of transmission power lines (Jones , 2001). Overgrown trees within the vicinity or in the 'danger zone' of transmission power lines can lead to short circuits. This interrupts the continuous power supply and causes blackouts. Danger zone refers to the area around the vegetation growth, which may cause flashover and subsequent power failures. The companies monitor vegetation growth regularly along the danger zone and eliminate them to avoid blackouts and economic losses. Many methods can be deployed to monitor the vegetation growth, and more importantly to estimate the height of the vegetation within the danger zone. Traditional methods such as manual line patrol or inspection by foot lack accuracy primarily due to human judgmental errors [Lotti, 1994]. Moreover, these traditional methods consume more time, and can be dangerous; essentially due to bad weather, hazardous terrain, or sometime exposes human to wild and vicious animal. Aerial inspection of power lines using a helicopter, or airborne imaging sensors are very expensive and trivially feasible in a non-uniform terrain (Jones , 2001). In comparison with the manual visual inspection methods, the aerial inspection can cover a larger area in a lesser time but incorporates excessive costs. However, aerial inspection is prone to error introduced by camera shaking, and target location ambiguity, especially for non-uniform terrain. Videography, or aerial multispectral imaging utilizing computer vision techniques, is better than the previous two methods. This method also uses a helicopter or a balloon or an airborne vehicle to capture the aerial images of vegetation. This method has a better accuracy as compared with visual or video surveillance. However, it is more time consuming due to the low altitude of the airborne vehicle and its accuracy is dependent upon the multispectral resolution. A different method

based on satellite stereo imaging can provide a cost effective solution including lesser involvement of human resources and manual judgment. The time required to monitor a particular danger zone is less, since the images are captured using satellite. The use of satellite stereo images has many advantages over visual inspection on foot and airplane based technique (Lotti, 1994). The satellite images cover a wide area. It is cost effective and can easily monitor restricted areas (Lotti, 1994).
This paper describes algorithms to process stereo images obtained via Pleiades satellite sensor. The algorithm can perform calculation to monitor the vegetation. It is followed by performance comparison with Graph-cut algorithm for disparity calculation, in terms of processing speed and accuracy. The result shows that the proposed Graph-Cut algorithm associated with each pixel for disparity calculation is more accurate as compared with dynamic programming algorithm. The background of depth estimation is discussed in Section II. The proposed technique is explained in Section III. Section 1V presents the simulation results. The conclusion and future research directions are presented in Section V.

2. RELATED WORK

Various stereo vision algorithms are available to compute stereo map and depth map. For example, the stereo vision system can determine depth of scene with help of two images, which are captured from the same or vantage points (Ghaffar, 2004). The stereo matching is the process of matching the pixel of left image to corresponding right image (Tomasi, 1998). The depth of the scene depends upon the disparity map/stereo matching. A good stereo matching calls for an accurate depth map; however, this task is very difficult and time consuming. The disparity assignment in stereo matching is difficult due to occlusion and existence of texture-less region (Tomasi, 1998). Sun et al.

---

* Corresponding author.

presented the fast cross correlation technique, and applied box filtering to measure stereo matching.

Stereo matching methods are generally categorized into two classes: local and global. The local methods are fast and efficient in computation based on area or windows (Cai,2006 ;2010 ; Cox, 1996). On the other hand, global methods based on specific energy function and are computationaly expensive (Boykov, 2001, 2004). However, stereo matching method demonstrates more noise when the smaller window in area based method is used. Upon increasing the window size, the noise is less affected, but the computational complexity increases with the increase in the window size. For the good construction of 3D, the surface should be continuous and fully textured. The variation in intensity is not covered for small window size, and if we increase the size, then occlusion and discontinuities in disparity occur.

Area based methods are used to measure similarity between two blocks using different types of window to measure disparity map from stereo images. The maximum similarity between two stereo images in stereo matching depends upon the cost/similarity function. The efficient designing of the cost/similarity function produces fast and robust stereo matching.

Global optimization algorithms like Graph-Cut and Belief propagation sometimes require extra parameters which are computationally more expensive (Boykov, 2001, 2004). These algorithms are not suitable for real time processing due to higher running time. These algorithms can be used for non-real time processing of data where higher accuracy is required. However, the Graph-Cut Algorithm is more accurate than Belief Propagation and dynamic programming algorithms (Sun, 2006). Therefore, Graph-Cut is suitable candidate for stereo matching for estimation of disparity maps or depth maps. The Graph-Cut produced new energy minimization algorithm and give good architecture for stereo matching problems. Boykov and Kolmogorov show graph-cut based energy minimization algorithms, which are faster by 2 or 5 times as compared to traditional push-reliable approaches (Scharstein, 2002). Graph-Cut for energy minimization using Potts model are used in segmentation, stereo, object recognition, shape reconstruction and augmented reality. The Boykov produces excellent algorithms that are expansion move and swap-move. These algorithms are based on pixel labelling for large pixel sets. Stereo matching based on multi-labelling problems and these labels are called disparities.

3. METHODOLOGY

3.1 Proposed framework

The Pleiades satellite was successfully launched with two sensors Pleiades-1A and Pleiades-1B sensor on 16 December 2011 and 1 December 2014. Pleiades-1(A&B) has the capability to acquire stereo imagery in one pass, with a few second differences. It also has ability to provide stereo-pairs color images with 20 km swath width and 70 cm resolution obtained with base-to height ratio from 0.15 to 2 (Lebègue, 2012). The Pleiades has been placed on the same sun-synchronous orbit at 694 km. It has been acquiring the panchromatic stereo images with resolution of 50 cm and multispectral images with resolution 200cm and also in bundle form 50 cm black and white and 200 cm multispectral (Lebègue, 2012). The Pleiades satellite has high resolution and low weight and also low cost for acquiring the images of small area. We have area of 10x10 km square in Sabah in east Malaysia, so that's why we choose high resolution, small

satellite sensor like Pleiades. This satellite has varieties probable various acquisition plans, such as a monoscopic cover up to 100x100 km or a stereoscopic instantaneous cover up to 60x60 km. The stereoscopic coverage is comprehended by only a single flyby of the area, which allows collection of a homogeneous product quickly. A classical forward and backward looking stereo pair provides the highest accuracy, but this combination is limited to areas with moderate terrain. A nadir and forward/backward looking stereo pair can be used in most kinds of terrain. The depth estimation was calculated on selected patches of imagery by employing the proposed dynamic programming and Graph-Cut algorithm.

The acquired data is first preprocessed and cropped. The two stereo images are then used to calculate the disparity maps. These disparity maps are then further used to find depth via disparity map algorithms. Furthermore, the depth maps are compared with previously recorded satellite data to find the area, where vegetation strikes the power transmission poles. Figure1 shows how our framework gets the desired information by using disparity maps and depth estimation technique blank line.

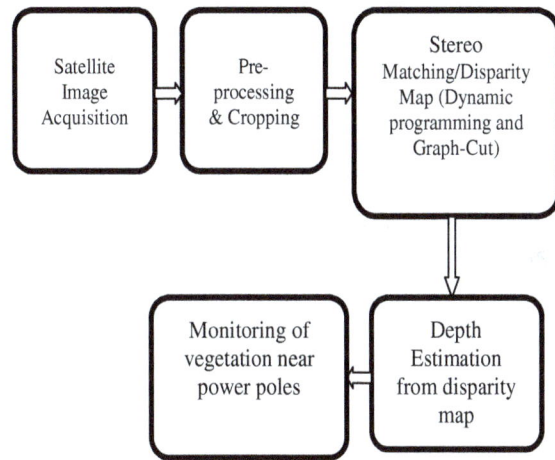

Figure 1. Proposed framework for monitoring of vegetation near power poles

3.2 Disparity Map Generation

Depth information is computed from a pair of stereo images by calculating the pixel wise distance between the location of a feature in one image and its corresponding location in the second image, hence generating a disparity map. Consequently, it gives a depth map because the pixels with larger disparities are closer to the camera, and those with smaller disparities are farther from the camera.

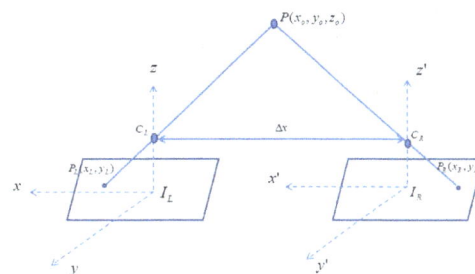

Figure 2. Stereo camera model (Boyer, 1988).

In the Figure 1, we have left and right camera images, where the left image have a center at 0 and right has a center at 0'. Therefore, we can calculate 3D depth point at coordinates (X0, Y0, Z0). We have the following relation from the above diagram (Boyer, 1988). Solving equation (4) and equation (5), we have the value of Zo.This value of z depends upon the value of the denominator factor which is called disparity value.

$$\frac{x_0}{x_L} = \frac{y_0}{y_L} = \frac{\lambda - Z_0}{\lambda} \qquad (1)$$

$$\frac{x_0 + \Delta x}{x_R + \Delta x} = \frac{y_0}{y_R} = \frac{\lambda - Z_0}{\lambda} \qquad (2)$$

Solving equation (1) and equation (2),we obtain equation (3).

$$Z_0 = \frac{\lambda + \lambda \Delta x}{x_L - (x_R + \Delta x)} \qquad (3)$$

The distance in pixels between the first and second image of the stereo pair is used to estimate the depth information and this information is called a disparity map. Pixels with smaller disparity are far from the camera and the pixels having large disparities are near to the camera. In other words, depth is inversely proportional to the disparity map as shown in the equation (3). We discussed Graph-Cut and dynamic programming Algorithms for stereo matching on plaids satellite stereo images.

3.3 Graph-Cut algorithms for stereo matching

Stereo matching is a classical vision problem, where graph based energy minimization method has been successfully applied. Three basic graph-based methods are used to solve stereo corresponding problems: pixel labelling with the Potts model, stereo matching with occlusion handling, and multicamera scene reconstruction. The multicamera scene reconstruction method is used for more than three stereo cameras. We are interested to handle the stereo matching with occlusion and also detect objects in stereo vision at textureless region. We used satellite stereo images that have low textures in some regions. In this paper, our work is closest to the formulation based on graph-Cut introduced by Kolmogorov & Zabih.They used symmetrical images in both stereo pair and used binary labels to pixel from each pair instead of assigning labels to individual pixel. If the pixel pair have the same correspondence in stereo pair, it assigns label '1' in the final disparity map, otherwise it is assigned '0' label. They further create a disparity map that imposes the uniqueness constraint. The Boykov introduced the similar work based on energy minimization using an expansion move algorithm. This algorithm minimizes the energy function in an iterative manner. It minimizes energy function by transforming into minimum cut problem on the graph and cuts the graph at each iteration to solve such problem at each iteration. The algorithm is run until convergence is achieved, and the result is a pretty strong local minimum of the energy function. The stereo correspondence algorithms based on graph cut discussed here endow with the base, from which innovative algorithms have emerged. The expansion-move algorithm [12] has the following chraractestics.

- Large number of pixels can change their labels simultaneously
- Finding an optimal move is computationally interactive
- It takes almost less than one minute to complete an execution as compared with other energy minimization algorithms like simulated annealing and iterated-

conditional model which take 19 hours to complete execution in early days.

- Finds local minimum of energy with respect to small "one-pixel" moves.
- Initialization is important practice. Theoretically, solution reaches the global minima.

Kolmogorov & Zabih introduced the energy function which comprises three terms: a data term, an occlusion term and a smoothness term penalizing neighboring pixels pairs for having different labels.Based on energy function f of Kolmogorov and zabih, different energy functions can be defined as

$$E(f) = E_{data}(f) + E_{occ}(f) + E_{smooth}(f) + E_{unique}(f) \qquad (4)$$

We can define these energy terms one by one as the following.

$E_{data}(f)$ define the matching cost of corresponding pixel and this matching cost can be calculated using four matching cost function given as

- Sum of absolute difference (SAD)
- Sum of Squared difference (SSD)
- Normalized cross-correlation (NCC)
- Zero-mean normalized cross-correlation (ZNCC)

The kolmogrov and zabih discussed squared difference of intensity values. We used sum of absolute difference which is easy and cost effective. The formula of the data cost function is given below.

$$E_{data}(f) = \sum_{\langle p,q \in B(f) \rangle} \left| I_{LeftInten\ ty}(p) - I_{rightInten\ sity}(q) \right|^a \qquad (5)$$

Where a is may be 1 for SAD and 2 for SSD.

$E_{occ}(f)$ adds a constant value to total energy function for each occluded pixel in the stereo corresponding of the stereo pair.

$$E_{occ}(f) = \sum_{p \in P} K_p . F\left(|U_p(f) = 0| \right) \qquad (6)$$

Where F evaluates 1 if its argument is true otherwise zero.

$E_{smooth}(f)$ If the neighboring pixels have different disparity this smooth energy function imposes the penalty and can be defined as

$$E_{smooth}(f) = \sum_{\{b_1,b_2 \in N1\}} U_{b_1,b_2} . F(f(b_1) \neq f(b_2)) \qquad (7)$$

The smoothness term will be zero if the assignment b_1 and b_2 have the same disparity in the $N1$ neighbourhood system for 4-neighbours in the input images otherwise it imposes penalty for different disparity of the neighbouring pixels. $E_{unique}(f)$ confines the possible solutions of the optimisation problem to unique solutions. If pixel is containing more than one value in the crossponding image in stereo pair then it assign penalty for infinite value otherwise null value assign.

This can be defined as

$$E_{Unique}(f) = \sum_{P \in p} F\big(\big|N_p(f)\rangle 1\big|\big).\infty. \qquad (8)$$

We introduced the ordering term in the above total energy function for calculating stereo matching.

$E_{order}(f)$ can be written as

$$E_{order}(f) = \sum_{\{b_1, b_2\} \in N_2} F(f(b_1) = f(b_2) = 1).\infty. \qquad (9)$$

Where N_2 is a neighbourhood system and can be explain as in such a way that $b_1 = \langle p, q \rangle$ and $b_2 = \langle p', q' \rangle$ are neighbours pixels. They must fillfull the order as if $\langle p_x \rangle p'_x \rangle$ and $\langle q_x \langle q'_x \rangle$ is true.

The final energy function can be written as

$$E(f) = E_{data}(f) + E_{occ}(f) + E_{smooth}(f) + E_{unique}(f) + E_{order}(f) \qquad (10)$$

The energy function minimized using Graph-Cut algorithm gives a general solution of the correspondence between stereo images.

3.4 Dynamic programming

It can produce good results in the lower contrast region and at object boundaries in the images. Dynamic programming also provides excellent computation between scan lines of the stereo images (Cai, 2010). On rectified stereo images, the dynamic programming exhibits the lowest matching cost of scan lines between first and second image of the stereo vision. It also illustrates the path has low cost in 2D grid form. The dynamic programming can be exploited in stereo matching to estimate the disparity map and calculate the depth map using satellite stereo images. We can apply dynamic programming in stereo matching to compute the disparity map. The dynamic programming search the best correspondence points between left and right stereo images and must enforce the ordering constraint between scan lines of both stereo images. This ordering constraint should be applied on rectified stereo images. The dynamic programming gives a smoothness disparity map due to strong correspondence between left and right image. Dynamic programming finds the minimum path from top left corner of the 2D matrix to the right-bottom corner. It finds the optimal path and matches the sequence in left scan-line in left image to the right scan-line in the right image optimally. If we assume ordering constraint, the best path can be computed to match the pixels belonging to left image and the right image, so the dynamic programming provides the best path on the 2D grid that satisfies the ordering constraints.

The simple dynamic programming cannot detect occlusion; the algorithm was introduced by J.cai for occlusion detection using dynamic programming (Cai, 2010). In proposed DP algorithm employs fixed occlusion parameter for detection of occlusion to calculate disparity map. The proposed algorithm not only produces high accuracy, but also it detects occlusion in the disparity map based on dynamic programming. We incorporate the following steps to calculate the minimum cost value for accurate disparity map and introduce some fixed occlusion cost

from the original source algorithm steps. The four steps to calculate the minimum cost path and minimum disparity value using dynamic programming as are as follows.
Set the initial value to each path cost and accumulative cost.

Step 1) Initialization

$$\beta(n, d) = C(n, d)$$
$$p(n, d) = 0$$

Calculate the minimum cost and occlusion detection in inter-scanline using dynamic programming.
$n = 0, 0 \leq d \leq 30$

Step 2) Recursion from n=1 to N-1

$$\beta(n,d) = Min \begin{cases} \beta(n-1,t) + C(n,d) + C_1 & if\ t < d \\ \beta(n-n,t) + 2C(n,d) & if\ t = d \\ \beta(n-t-d-1,t) + 2C(n,d) + C_2 & if\ t > d \end{cases}$$

Step 3) Number of paths to determine

$$\beta(n, d) = t;$$

Save minimum Cost path and track the index of minimum path in 2D matrix.

$$d_{min} = min\{\beta(N-1, d)\}$$

Step 4) Path backtracking from time n=N-2 to 0

$$d_{min} = C(n+1, d_{n+1})$$
$$n = n - (d_{n+1} - d_{min}) - 1$$
$$d_n = d_{min}$$

Where n is the pixel index of the second scan line, $\beta(n, d)$ is the accumulated matching cost at nth pixel. The disparity range is from 0 to 30, $C(n, d)$ is the matching cost, C1 and C2 are matching costs of the left and right occlusion respectively. This disparity produces good smoothness and detects objects at depth discontinuous. Dynamic programming produces fine results as compared to block matching on satellite stereo images. Our proposed dynamic programming algorithm performs better in terms of accuracy for satellite stereo images and also consumes less computational time as compared to block matching algorithm with energy minimization. It detects occlusion, which produces error in stereo correspondences between left and right stereo image.

4. EXPERIMENTAL RESULTS

We have applied two different stereo matching algorithms (dynamic programming and Graph-Cut) to estimate the disparity map from Pleiades satellite stereo images (Fig. 3,4) and results are shown in (Figure. 5,6). The disparity map that is produced by dynamic programming on Figure. 3 and 4 carries noise and no smoothness is observed as shown in Figure. 5. It accurately detects occlusion, and cannot handle noise in the satellite stereo images. By means of visual inspection, it can be seen that the three power poles are detected and the fourth is missing. It also detects the vegetation near the third power pole as shown in the disparity map. It cannot handle textureless area smoothly as shown in the disparity map.

The disparity map produced by using Graph-Cut algorithm on Figure. 3 and 4 has smoothness, which implies less-noise and good accuracy as shown in Figure. 6. Graph-Cut produces accurate disparity map as compared with dynamic programming as shown in Figure. 6.

Three important and difficult areas of input images were discussed: occluded areas, low-textured areas and noisy areas. Graph-Cut algorithm is successfully applied to Pleiades satellite stereo images. We introduced ordering constraint for measurement disparity map using Graph-Cut Algorithm and see from the results, In occluded areas the introduction of the ordering constraint was successful. In the disparity maps, no visual differences can be seen, and their error measurements are also very similar in both dynamic programming and Graph-Cut Algorithms.

We can conclude that the ordering constraints cooperate in Graph Cut Algorithm helps in occluded area and also minimize the noise as shown in disparity map. Our satellite images have very low textured, so there is no big difference in textureless areas using ordering constraint in Graph-Cut. In the occluded areas the use of the ordering constraint was successful in that the modelled occluded regions were much clearer from spurious mismatches; a fact that could be seen visually in the disparity maps, and verified by the calculated error percentages. The ordering constraint assisted in some areas, but not in texturless areas, and hence did not affect overall performance of the algorithm. We certainly need another solution for textueless region, which is very important in satellite images. Based on the results presented here, the use of the ordering constraint in a global method looks encouraging. Although the results are not amazing, they show that the significance of the ordering constraint will increase with images taking larger disparity ranges and more noise.

The Incorporation of ordering constraint in Graph-Cut is not sufficient to solve the texturless areas in satellite image. This is not that amazing, since the problem of these areas is their low signal to noise ratio, and the only way to increase this ratio is to include more pixels in the calculation of the matching cost. Therefore, the problem can be fixed to combine local area method based on correlation with Graph-Cut algorithm for stereo matching into one algorithm.

The introduction of the ordering constraint was not sufficient to solve the problems of textureless areas. This was not that amazing, since the problem of these areas is their low signal to noise ratio, and the only way to increase this ratio is to include more pixels in the calculation of the matching cost. Therefore, it would be of great interest if one could combine the robustness of area based correlation matching with the perfectionism of global graph cut based matching into one algorithm.

Another problem is that the conversion of matching problems into Graph-Cut problem necessitates that the matching problem is seen as a labelling problem. For labelling problem, we require a discrete disparity map which has less capability to capture the shape of the objects. We need fronto-parallel surfaces for this purpose. To calculate the discreet disparity map, we need a solution to solve the problem iteratively on segmented image and separate the image as foreground.The structural similarity matrix (SSIM) index can be used to calculate the similarity between reference and calculated images. The value of SSIM should be one for accurate similarity. The SSIM value of the proposed algorithm is near to accuracy as compared to the existing algorithm as shown in table. 1. The accuracy of graph-Cut is 80% and accuracy of dynamic programming algorithm is 77% according to SSIM value.

Figure 3.   Reference satellite Pleiades satellite stereo image contains four power poles and vegetation near power two poles.

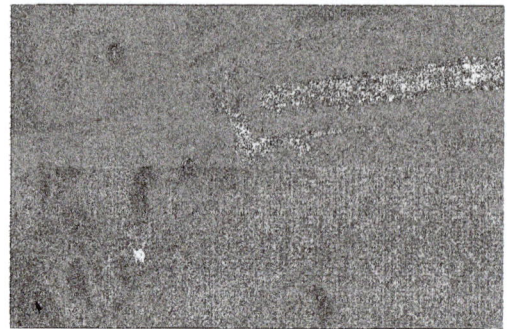

Figure 4.   Right Pleiades satellite stereo image contains four power poles and vegetation near power two poles

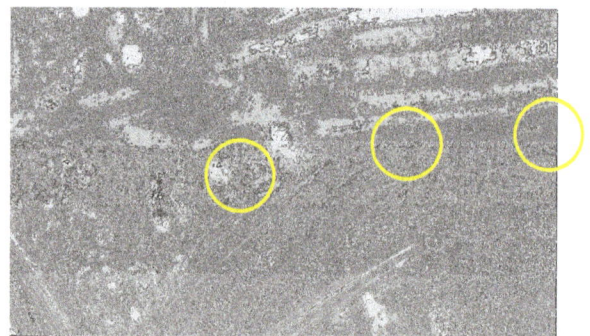

Figure 5.   Disparity map using dynamic programming algorithm (Detect poles and power lines seems some extent and also the vegetation in white color).

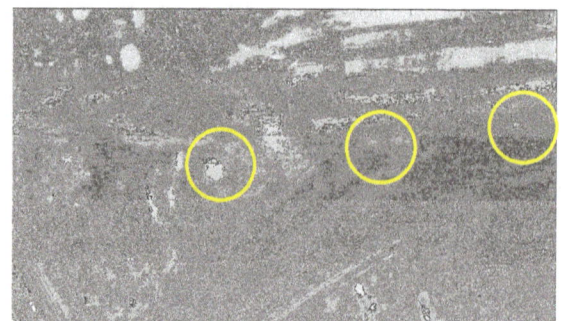

Figure 6.   Disparity Map using Graph-Cut algorithm (Detect poles and power lines seems some extent and also the vegetation in white color, well smooth, less noise).

Algorithms	Proposed	Existing
Dynamic Programming	77%	70%
Graph-Cut	80%	72%

Table 1.Comparison of accuracy using SSIM between proposed and existing algorithm

5. CONCLUSION AND FUTURE DIRECTION

In this paper, it is a novel technique is proposed to monitor vegetation/trees near or under the power lines using satellite stereo images. The proposed method employs Graph-Cut and dynamic programming algorithms to measure height of vegetation from stereo satellite images. The proposed technique was employed imagery from Pleiades satellite images. The experimental results illustrate that Graph-Cut algorithm based disparity measurement technique outperforms Dynamic Programming based technique in accuracy. The Graph-Cut shows an accuracy of 80%, while Dynamic Programming algorithm has 77% using SSIM value.

In future, we can compare two-way-dynamic programming algorithm with graph cut to estimate the disparity map for this application in order to improve the ordering constraint of Grap-Cut. The two-way dynamic programming is used to compute the best optimal cost within inter-scan line between stereo images.

ACKNOWLEDGEMENTS

This research is sponsored by Ministry of Energy, Green Technology and Water (KeTTHA cost No: 153AB-G07), Malaysia.

REFERENCES

Boyer, K, L. and Kak , A, C., 1988. Structural stereopsis for 3-D vision, IEEE Trans. Pattern Anal. Machine Intell. vol.10, no. 2, pp. 44-166.

Boykov,Y. and Zabih,R., 2001.Fast approximate energy minimization via graph cuts, IEEE Trans. Pattern Anal. Mach. Intell., vol.23, no.11, pp. 1222–1239.

Bank, J., Bennamoun ,M., Corke, P.,1997. Non-parametric techniques for fast and robust stereo matching, IEEE Tencon Speech and Image Technologies for Computing and Telecommunications, pp. 365–368.

Boykov.Y. and Kolmogorov,V., 2004. An experimental comparison of mincut/max-flow algorithms for energy minimization in vision, IEEE Trans. Pattern Anal. Mach. Intell., vol. 26, no. 9, pp. 1124–1137.

Cai, J., 2007. Fast stereo matching: coarser to finer with selective updating. International Conference on Image and Vision Computing, New Zealand, pp. 266–270.

Cox, S., Maggs, b, M ., 1996. A maximum likelihood stereo algorithm, vol. 63, no.3, pp. 542–567.

Krauss, T., Lehner, M.,Reinartz, P.and Stilla.J.,2006. Comparison of DSM generation methods on IKONOS images, Urban Remote Sensing – Photogrammetrie – Fernerkundung – Geo information special issue, 04/2006.

Cai,J. and Walker,R., 2010. Height estimation from monocular image sequences using dynamic programming with explicit occlusions,IET Computer Vision, vol.4, pp.149–161.

Ghaffar, R., Jafri, N. and Khan, S, A., 2004. Depth extraction system using stereo pairs, Image Processing & Computer Vision (IPVC), pp.512–519.

Jones, D, I and Earp, G, K., 2001. Camera sightline pointing requirements for aerial inspection of overhead power lines, Electric Power Systems Research (EPSR), vol.57, pp.73-82.

Kobayashi, Y., Karady, G and Olsen, R, G., 2009. The utilization of satellite images to identify trees endangering transmission lines, IEEE Transactions on Power Delivery, pp. 1703–1709.

Kolmogorov,V.and Zabih,R.,2001. Computing visual correspondence with occlusions via graph cuts, In International Conference on Computer Vision.

Lebègue, L., Greslou, D., deLussy, F., Fourest, S., Blanchet, G., Latry, C., Lachérade, S., Delvit, J.-M., Kubik, P., Déchoz, C., Amberg, V., Porez-Nadal, F., 2012, "Pleiades-HR image quality commissioning. Int.", Archives of Photogrammetry, Remote Sensing and Spatial Information Sciences, Vol. 39(1), pp. 561-566. XXII ISPRS Congress, Melbourne, Australia.

Lotti, J, L. and Giraudon, G., 1994. Correlation algorithm with adaptive window for aerial image in stereo vision, European Symposium on Satellite Remote Sensing (EUROPTO), (Rome, Italy), pp. 2315–10.

Scharstein, R.and Szeliski, R., 2002 A taxonomy and evaluation of dense two frame stereo correspondence algorithms, International Journal of Computer Vision,vol.47,no.1-3, pp. 7–42.

Sun, C ., Jones, R., Talbot, H., Wu. X. Cheong, K., Beare ,R., Buckley, M. and Berman,M.,2006. Measuring the distance of vegetation from power lines using stereo vision, ISPRS Journal of Photogrammetry & Remote Sensing, vol.60, pp. 269–283.

Sun, C., 2002. Fast stereo matching using rectangular subregioning and 3d maximum-surface techniques, International Journal of Computer Vision, vol. 47, no.1-3, pp. 99–117.

Sun, J., Yeung Shum, H. and Ning Zheng,ZHENG ,N., 2003. Stereo matching using belief propagation, IEEE Trans. Pattern Anal. Mach. Intell , vol.25, pp. 787–800.

Tomasi , C., 1998. A pixel dissimilarity measure that is insensitive to image sampling, IEEE Trans. Pattern Anal. Mach. Intell, vol.20, no. 4, pp. 401–406.

Young, M.,1989. The Technical Writer's Handbook. Mill Valley, CA: University Science.

17

RELATIONSHIPS BETWEEN PRIMARY PRODUCTION AND CROP YIELDS IN SEMI-ARID AND ARID IRRIGATED AGRO-ECOSYSTEMS

H. H. Jaafar [a,*], F. A. Ahmad [a],

[a] Faculty of Agriculture and Food Sciences, American University of Beirut - (hj01,fa76)@aub.edu.lb

KEY WORDS: Agriculture, Crop, GIS, Analysis, Correlation, Satellite, Spatial.

ABSTRACT:

In semi-arid areas within the MENA region, food security problems are the main problematic imposed. Remote sensing can be a promising too early diagnose food shortages and further prevent the population from famine risks. This study is aimed at examining the possibility of forecasting yield before harvest from remotely sensed MODIS-derived Enhanced Vegetation Index (EVI), Net photosynthesis (net PSN), and Gross Primary Production (GPP) in semi-arid and arid irrigated agro-ecosystems within the conflict affected country of Syria. Relationships between summer yield and remotely sensed indices were derived and analyzed. Simple regression spatially-based models were developed to predict summer crop production. The validation of these models was tested during conflict years. A significant correlation ($p<0.05$) was found between summer crop yield and EVI, GPP and net PSN. Results indicate the efficiency of remotely sensed-based models in predicting summer yield, mostly for cotton yields and vegetables. Cumulative summer EVI-based model can predict summer crop yield during crisis period, with deviation less than 20% where vegetables are the major yield. This approach prompts to an early assessment of food shortages and lead to a real time management and decision making, especially in periods of crisis such as wars and drought.

1. INTRODUCTION

In the last decade, remote sensing techniques were one of the main components that contributed to a shift toward increased precision in crop management (Jones & Vaughan, 2010). Such techniques provide spatially and temporally distributed information, leading to a real time management and decision making. Early diagnosis and estimation of yield is a must when early intervention is needed, mainly in the case when yield deficit threatens food security. The importance of these techniques lies in regions where yield data is either unreliable or non-existent, the case in war affected countries. Many image-based parameters and models that monitor agricultural performance exist in literature. However, verification and validation of such models remain a challenge. Primary production and vegetation indices were used to assess and predict crop yields (Running, et al., 2004; Reeves, Zhao, & Running, 2005). The Normalized Difference Vegetation Index (NDVI) was frequently used in crop forecasting and to detect crop areas (Domenikiotis, Spiliotopoulos, Tsiros, & Dalezios, 2004; Mkhabela, Mkhabela, & Mashinini, 2005). Quarmby et al. (1993) demonstrated that NDVI is an accurate early warning indicator for years with poor yield. NDVI was a good predictor of wheat, cotton and rice yields in northern Greece (Quarmby, Milnes, Hindle, & Silleos, 1993). NDVI was also used in crop discrimination in Northern China (Mingwei, et al., 2008). Maize and cotton fields were discriminated using MODIS derived NDVI and results were well correlated with statistical data at regional spatial scales. The Enhanced Vegetation Index (EVI), an improved index that accounts for soil reflectance, was rarely assessed in literature. EVI has similar potential as NDVI in estimating yield of many crops (corn, wheat, alfalfa, sorghum, soybeans) (Wardlow, Egbert, & Kastens, 2007). EVI was also used in the estimation of wheat area in China (Pan, et al., 2012). The Moderate Resolution Imaging Spectroradiometer (MODIS) produces both EVI and NDVI in addition to the primary production parameters such as the Gross Primary Production (GPP), Net Primary Production (NPP) and Net Photosynthesis (net PSN). Data acquired from MODIS allow an accurate monitoring of crop due to its frequent acquisitions of remote sensing data and the rapid availability of data over large regions (Running, et al., 2004; Zhao, Heinsch, Nemani, & Running, 2004). MODIS GPP is the result of combining MODIS data with meteorological inputs in a plant growth algorithm. Few studies had used this parameter in yield estimation. Reeves et al. (2005) converted GPP to biomass through a conversion equation of carbon to yield. This conversion was sufficiently accurate at state level but not at county level nor at climate district.

In this paper a methodology is introduced for summer crop-yield prediction using MODIS vegetation and productivity indices. The remote sensing algorithm capitalizes on MODIS historical archive of these indices. The study is useful in conflict affected areas where reported data are unavailable or discrepant, or where access to agricultural areas is not possible due to security situations. Summer crop production in selected governorates of Syria was estimated from MODIS derived indices (GPP, net PSN and EVI). Regression models were built during pre-conflict years (2000-2011) and simulated during years of conflict (2012-2013).

2. METHODOLOGY

2.1 Study area

The conflict affected country of Syria is chosen as a case study. Syria produces 3.3 million tons of crops and vegetables in summer season. The vast majority of summer production relies on irrigation (from wells, rivers, and governmental irrigation projects). Most of irrigated land is located around Euphrates and Orontes River and their streams. Cotton is the most important summer crop. Syria is ranked by FAO as the 10th in the world in cotton production. The annual average production of cotton in the last decade amounts to 740,000 tons equivalent to 3.75 tons/ha. The north-eastern region of Syria (Al-Hassake, Deir-Ezzor, Al-Raqqa and Aleppo) is reputed in cotton and maize culture in summer. The other governorates produce in summer mainly vegetables, in addition to cotton and tobacco. Winter and spring agricultural production is not within the scope of this study.

2.2 Data analysis

Three remotely sensed parameters were analyzed and compared to administrative statistics. The Gross Primary Production (GPP), Net Photosynthesis (net PSN) and Enhanced Vegetation Index (EVI) were extracted from MODIS datasets published by NASA and improved by the Numerical Terradynamic Simulation Group (NTSG) at the University of Montana. Those datasets are available from 2000 until present, on monthly and yearly basis, at 1-km spatial resolution (NASA LPDAAC, 2014; NTSG, 2014). The photo-synthetically active radiation (PAR), meteorological data, the estimated growth and maintenance respiration are the main parameters to obtain primary production. The EVI is an improved form of NDVI (Normalized Difference Vegetation Index) where vegetation conditions are compared in a spatio-temporal horizon. EVI is defined as per equation (1).

$$EVI = G \times \frac{\rho_{NIR} - \rho_{Red}}{\rho_{NIR} + (C_1 \times \rho_{Red} - C_2 \times \rho_{blue}) + L} \qquad (1)$$

Where ρ are atmospherically corrected or partially atmosphere corrected surface reflectance, L is the canopy background adjustment that addresses nonlinear, differential NIR and red radiant transfer through a canopy, and C1, C2 are the coefficients of the aerosol resistance term, which uses the blue band to correct for aerosol influences in the red band (Huete, et al., 2002).

Delineation of summer irrigated lands was relatively easy using the EVI. The EVI gives a good first approach to these lands. The mean (2000-2011) summer EVI raster was used to delineate summer irrigated lands. The resulting shape file was then compared with high resolution Google Earth imagery available for the summer seasons between 2000 and 2011, and fine-tuned where necessary. The delineated irrigated lands were comparable with the average of total irrigated lands stated by the MOAAR over the period 2000-2011 in each political unit. The statistical department in the Syrian Ministry of Agriculture and Agrarian Reform (MOAAR) published, since 2000, annual reports about summer crops and vegetables production and areas in all Syrian governorates (Ministry of Agriculture & Agrarian Reform in Syria, 2014). Two years of record are excluded from analysis (2005 and 2006) due to missing/discrepant data.

Monthly GPP, net PSN and EVI rasters were summed for the months of June, July and August to obtain summer indices between 2000 and 2011. Zonal statistics analysis was performed over delineated irrigated lands. Means of the cumulative sums for summer GPP, summer net PSN and summer EVI were obtained in the irrigated lands of each Syrian governorate. Linear regression was applied to relate summer crop production and remotely sensed indices in all governorates during the pre-conflict period (2000-2011). An F-test statistic was used to assess the significance of the linear regression parameters. The significance of analysis was evaluated using F-statistics at p < 0.05. The derived relationships were simulated to estimate the summer crop production during conflict years (2012-2013).

To evaluate the spatial variation of production, the standardized summer EVI was calculated pixel-by-pixel as in equation (2).

$$SEVI_i = \frac{EVI_i - \overline{EVI}}{S_n} \qquad (2)$$

Where $SEVI_i$ is the standardized EVI in year i; EVI_i is the sum of summer EVI in year i; $\overline{EVI}$ is the mean of summer EVI in n years; S_n is the standard deviation of summer EVI in n years

Pixel-by-pixel statistics were derived from multiple rasters. The spatial mean and spatial standard deviation of EVI were calculated for the period of record 2001 and 2011. The spatial summer SEVI was derived by applying the Equation 1 on a cell-by-cell basis using a GIS-based raster calculator.

3. RESULTS AND DISCUSSIONS

The regression was significant where cotton and vegetables were highly produced. Cotton is produced in Al-Hassake (85%), Al-Raqqa (64%), Ghab (52%), Deir-Ezzor (49%), and Aleppo (42%). More than the half of production in Hama, Dar'a and Homs is of vegetables.

Summer EVI was regressed against summer crops ($r^2 > 0.5$) in major irrigated lands (Hama, Ghab, Homs, Al-Hassake, Deir-Ezzor and Al-Raqqa) with high significance. The regression was noticed to be significantly negative in Deir-Ezzor and Al-Raqqa. The agricultural lands of the latter governorates surround Euphrates River within a narrow strip in a hyper-arid landscape. The second major crop in these governorates is maize. Summer GPP was a significant estimator of summer crops in Hama, Al-Hassake, Deir-Ezzor, Ghab and Dar'a. Summer crops can be predicted from net PSN in four governorates (Hama, Ghab area within Hama, Al-Hassake and Dar'a) but with less significance than the other parameters.

Governorates	GPP		PSN net		ΣJJA EVI	
	2012	2013	2012	2013	2012	2013
Hama	-8	9	-9	22	3	-8
Ghab	-10	-4	-11	24	-11	-63
Al-Hassake	26	164	33	237	-7	-73

Table 1- Deviation of predicted total summer yield from reported (%) for conflict years of 2012 and 2013

Following examination of significant relationships between summer crops and summer EVI, GPP and net PSN, regression models were derived for the period 2000-2011 and tested over the period 2012-2013. Table 1 show the deviation of predicted summer yield from reported during conflict years 2012-2013. The EVI based model performed well in 2012 in Hama, Ghab, Al-Hassake and Al-Raqqa. In 2013, the regression model did not conform well to the reported crop yields, with the exception of Hama governorate where the major crops are cotton and vegetables. Hama also is almost entirely under full government control. In governorates were maize was planted, no significant relationship between the tested indices and crop yield could be derived. It was noted that summer EVI was inversely correlated with netPSN is such areas (cotton and maize). It is well-worth noting that reported yield cannot be verified, as many of the areas are combat grounds, and hence the reliability of reported yields by the Syrian government in such areas is questionable. For example, heavy battles occurred in the Deir-Ezzor and Al-

Raqqa governorates during 2012 and 2013. This may have prevented the government employees to access farmers' lands to conduct farm surveys and/or to collect crop yields.

Two main reasons could be arbitrated: 1) the unreliability of government reported data in conflict years (Endowment, 2014); 2) the sharp drop in production due to damages in irrigation systems and farmers displacement (Swiss Agency for Development and Cooperation SADC, 2014).

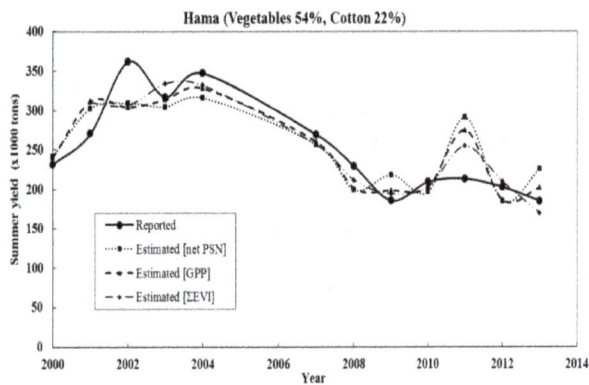

Figure 1- Time series of reported and simulated summer crop yields for Hama governorate (2000-2013).

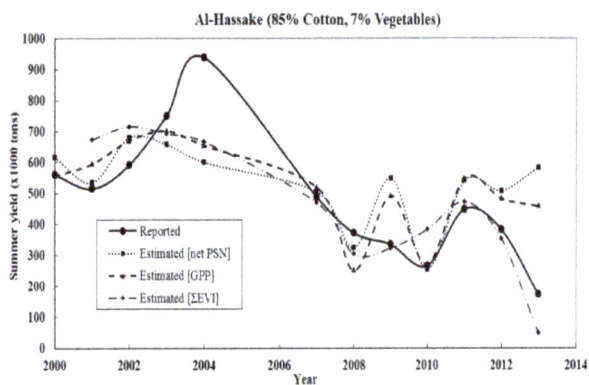

Figure 2- Time series of reported and simulated summer crop yields for Al-Hassake governorate (2000-2013).

To compare different models, the work focuses on two governorates: Hama (vegetables main producer) and Al-Hassake (cotton main producer). Figure 1 and 2 illustrates the time series of reported and simulated summer crops and vegetables production in Hama and Al-Hassake. In both governorates, a rise in production is noticed between 2000 and 2004 where irrigation projects were enhanced (Hole & Smith, 2012). Since 2006, Syria had faced a severe drought that contributed to water shortages allowing for decreasing irrigated lands and a decrease in production (FAO, 2009). In 2011, the production slightly increased. The Syrian conflict started in 2011, and peaked during 2012 and 2013 where a decline in production is noticed both from reported yield and from cumulative summer EVI. The EVI predictor fits well the reported production in Hama most years, with less than 20% of error. In Al-Hassake, more fluctuations were observed. Models underestimated the high production in 2004 (-30%). The more suitable model during crisis years is the EVI model.

To evaluate the spatial variation of production, the SEVI was calculated pixel-by-pixel. Figure 3 and 4 shows the summer SEVI of 2012 against the mean and the standard deviation of pre-conflict years (2001-2011). Most regions in Hama and Al-Hassake faced a drop in EVI during 2012. Interior irrigated lands in Hama was not affected in 2012 nor the western regions. Ghab plain faced a significant drop of summer EVI in 2012. In the Kurdish area, northeastern Al-Hassake, summer EVI was similar to pre-conflict years. In other regions, summer SEVI was significantly less than before crisis.

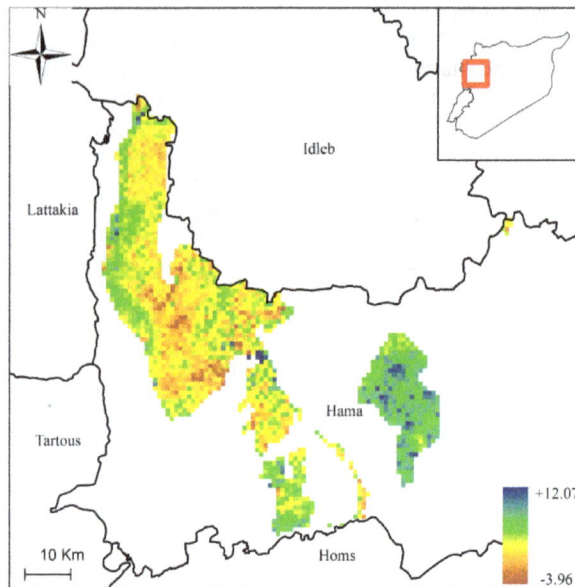

Figure 3- Standardized cumulative summer EVI for 2012 Hama.

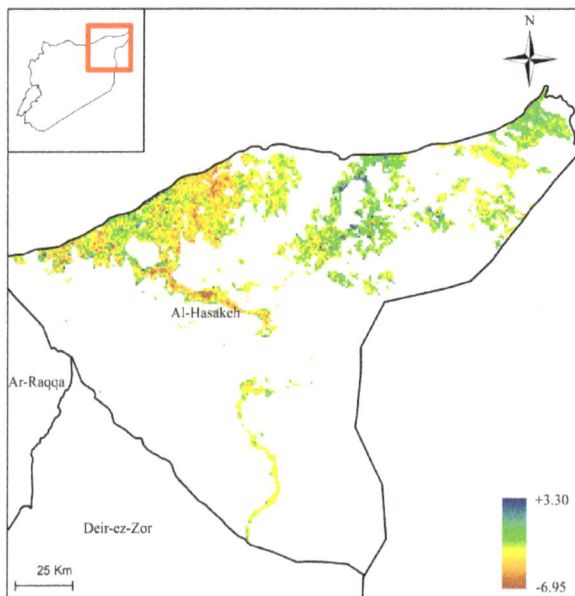

Figure 4- Standardized cumulative summer EVI for 2012 in Al-Hassake.

4.   CONCLUSION

The analysis of MODIS-derived EVI and primary production indices indicate a high correlation with reported summer yield in major irrigated agriculture in the pilot area within Syria. Simple regression spatially-based models were developed to predict summer crop production, found mostly effective for predicting cotton yields and vegetables. The regression models can be used as an indicator to predict summer crop yields during conflict years, and are able to show incidents were reported data could be questionable. Cumulative summer EVI-based model was the most effective among other parameters in predicting summer yields. The approach can contribute to an early diagnosis of food shortages and help decision makers' to focus relief efforts, especially in wars and periods of drought.

5.   REFERENCES

Domenikiotis, C., Spiliotopoulos, M., Tsiros, E., & Dalezios, N. (2004). Early cotton yield assessment by the use of the NOAA/AVHRR derived Vegetation Condition Index (VCI) in Greece. *International Journal of Remote Sensing, 25*, 2807-2819.

Endowment, C. (2014, April 18). *Drought, Corruption, and War: Syria's Agricultural Crisis.* Retrieved from Syrian Economic Forum: http://www.syrianef.org/En/?p=3361

FAO. (2009). *FAO's role in the Syria drought response plan.* Damascus, Syria: Food and Agriculture Organization of the United Nations (FAO).

Hole, F., & Smith, R. (2012). Arid land agriculture in northeastern Syria. *Land Change Science*, 213-226. doi:10.1007/978-1-4020-2562-4

Huete, A., Didan, K., Miura, T., Rodriguez, E. P., Gao, X., & Ferreira, L. G. (2002). Overview of the radiometric and biophysical performance of the MODIS vegetation indices. *Remote Sensing of ENvironment, 83*, 195-213.

Jones, H. G., & Vaughan, R. A. (2010). *Remote sensing of vegetation.* New York: Oxford University Press Inc.

Mingwei, Z., Qingbo, Z., Zhongxin, C., Jia, L., Yong, Z., & Chongfa, C. (2008). Crop discrimination in Northern China with double cropping systems using Fourier analysis of time-series MODIS data. *International Journal of Applied Earth Observation and Geoinformation, 10*, 476-485.

Ministry of Agriculture & Agrarian Reform in Syria. (2014). *Statistical datasets.* Retrieved from http://moaar.gov.sy

Mkhabela, M. S., Mkhabela, M. S., & Mashinini, N. N. (2005). Early maize yield forecasting in the four agro-ecological regions of Swaziland using NDVI data derived from NOAA's-AVHRR. *Agricultural and Forest Meteorology*(129), 1-9.

NASA LPDAAC. (2014). *NASA Land Processes Distributed Active Archive Center (LP DAAC), USGS/Earth Resources Observation and Science (EROS) Center.* Retrieved from https://lpdaac.usgs.gov/products

NTSG. (2014). *Modis Proucts.* Retrieved from the Numerical Terradynamic Simulation Group: http://www.ntsg.umt.edu/project/mod17

Pan, Y., Li, L., Zhang, J., Liang, S., Zhu, X., & Sulla-Menashe, D. (2012). Winter wheat area estimation from MODIS-EVI time series data using the Crop Proportion Phenology Index. *Remote Sensing of Environment, 119*, 232-242.

Quarmby, N. A., Milnes, M., Hindle, T. L., & Silleos, N. (1993). The use of mutli-temporal NDVI measurements from AVHRR data for crop yield estimation and prediction. *International Journal of Remote Sensing, 14*, 199-210.

Reeves, M. C., Zhao, M., & Running, W. (2005). Usefulness and limits of MODIS GPP for estimating wheat yield. *International journal of remote sensing, 26*(7), 1403-1421.

Running, S. W., Nemani, R. R., Heinsch, F. A., Zhao, M., Reeves, M., & Hashimoto, H. (2004). A continuous satellite-derived measure of global terrestial primary production. *BioScience*, 547-560.

Swiss Agency for Development and Cooperation SADC. (2014). *Syria: The imapct of the conflict on population diplacement, water and agriculture in the Orontes River basin.* Geneva: Swiss Agency for Development and Cooperation SDC.

Wardlow, B. D., Egbert, S. L., & Kastens, J. H. (2007). Analysis of time-series MODIS 250 m vegetation index data for crop classification in the U.S. Central Great Plains. *Remote Sensing of Environment*, 290-310.

Zhao, M., Heinsch, F. A., Nemani, R. R., & Running, S. W. (2004). Improvements of the MODIS terrestrial gross and net primary production global dataset. *Remote Sensing of Environment*, 164-176.

18

DRAWING AND LANDSCAPE SIMULATION FOR JAPANESE GARDEN BY USING TERRESTRIAL LASER SCANNER

R. Kumazaki [a, *], Y. Kunii[a]

[a] ITU, Department of Landscape Architecture Science, Tokyo University of Aguriculture,
1-1-1 Sakuragaoka, Setagaya, Tokyo, 156-8502, Japan – (45715004, y3kunii)@nodai.ac.jp

Commission V/WG 4

KEY WORDS: terrestrial laser survey, drawing of ground plan, landscape simulation, extract contour lines

ABSTRACT:

Recently, many laser scanners are applied for various measurement fields. This paper investigates that it was useful to use the terrestrial laser scanner in the field of landscape architecture and examined a usage in Japanese garden. As for the use of 3D point cloud data in the Japanese garden, it is the visual use such as the animations. Therefore, some applications of the 3D point cloud data was investigated that are as follows. Firstly, ortho image of the Japanese garden could be outputted for the 3D point cloud data. Secondly, contour lines of the Japanese garden also could be extracted, and drawing was became possible. Consequently, drawing of Japanese garden was realized more efficiency due to achievement of laborsaving. Moreover, operation of the measurement and drawing could be performed without technical skills, and any observers can be operated. Furthermore, 3D point cloud data could be edited, and some landscape simulations that extraction and placement of tree or some objects were became possible. As a result, it can be said that the terrestrial laser scanner will be applied in landscape architecture field more widely.

1. INTRODUCTION

3D Laser scanner has various kinds. Terrestrial 3D Laser Scanner measure products of the district by install it in the ground. Airborne Laser Survey System checks a shape of the ground altitude and topography precisely by Positional information of a plane distance to the point that a laser reflected back and information of the angle from the GPS, IMU by fire a laser to the ground from the laser scanner which it was equipped by the plane. MMS (Mobile Mapping System) measures a road and neighbouring mosaics 3D point cloud data by equipped with a digital camera and the laser scanner which were carried by a vehicle. (Kataoka, 2014)(Ishioka, 2009)
As described above, terrestrial laser survey using terrestrial 3D Laser Scanner attracts attention as surveying machinery and tools measuring a shape easily in contract with the objects of the comparative short distance. As a reason for that, it can acquire 3D dimensional shape data for an object because of contactless and high speed (Kunii, 2010). Following are some examples, the extraction of the part of the historic structure. (Ajioka, 2013) The highly precise 3D modeling of the city. The modeling of the 3D printer with the use of 3D point cloud data.
Recently, use of terrestrial 3D laser survey advances mainly in a field of civil engineering, but use study is expected in a field of landscape architecture that perform to record by 3D dimensional shape data. Specifically, it is expected that the use of terrestrial 3D laser survey is effective for the making of the ground plan of the garden, because the surveying by Total Station needs great labour (Awano, 2013).
I left need to find utilization other than visual technique such as computer animation, so as to examine the applicability of terrestrial 3D laser scanner in the garden investigation. In this research, I examined that studied various usefulness of

Terrestrial 3D Laser Scanner in the garden by perform that making of the ground plan with garden from 3D point cloud data.

2. JAPANESE GAREDENS FOR MEASUREMENTS

In this research, four Japanese gardens are adopted as measurement sites which are as follows; Taikan Yokoyama Memorial Garden (July, 2013: Taito-ku Tokoy), Mitsuaki Tanaka Old House Garden (August, 2013: Odawara Kanagawa), Saito Family Garden (August, 2014: Ishinomaki Miyagi) and Gokichi Matsumoto Old House Garden (August 2014: Odawara Kanagawa).
Laser measurements for these gardens were performed, and 3D point cloud data was acquired respectively. In addition, these laser measurements were included each own observation which were requested by each owner or local government. The purposes of these observation were maintenance of garden, grasping the present situation and create a new ground plan from past situation.

3. DRAWING FOR JAPANESE GARDEN BY USING 3D POINT CLOUD DATA

The point cloud data of Japanese gardens can be obtained by the laser measurements. Especially, in order to drawing of these gardens, the point cloud for ground level are as an important data. However, ground surface of the data were difficult to recognize from above view due to tree crowns (figure 1). Therefore, 3D point cloud data were deleted more than 1m and make it possible visibility of ground surface on the state of clouded by tree (figure 2).

---

* Corresponding author. This is useful to know for communication with the appropriate person in cases with more than one author.

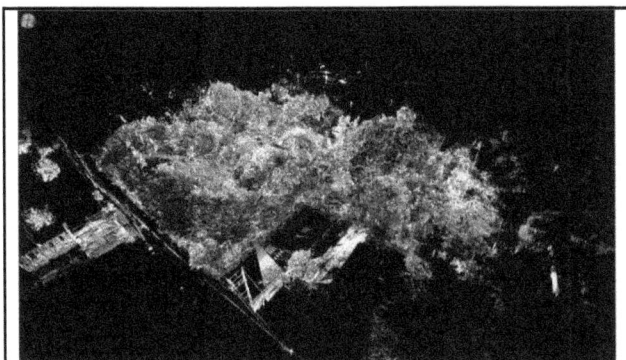

Figure 1. 3D point cloud data (whole data)

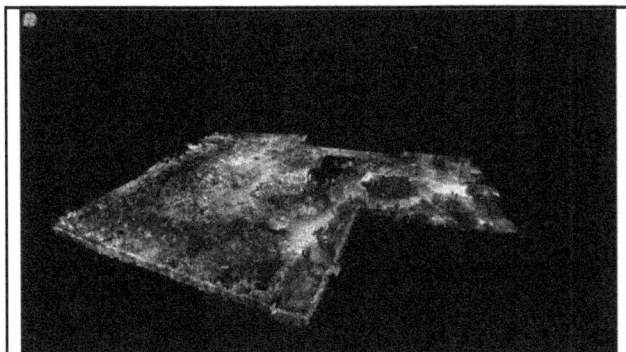

Figure 2. 3D point cloud data (delete more than 1m data)

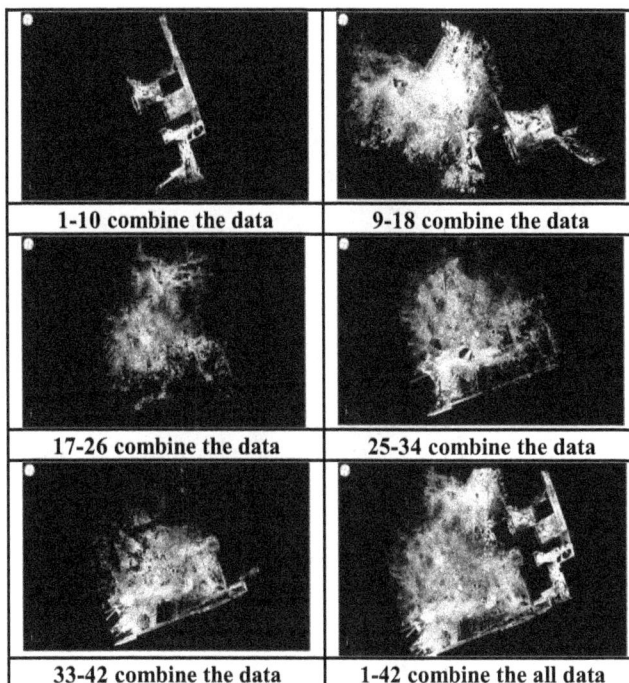

1-10 combine the data	9-18 combine the data
17-26 combine the data	25-34 combine the data
33-42 combine the data	1-42 combine the all data

Figure 3. The point density map of each group

3.1 Create of the Point Density Map

In order to obtain detail measurement data for ground level, the high density point cloud is required. However, the 3D point cloud data has huge volume due to include useless data, and smooth operations of the point cloud data processing are so difficult. Therefore, 3D point cloud data of each equipment point severally were combined by divide group of various places. After that, all group data were combined, and detail 3D dimensional shape data could be created. These procedures are shown in figure 3.

3.2 Output as DXF Format

DXF format is supported many CAD software, and feature points of the format are easy understanding structure and versatility. Then, point cloud data that deleting useless point was outputted in DXF, and point cloud of ground level became possible to show as ortho image. The ortho image by DXF is shown in figure 4.

3.3 The Point Density and Scale

The visualization of the point cloud data is influences the density. Then, suitable point density was considered for scale. Figure 5 shows ground plan of the most high point density.

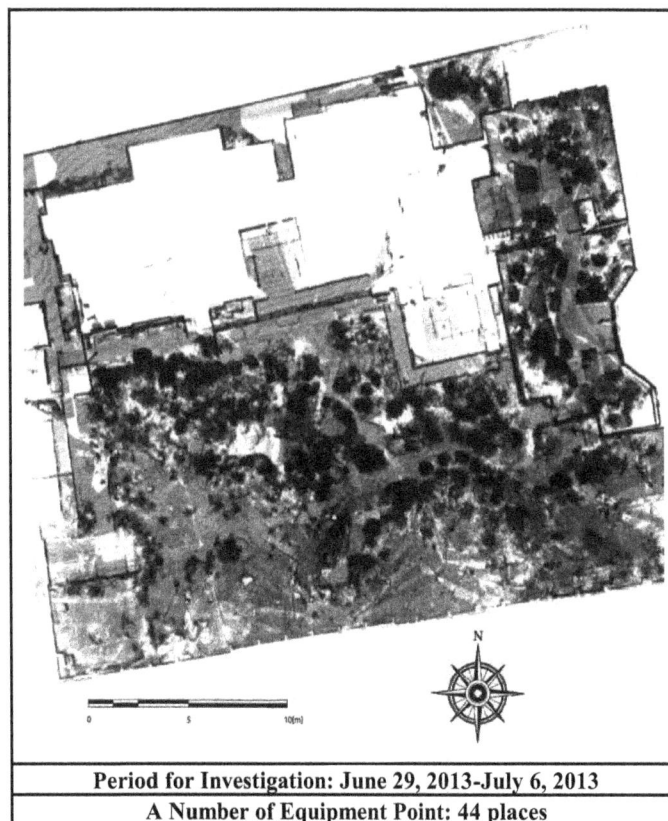

Period for Investigation: June 29, 2013-July 6, 2013
A Number of Equipment Point: 44 places
(a) Yokoyama Taikan Memorial Garden (Taito-ku, Tokyo)

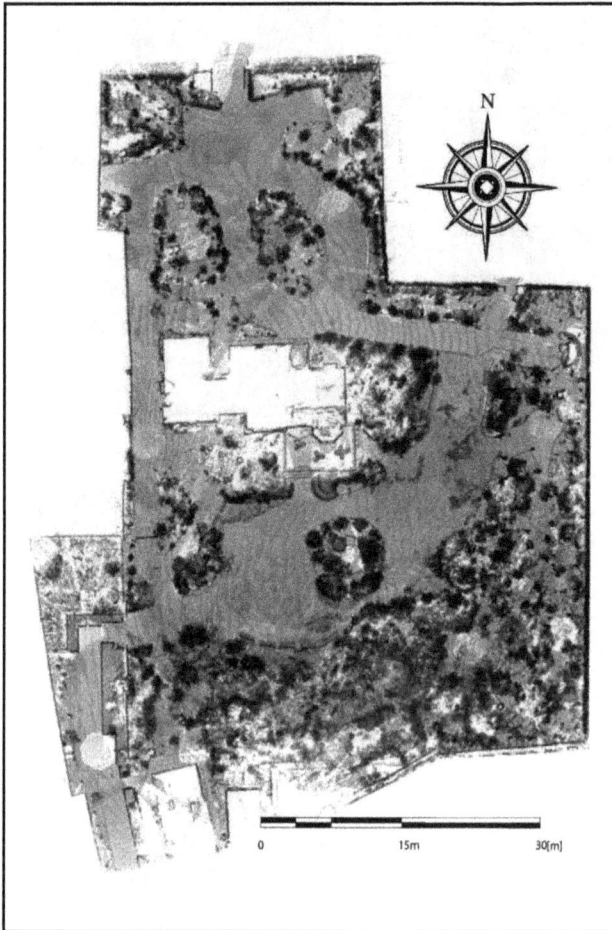

Period for Investigation: August 6-10, 2013

A Number of Equipment Point: 52 places

(b) Tanaka Mitsuaki Old House Garden (Odawara, Kanagawa)

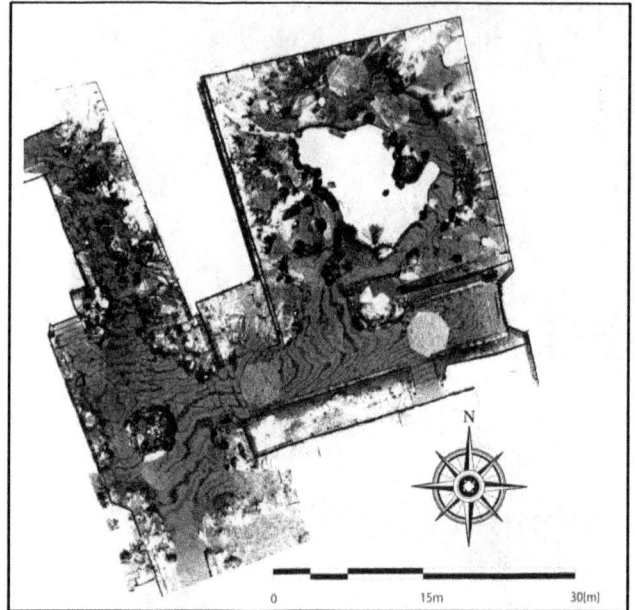

Period of Investigation: August 6-8 2014

A Number of Equipment Point: 27 places

(d) Saito Family Garden (Ishinomaki Miyagi)

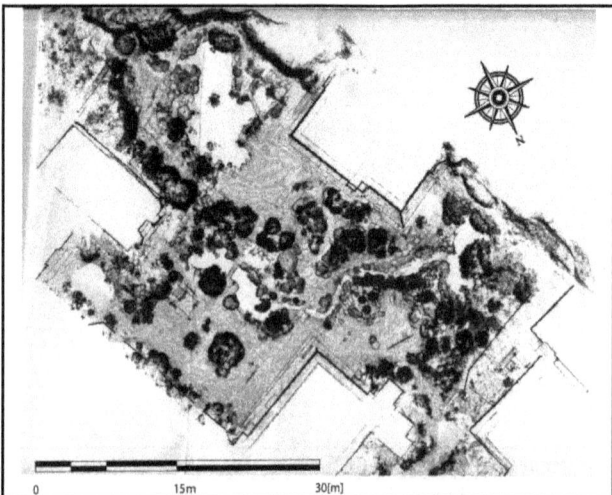

Period of investigation: May 16-18 2014

A Number of Equipment Point: 32 places

(c) Saito Family Garden (Ishinomaki, Miyagi)

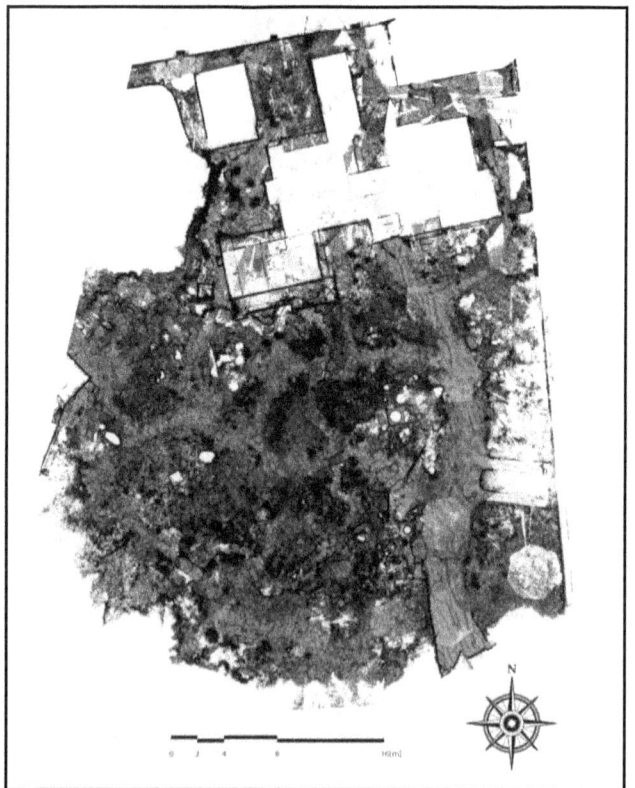

Period of Investigation: August 25-29

A Number of Equipment Point: 42 places

(e) Matsumoto Gokichi Old House Garden (Odawara Kanagawa)

Figure 4. A ground plan of point cloud by outputting orthochromatic images

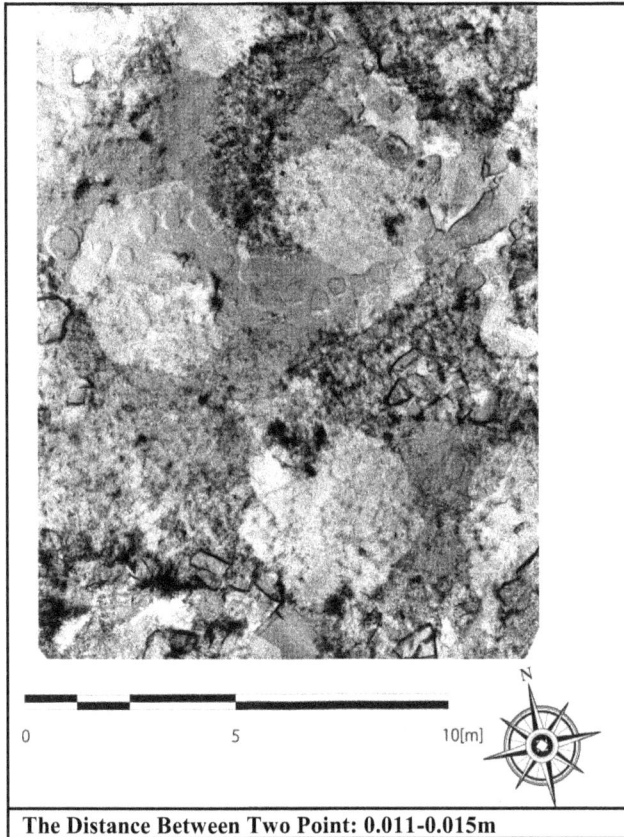

The Distance Between Two Point: 0.011-0.015m

Figure 5. Ground plan of the high point density

3.4 Drawing of Contour Line

3.4.1 Classification of height and chose original point: 3D point cloud data include height information, and it becomes possible to divide each height information. In addition, it is possible to create data that divide height by using the DXF data. In order to draw contour line of each garden, the each height data of Yokoyama Taikan Memorial Garden and Matsumoto Gokichi Old House Garden was extracted by following procedure. Firstly, temporary bench mark was chosen in the field, and the height of temporary bench mark was set as ±0. In this study, position of temporary bench mark was chosen on the end of washbasin (figure 6). Secondly, the height classification data of point cloud was outputted. The height data of Matsumoto Gokichi Old House Garden is shown in figure 7.

Drawing of Contour Line and Crown Projection: From the above procedure, the height classification data were converted layers. In that case, the overlay for each layer was performed by using the temporary bench mark. After that, contour lines were drawn (figure 8). Furthermore, crown projection area of each tree was surveyed in the actual gardens, and these were drawn in ground plan. Figure 9 shows the ground plan of Matsumoto Gokichi Old House Garden by using these information.

Figure 6. Original point and temporary bench mark

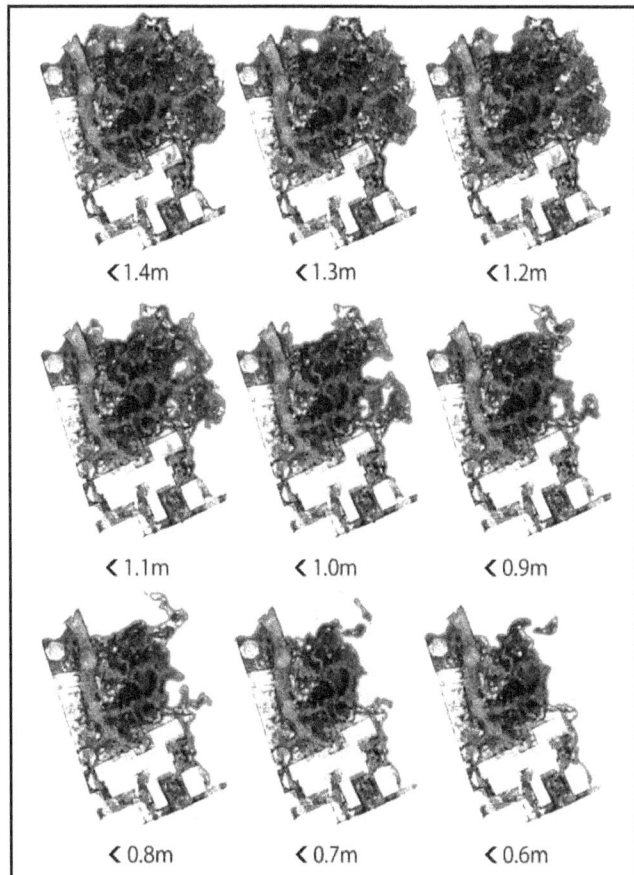

Figure 7. The height classification of Matsumoto Gokichi Old House Garden

Figure 8. Drawing of contour line

Before

After

Figure 10. Landscape Simulation of Cutting Trees. (Saito Family Garden)

4.2 Triangular Mesh Forming and Polygon Model Creation

3D shape data by measurement are constituted by point cloud data, and objects are shown to transmit. Therefore, the exterior wall of architecture and the trunk of a tree in the garden were desirable to apply constitute the triangular mesh. By doing so, it is become possible to create 3D model which were much closer to reality (Figure 11).

4.3 Available Polygon Model

Figure 11 shows the polygon model of trees that in order to practice landscape simulation. Perform to insert polygon model of Japanese red pine for virtual field of figure 10. Figure 12 shows landscape simulation that insert Japanese red pine of 3D point cloud.

5. CONCLUSION

In this study, carried out drawing ground plan at the end series using 3D point cloud data. As a consequence, it is thought that showed one approach describes a method until created ground plan. For example, create ground plan of high point cloud density and extract contour line data of 3D point cloud data by create layer that has been divide height. Also, 3D model of garden consisted by 3D point cloud data can confirm that simple structure however, it was impossible to confirm the trunk of trees and trees overhang in detail. And with that, attempt to part extract 3D point cloud data of tree, and it is presumed that lead to utilization of create polygon model by using this data.

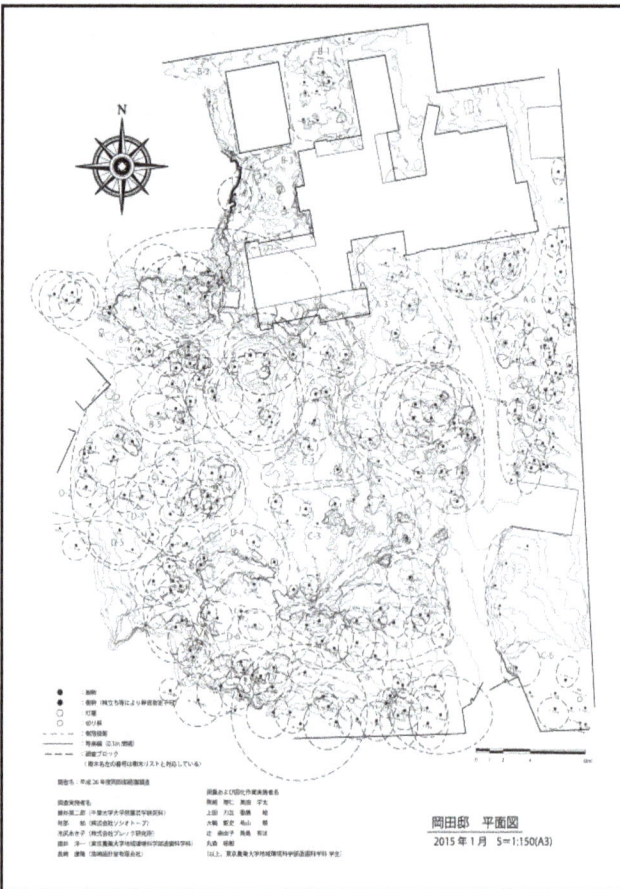

Figure 9. Ground Plan of Matsumoto Gokichi Old House Garden (This is created as part of the surveying project garden)

4. DATA POCESSING FOR GIVE LANDSCAPE SIMULATION

4.1 Landscape Simulation of Cut down Tree

3D point cloud data by survey the field can be used for landscape simulation. In order to simulate, it needs to be delete useless point cloud data that cut down tree of point cloud by virtual 3D visual world. The comparison of before and after of the simulations are shown in figure 10.

Red pine woods

Zelkova

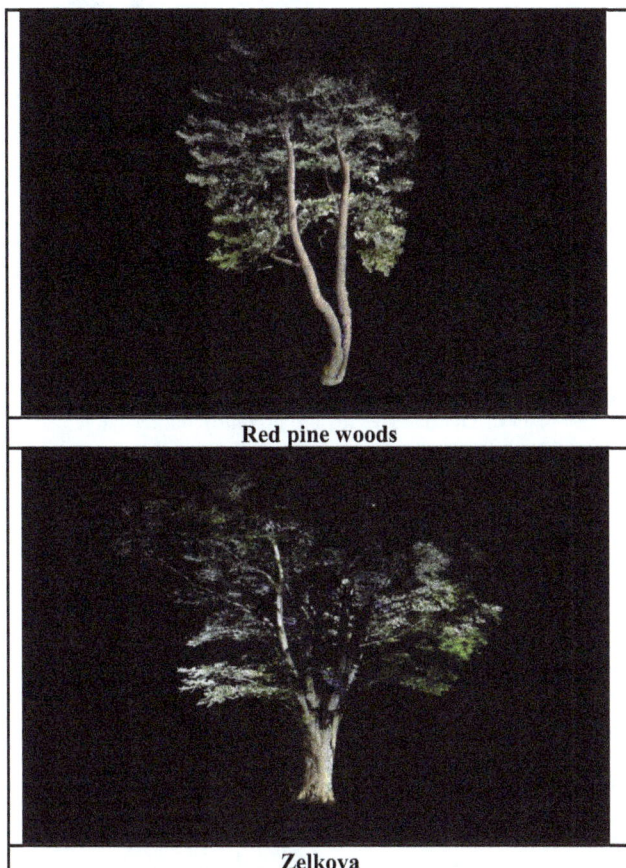

Figure 11. 3D model by using point cloud and triangular mesh

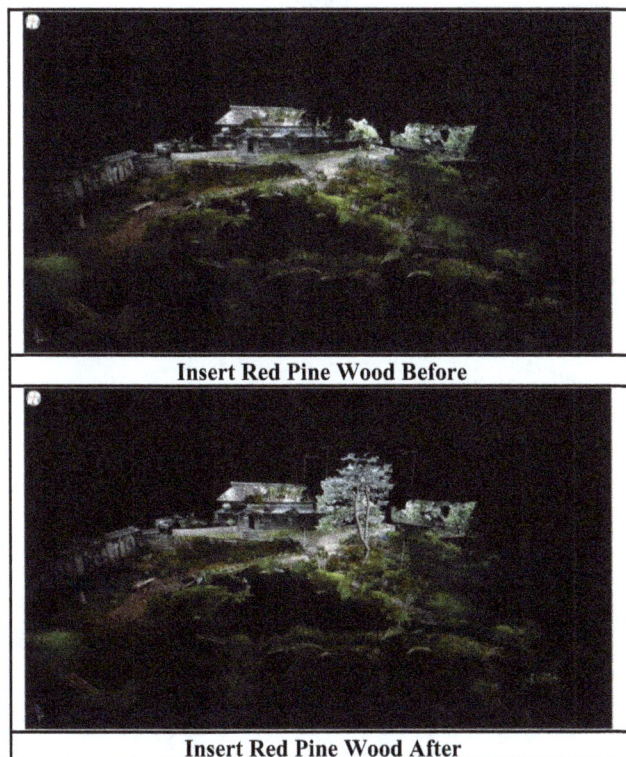

Insert Red Pine Wood Before

Insert Red Pine Wood After

Figure 12. Insert Red Pine Wood on the Saito Family Garden.

5.1 References

Kataoka, K., Nakagawa, M., 2014.
Evaluation of Projection Model for Random Point Cloud

The 35th Asian Conference on Remote Sensing (ACRS2014), Naypyitaw, Myanmar
Session G1, OS-239, pp.6, CD-ROM, 2014/10/30

shikawa, K., Amano, Y., Hashizume, T., Takiguchi, J., and Shimizu, S., 2009. City Space 3D Modeling Using a Mobile Mapping System *SICE Journal of Control, Measurement, and System Integration* Vol.8, No.17, pp.132-139

Kunii, Y., Yanagi, T., Yamazaki, M., 2010.
Grasping of Landscape in Campus by using Terrestorial Laser Scanner and its Application Journal of Agriculture Science, Tokyo University of Agriculture Vol. 55 No.2: pp.199-204 September 2010 Setagaya Tokyo, 156-8502, JAPAN

Awano, T., Kunii Y., 2013
Survey Technique of Historic Garden by Using 3D and Direct Measurements in the Case Study of Seikan-Tei Garden.
JILA Technical Reports of Landscape Architecture No.7: pp. 126-129

Ajioka, O., Watanabe, H., 2013. Study about possibility of application for making drwings of historical buildings using extracting ridge lines from point cloud
Summaries of technical papers of annual meeting pp. 639-640

5.2 Acknowledgements

In this study, really appreciate your kindness for Prof: Makoto Suzuki and Prof: Takashi Awano, each municipality and related organization in illustration of ground plan and field survey. I would like to express my heartfelt thanks.

19

SATELLITE-BASED ASSESSMENT OF GRASSLAND YIELDS

K. Grant [a, *], R. Siegmund[b], M. Wagner[b], S. Hartmann[a]

[a] Bavarian State Research Center for Agriculture (LfL), Institute for Crop Science and Plant Breeding, 85354 Freising, Germany – (Kerstin.Grant, Stephan.Hartmann)@lfl.bayern.de
[b] GAF AG, 80634 Munich, Germany – (Melanie.Wagner, Robert.Siegmund)@gaf.de

KEY WORDS: Radar, Change detection, Sentinel-1, Grassland, Cutting date, Forage yield model

ABSTRACT:

Cutting date and frequency are important parameters determining grassland yields in addition to the effects of weather, soil conditions, plant composition and fertilisation. Because accurate and area-wide data of grassland yields are currently not available, cutting frequency can be used to estimate yields. In this project, a method to detect cutting dates via surface changes in radar images is developed. The combination of this method with a grassland yield model will result in more reliable and regional-wide numbers of grassland yields. For the test-phase of the monitoring project, a study area situated southeast of Munich, Germany, was chosen due to its high density of managed grassland. For determining grassland cutting robust amplitude change detection techniques are used evaluating radar amplitude or backscatter statistics before and after the cutting event. CosmoSkyMed and Sentinel-1A data were analysed. All detected cuts were verified according to *in-situ* measurements recorded in a GIS database. Although the SAR systems had various acquisition geometries, the amount of detected grassland cut was quite similar. Of 154 tested grassland plots, covering in total 436 ha, 116 and 111 cuts were detected using CosmoSkyMed and Sentinel-1A radar data, respectively. Further improvement of radar data processes as well as additional analyses with higher sample number and wider land surface coverage will follow for optimisation of the method and for validation and generalisation of the results of this feasibility study. The automation of this method will than allow for an area-wide and cost efficient cutting date detection service improving grassland yield models.

1. INTRODUCTION

1.1 Importance of cutting dates for grassland yield

Grassland ecosystems support human, fauna and flora populations worldwide by providing numerous goods and services such as provision of forage for livestock, wildlife habitats, and biodiversity conservation (White et al., 2000). In Bavaria (Germany), grassland covers 34% of the agricultural land area and contributes primarily to the production of forage for the dairy and meat industry (StMELF, 2014). Despite the economic importance of Bavarian grassland, actual and area-wide data of grassland yield is not available (Diepolder et al., 2013). Detailed grassland harvest quantification is missing because, in contrast to other agricultural products (e.g. grain, maize, sugar beets), grassland forage does not enter the market and usually remains on the farm. Therefore, quantification techniques such as scales or sensors are generally not in practice. Cutting date and frequency are important parameters, which determine grassland yield, in addition to the effects on yield of climate, soil, plant composition and fertilisation. These parameters would be required for yield modelling (e.g. see yield model in Herrmann et al., 2005). Furthermore, in the absence of yield data, cutting frequency is currently used to estimate amounts of fertiliser for managed grassland (Wendland et al., 2012). This is necessary because, by state regulation, fertilisation must be oriented on the anticipated nutrient requirements based on actual nutrient discharge via harvest. Thus, information on cutting frequencies might also assist in plausibility tests for fertilizer use. However, a method to record dates when grass is harvested over large areas would not only improve estimations of grassland yields and sustainable use of fertiliser, but could also be relevant for questions of nature conservation (Herben and Huber-Sannwald, 2002). Thus, it is necessary to find a cost- and time-efficient method to detect cutting dates for whole regions or countries.

1.2 Sentinel-1

Remote sensing techniques are useful to monitor surface changes for large areas. So far, it has been very expensive to get the necessary satellite images with high time resolutions for wide areas. The new European earth observation programme Copernicus has developed a set of satellites (called Sentinel), which will cover the entire world's land masses at least on a bi-weekly basis. The European Space Agency and the European Commission provide the data obtained with Sentinels on an open and free basis (DLR, 2014). The first Copernicus satellite, Sentinel-1A, carrying a C-band radar system, was launched in April 2014 and radar images are now available routinely every 12 days and systematically for land monitoring (ESA, 2014). Together with the identically constructed Sentinel-1B (launch 2016) the revisit time of each point will be shortened to 6 days. Thus, this study aims to investigate the applicability of Sentinel-1A radar data for the derivation of agricultural information as they offer a great potential for research across regions.

1.3 Radar & grassland cuts

This study focusses on the detection of cutting dates in grasslands as changes in the radar backscatter. Cutting of grass significantly affects the vegetation structure and surface of the grassland (height, density, shape) and therefore results in changes of the backscatter intensity of the radar signals. By comparing the reflection signals over a set of radar images acquired at a high temporal sampling frequency or with short time interval, cuts are expected to be detectable using change detection techniques. The application of change detection methods is therefore promising since the cuts and harvest events are temporally sampled in an adequate way.

---

* Corresponding author.

2. TEST-PHASE – GRASSLAND CUT DETECTION

2.1 Acquisition plan

In order to achieve a high monitoring potential a variety of operational SAR systems are used. For the test-phase, specifically to increase the temporal acquisition frequency, the high resolution X-band systems COSMO-SkyMed (CSK) and TerraSAR-X (TSX) are utilised. Future operational monitoring will exploit the C-band system Sentinel-1 (A and B) plus optional X-band acquisitions. During the test-phase a multimission acquisition plan (Table 1) will be used in a way that for given harvest periods of three weeks the subsequent scheme will be applied. Begin and end of these periods are determined according to reports of contract farmers. Validation of cutting dates is done by *in-situ* measurements and reports of date and location of grassland cuts in the study area.

Table 1. Acquisition plan for one harvest period using TerraSAR-X (TSX), Sentinel-1A (S1) and COSMO-SkyMed (CSK) system; study area can be acquired with CSK using incidence angles h4-09 = 35°, h4-14 = 41°, h4-18 = 46°, h4-21 = 49°)

Day	S1	TSX	CSK	CSK	S1	TSX	CSK	CSK
		ascending				descending		
1							h4-09	
2		h4-14**					h4-09	
3								h4-18
4	IW*			h4-21				
5							h4-09	
6								
7								
8		11**		h4-21	IW			
9	IW						h4-09	
10			h4-14					
11			h4-14			10		h4-18
12								h4-18
13					IW			
14			h4-14					
15								h4-18
16	IW			h4-21				
17				h4-21			h4-09	
18			h4-14				h4-09	
19		11						h4-18

*IW= Interferometric Wide Swath   **Beam Mode

A feasibility study was done using CSK and Sentinel-1 radar images from October 2014 during the last grassland cuts in a study area south-east of Munich, Germany.

2.2 Feasibility study

2.2.1 Incoherent monitoring

The acquisition capacity of a multi-mission exercise provides a flexible basis to map and monitor both continuous, as well as episodic, events with a high degree of flexibility and agility plus a high unmatched revisit rate with multiple data take opportunities (DTO) for any location. Due to the robustness of the monitoring approach change detection is applied for the

same systems and similar acquisition geometries only. Therefore only comparisons of the same beam modes and sensors are made. The approach continues recent investigations to apply SAR-satellite monitoring to the European subsidy control system (INVEKOS), in which the potential of COSMO-SkyMed data were evaluated for its integration in operational INVEKOS tasks or in precision farming support systems in study areas in Germany (Britti et al., 2011, Cesarano et al., 2011, Strehl, 2012, Wagner, 2014). In the present case changes in backscatter – resulting from cutting of the grass layer – is derived from robust amplitude change detection methods, i.e. amplitude ratios. It is assumed that the high acquisition rate samples grassland cutting events sufficiently in a way that changes in density of the vegetation are mapped prior to significant /full regrowth. During our test phase the change detection technique and relevant optimisation methods, such as filter settings, are validated. The determination of changes is performed based on a plot level (field boundaries) using spatial statistics over the amplitude variation and is indicated in the field layer. The resulting change map is verified using all available *in-situ* data.

2.2.2 COSMO-SkyMed data

COSMO-SkyMed Images of the 3rd and 15th October 2014 were evaluated. All images were HH polarized X-band full-resolution data acquired in HImage mode. The SAR data was in Single Look Slant Range Complex format in product level 1A. The average ground range and azimuth resolution of the imagery are 2 m and 2.8 m, respectively. Both acquisitions were recorded at 16:58 (UTC) with a scene centre incidence angle of 46°.

2.2.3 Sentinel-1A data

Sentinel-1 data for 5th and 17th October 2014 were retrieved from the ESA archive and evaluated. Both images were VV/VH polarized C-band high-resolution data acquired in the interferometric wide swath mode. The data was in ground range detected (GRD) format and product level 1. The average ground range and azimuth resolution of the imagery are 20 m and 22 m, respectively. Both acquisitions were recorded at 05:26 (UTC).

2.2.4 Field data – grassland cuts

Grassland plots in the study area were surveyed *in-situ* between 30th September and 13th October 2014 for cutting activities. Dates and location of grassland cuts were noted on a map. In a next step, this information was digitalised resulting in a shapefile for further analysis in a GIS environment. During this time period, cuts on 154 grassland plots were detected.

2.2.5 SAR data processing

COSMO-SkyMed
The CSK images were georeferenced using a digital terrain model (Range Doppler Terrain Correction, SRTM) and reprojected to the coordination system 3-degree-Gauss-Krüger zone 4. For analysis and comparison the radar images were radiometrically calibrated. The corrected amplitude data was resampled to 3 m and transformed to the logarithmic scale (unit dB). Speckle was reduced with an adaptive Frost filter (window size 7*7). All image processing was performed with SARscape (ENVI) and ERDAS Imagine. Radar data was exported as GeoTIFF (unsigned 8bit) to visualize and analyse data in a GIS environment.

Sentinel-1

Sentinel-1 data was pre-processed using data conversion and orthorectification processes implemented in SARscape. Data was georeferenced using a digital terrain model (Range Doppler Terrain Correction, SRTM) and reprojected to the coordination system 3-degree-Gauss-Krüger zone 4. For analysis and comparison the radar images were co-registered. For each image, both polarisation channels (VV and VH) were radiometrically corrected and exported as sigma0 (σ^0) values (unit dB).

Grey level statistic

Radar data was overlaid with a shapefile including the grassland plots with cutting dates from *in-situ* measurements. The grey values of each image represent the strength of the radar return. For a qualitative comparison grey level statistics were calculated for each grassland plot illustrating the backscatter or intensity change before and after an area of grassland had been harvested.

2.3 Results

Alterations in the grey values of radar images of both systems (CSK, Sentinel-1) showed modified radar backscatter signals due to surfaces changes in the test area (example see Figure 1). These changes were caused by grassland cuttings, which could be verified by in situ measurements. In order to estimate the separability of cut and uncut grassland, mean grey values of each image were extracted and compared (example in Table 2).

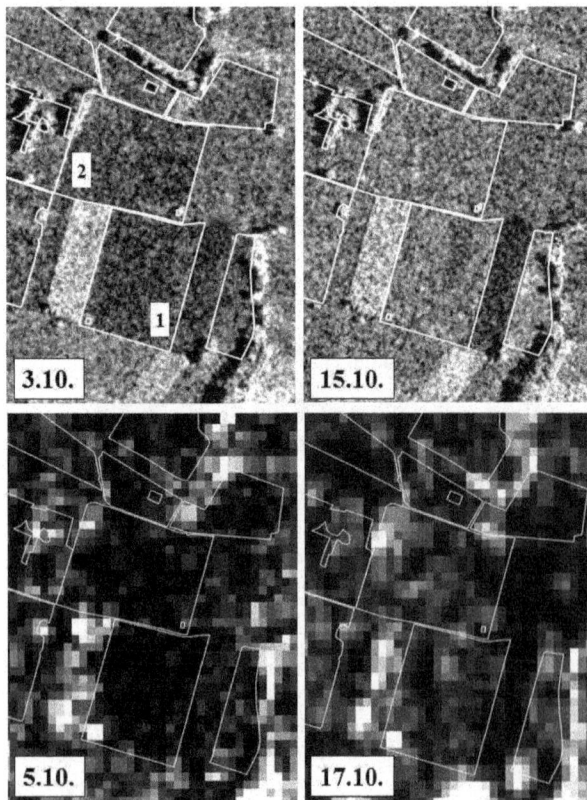

Figure 1. Alteration in radar backscatter signal/grey value by grassland cuts in a part of the study area; images: CosmoSkyMed: 3rd and 15th October 2014; Sentinel-1A: 5th and 17th October 2014 VV polarized; in situ measurements detected grassland cut on plot 1 on 6th and plot 2 between 10th)

Of 154 tested plots (covering in total 436 ha) 116 and 111 cuts were detected with CSK and Sentinel-1 data comparison, respectively. Neither CSK nor S1 grey value analyses resulted in the detection of cuts on 25 grassland plots. Furthermore, 18 cuts were missed by Sentinel-1 but not by CSK. In contrast, 13 cuts were detected by Sentinel-1 but not by CSK.

Table 2. Grey value comparison for the two grassland plot samples (see Figure 1); given are in-situ detected cutting date, grey value mean and differences extracted from CosmoSkyMed images of 3rd and 15th October 2014 as well as from Sentinel-1A data of 5th and 17th October; negative value in grey value and sigma0 (for Sentinel-1A) difference indicates grassland cuts

Plot	Cutting date	Grey value mean						sigma0
		CosmoSkyMed*			Sentinel-1A			
		3.10.	15.10.	difference	5.10.	17.10.	difference	difference
1	6.10.	80	127	-47	16	27	-11	-27.1
2	10.10.	90	133	-43	17	30	-13	-32.6

*after adaptive Frost-filtering

According to the sigma0-values (VV polarization), 151 plots showed an increase in the radar backscatter amplitude. Therefore, the difference between 5th and 17th October were negative, also indicating the grassland cuts on these plots (example in Table 2).

2.4 Discussion

This feasibility study showed that grassland cuts can be detected using radar images of chronologically close dates. So far, change detection with both SAR systems with a revisit time of 12 days resulted in the retrieval of about 74% known grassland cuts. Although the systems had various acquisition geometries (e.g. differences in polarisation, wave length, spatial resolution), the amount of detected grassland cut was quite similar. Further analysis has to show, if specific field conditions such as low grass height difference, environmental condition (wind, soil type, moisture, trees) or management practices let to the missing of 25 grassland cuts by both SAR systems (Bouman & van Kasteren, 1990a, b). The 18 cuts only missed by detection with Sentinel-1A data (S1) might be caused by the coarse resolution or differences in wave length. Zoughi et al., (1987) reported that in contrast to X-Band, soil and its properties was the dominant factor in backscatter from grass canopy at C-Band. Reason for the detection of 13 cuts by S1 but not by CSK might be the VV polarisation. Horizontal polarisation seems to be more sensitive to horizontally oriented target components such as soil surface, therefore vertical polarisation might result in stronger backscatter from the upper portion of grass canopy (Zoughi et al., 1987). However, due to the coarse resolution of S1 data, the grey value mean results from fewer pixels per plot compared to CSK data. Therefore, the statistics of S1 data are not as reliable as of CSK data. Further analysis with a higher sample number and wider land surface coverage is necessary for validation and generalisation of the results of this feasibility study. The detection rate of 74% is not yet sufficient for reliable estimation of cutting frequencies. However, the radar data processing could be further optimized for instance by filtering or using coherence information. Coherent monitoring was out of scope for the recent study conclusion, though the applicability of those techniques will be conducted during further steps of the test phase. Grassland is a dynamic system and plants are continuously growing (up to 10-20 cm in a week at optimal conditions). Enhancing the revisit times (e.g. 6 days instead of 12 with additional Sentinel-1B data) and thus using a set of radar images acquired at a higher temporal sampling frequency, changes in grassland growth might be better incorporated. In this context, it should be mentioned that, in October, the height difference in grassland before and after the cut is usually not very pronounced (10-15 cm). Therefore, the detection rate of this feasibility study is reasonable. For annual grassland yield the first cut in the year is the most important, usually contributing 25-40% depending on the management intensity. Then, pre- to post cut height difference is generally about 30 to 50 cm. Thus, a higher detection rate for the first cut can be expected.

3. OUTLOOK – YIELD ESTIMATION

The improved and automated method for grassland cutting date detection based on Sentinel-1 radar data will be integrated in a grassland yield and quality model such as used by Herrmann et al., (2005). The main model parameters are the satellite-based cutting date information in addition to climate and site specific data retrieved from state or national geographic information systems (Figure 2). Then, yield and quality estimations can be calculated area-wide

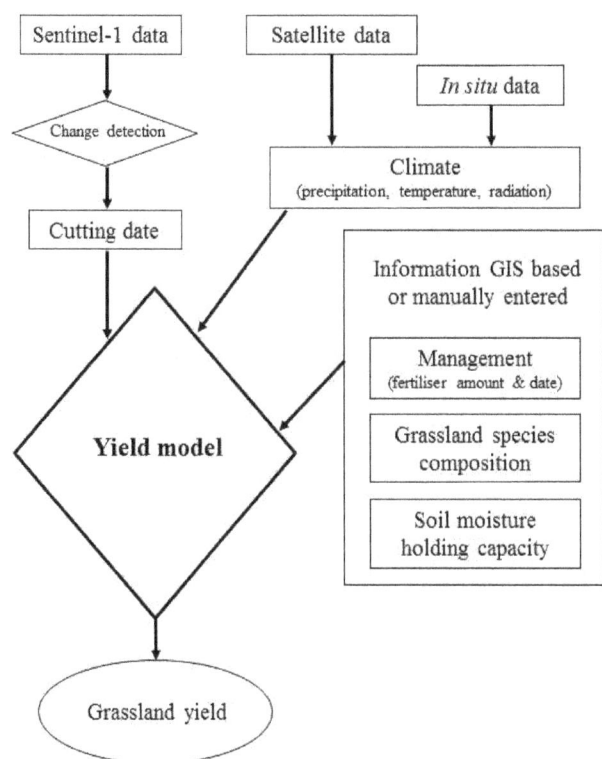

Figure 2. Outline of improved and regionalized grassland yield model

In addition, a yield prognosis model with a web interface might be developed using this model to predict the best date (in terms of yield and quality) for the next cutting for single grassland plots. Radar information could be used to set the starting point for the model run. If known, farmers might add more detailed information of their grasslands to further improve the estimation.

ACKNOWLEDGEMENTS

This project was funded by the German Federal Ministry of Economic Affairs and Energy as part of the program "GMES – Dienste für den öffentlichen Bedarf" (FKZ 50EE1318). We thank all farmers in the study area for their assistance.

REFERENCES

Bouman, B., van Kasteren, H., 1990a. Ground-Based X-Band (3-cm Wave) Radar Backscattering of Agricultural Crops. I. Sugar Beet and Potato; Backscattering and Crop Growth, *Remote Sensing of Environment*, 34, pp. 93-105

Bouman, B., van Kasteren, H., 1990b. Ground-Based X-Band (3-cm Wave) Radar Backscattering of Agricultural Crops. II. Wheat, Barley, and Oats; The Impact of Canopy Structure. *Remote Sensing of Environment*, 34, pp. 107–119.

Britti, F., Ligi, R., Rossi, L. Monaldi, G., 2011. New GAEC controls through high revisit SAR: COSMO-SkyMed in support to agro-environmental analysis and monitoring, *GEOCAP Conference*, Tallin

Cesarano, L., Pietranera, L., Britti, F., Gentile, V., 2012. Time and resolution: COSMO-SkyMed VHR data in support to precision farming applications – North Germany Pilot Project, *GEOCAP Conference*, Tallin

Diepolder, M., Raschbacher, S., Heinz, S., Kuhn, G., 2013. Rohproteinerträge und –gehalte bayerischer Grünlandflächen In: Mehr Eiweiß vom Grünland und Feldfutterbau Potenziale, Chancen und Risiken, *Schriftenreihe der Bayerischen Landesanstalt für Landwirtschaft*, 6/2013, pp. 136-140

Herrmann, A., Kelm, M., Kornher, A., Taube, F., 2005. Performance of grassland under different cutting regimes as affected by sward composition, nitrogen input, soil conditions and weather—a simulation study, *European Journal of Agronomy* 22, pp. 141-158

StMELF (Bayerisches Staatsministerium für Ernährung, Landwirtschaft und Forsten) 2014. Bayerischer Agrarbericht 2014, http://www.agrarbericht-2014.bayern.de/landwirtschaft-laendliche-entwicklung/index.html (25.03.2015)

Strehl, F., 2012. *Monitoring of agricultural fields using very high resolution SAR-data*, Bachelor thesis, University of Applied Sciences Munich

Wagner, M., 2014. *Potential analysis of COSMO-SkyMed VHR data for substituting optical data in remote sensing based agricultural subsidy control*, Master thesis, HafenCity University Hamburg.

Wendland, M., Diepolder, M., Capriel, P., 2012. *Leitfaden für die Düngung von Acker- und Grünland – Gelbes Heft*, LfL-Information, Bayerische Landesanstalt für Landwirtschaft, Freising, pp. 1-97

White, R., Murray, S., Rohweder, M., 2000. *Pilot analysis of global ecosystems: grasslands ecosystems*, World Resources Institute, Washington, DC, pp. 1-69

Zoughi, R., Bredow, J., Moore, R., 1987. Evaluation and comparison of dominant backscattering sources at 10 GHz in two treatments of tall-grass prairie, *Remote Sensing of Environment*, 22, pp. 395-412

20

DISCUSS ON SATELLITE-BASED PARTICULATE MATTER MONITORING TECHNIQUE

Bicen Li [a], *, Lizhou Hou [b]

[a] Beijing Institute of Space Mechanics and Electricity (BISME), Beijing, China - mou_lbc@163.com
[b] Beijing Institute of Space Mechanics and Electricity (BISME), Beijing, China - houlizhou@126.com

KEY WORDS: Fog and haze, PM2.5 concentration, Aerosol optical depth, Satellite remote sensing, Multi-angle polarimetric imager

ABSTRACT:

Satellite measurements for atmospheric pollutants monitoring provide full mapping, large spatial coverage, and high spatial resolution. Retrieved aerosol optical depth (AOD) from satellite data as the key parameter has been used in the study on particulate matter (PM) distributions which is complementary to ground-based measurements. Based on the empirical relations between aerosol optical properties and PM10 or PM2.5 concentration and its influencing factors, combining the Beijing-Tianjin-Hebei regional PM distribution feature, the specifications including bands, viewing angles and polarization measuring requirements of the on-orbit PM monitoring instrument are discussed. The instrument is designed to obtain the data for retrieving atmospheric AOD, shape and size of particles, refractive index, aerosol single scattering albedo (SSA), etc. The major pollutant PM2.5 concentration can be retrieved from the remote sensing data especially, and its global distribution can be mapped as well. The progress of conventional aerosol retrieval method using visible data is presented along with its limitation for serious haze-fog condition. Adding ultra violet (UV) bands to obtain UV aerosol index (UVAI) is useful for monitoring the main constituents of haze.

1. INTRODUCTION

With the rapid growth of China's economy, the problem of air quality is more and more serious. Because of the national energy structure adjustment and the increasing motor vehicle emissions, the current pollutions characteristics of ambient air quality have transformed to mixed pollutant from smoke pollutant. The dust haze pollution which mainly consists of atmospheric fine particulate matter happens frequently in the relevant developed areas such as Beijing-Tianjin-Hebei, Yangtze River Delta and Pearl River Delta (He, et al., 2002), which causes widespread public concern. Atmospheric fine particulate matter PM2.5 (the particulate matter whose aerodynamic equivalent diameter is less than or equal to 2.5 micron) which is the critical factor generating the pollution characteristics will reduce air visibility and affect health of human respiratory system and cardiovascular system. In recent years, research on PM monitoring have been developed rapidly along with the improving public attention to the air pollution.

The monitoring methods include satellite measurements and ground-based measurements. The time resolution and accuracy of ground based measurements are better but the effective observation range is limited by the fixed location of the instruments. Polar orbit satellite can obtain data of aerosol optical properties over very large spatial scale. Due to full mapping, large spatial coverage, and high spatial resolution, retrieved AOD from satellite data has been applied in the study on PM distributions. Although they are less precise than ground observations, satellite measurements can be very useful to improve the understanding of regional PM distributions and thus be complementary to ground-based measurements. Empirical relations between aerosol optical properties and PM10 or PM2.5 measurements have been studied for different parts of the world (Kacenelenbogen, et al., 2002; Hutchison, 2003; Wang and Christopher, 2003; Zhang and Li, 2013; Zhao,

et al., 2013). The capacity for monitoring and estimating fine particulates near the surface by satellite remote sensing will be improved based on these empirical results.

In this study, we discuss the specifications including bands, viewing angles and polarization measurement requirements of the on-orbit PM monitoring instrument based on the empirical relations between aerosol optical properties and PM concentration and the PM distribution features in Beijing-Tianjin-Hebei area. And some advices on development of the remote sensing instrument for fine PM monitoring are present.

2. REGIONAL FEATURES OF PM POLLUTION IN BEIJING-TIANJIN-HEBEI AREA

For a long time, pollution data formally released by China Ministry of Environment until 1st January 2013 were about PM10, and research on the relationship between pollutant and meteorological conditions mostly focused on PM10 (Wang and Li, 2003; Sui, et al., 2007). According to ambient air quality standard GB2095-2012 promulgated by Ministry of Environment in 2012, PM2.5 has been brought into the scope of general air quality monitoring and evaluation system along with O_3 and CO. Thus six kinds of pollutants including PM10, PM2.5, O_3, CO, SO_2 and NO_2 are monitored and evaluated comprehensively according to the standard. During January and February 2013, there were 51 days that PM2.5 was the critical pollutant in Beijing according to the statistical results based on the measurements of environment departments. PM2.5 is becoming the focus of public concern.

By studying the regional features of PM2.5, spatial distributions of PM2.5 in a broad scale has been learned. The airborne measurements find that the pollutants of Beijing area mainly come from Gobi desert in northwest China, south of Beijing and local emission sources (Zhang, et al., 2006). Almost 34% of

---

particulate matters in summer spread from the surrounding provinces and cities near Beijing (Streets, et al., 2007). By using PM2.5 concentration and aerosol scattering coefficient measured in urban and rural environment in Beijing area in autumn 2011, PM2.5 pollution and aerosol optical properties in fog-haze days are discussed (Zhao, et al., 2013).

As a new method to study atmospheric aerosol and particulate matter, satellite remote sensing can obtain pollution distributions of large scale and so provide more information to improve the insight in regional PM distributions. Global satellite-derived PM2.5 averaged over 2001-2006 is mapped by using Terra and Aqua data of NASA which is present in Figure 1 (van Donkelaar, et al., 2010). This figure show that PM2.5

concentrations is much higher in the regions including north Africa, west Asia and east Asia, especially over eastern China which exist long-term serious PM2.5 pollution. The fog and haze pollution situation in Beijing region are analyzed base on the AOD retrieved from data in 14th and 17th January 2013 obtained by HJ-1A and HJ-1B satellites (Figure 2). Figure 2(a) shows the heavy polluted weather in this fog-haze period that the air quality index (AQI) in 14th January was 331 which was measured at Olympic Sports Centre (OSC) observation station. Figure 2(b) shows the clean weather after fog-haze disappeared that the AQI in 17th January was 99 which also was measured at OSC station (Xu, et al., 2013). These retrieved AOD maps show many spatial distribution details of the fog and haze.

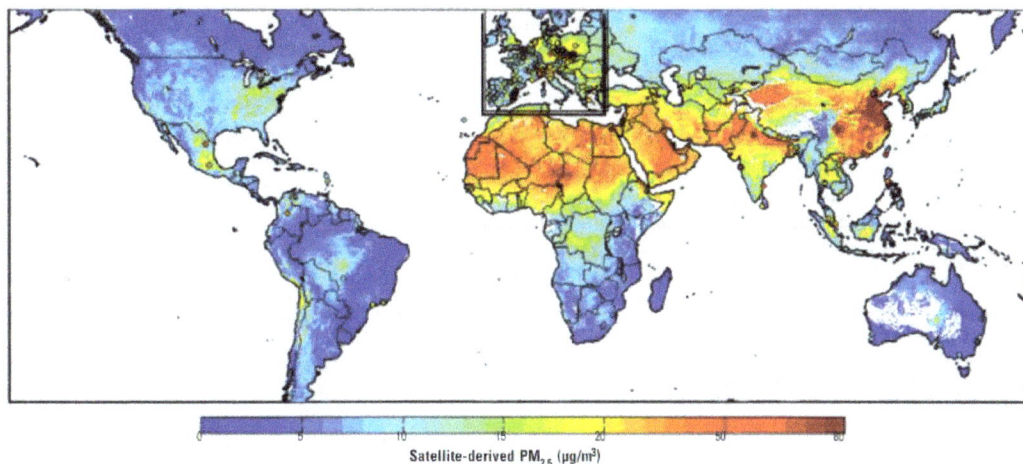

Figure 1. Global satellite-derived PM2.5 concentration

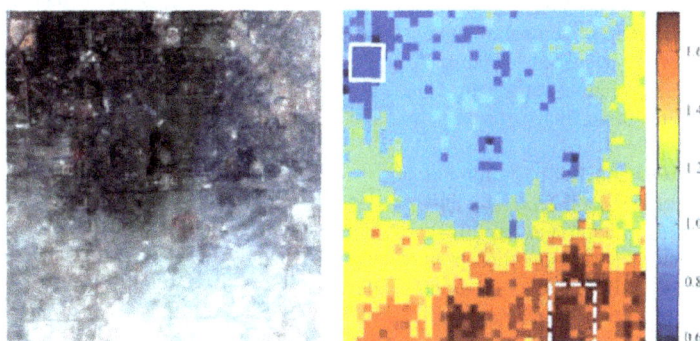

(a) The color composite image (left) and retrieved AOD map in 14th Jan 2013 in Beijing region

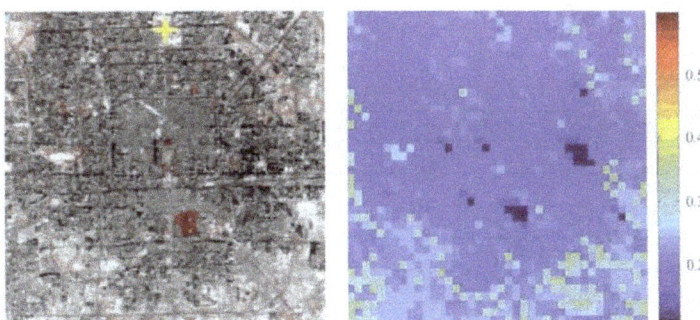

(b) The color composite image (left) and retrieved AOD map in 17th Jan 2013 in Beijing region

Figure 2. Image and AOD map of China HJ-1 satellite

The measurements of PM concentration and distribution near the surface by satellite will be more and more important for the fog-haze monitoring and treatments. Meanwhile, the study on correlation between AOD and PM2.5 concentration near surface is theoretical basis of applying satellite remote sensing to PM monitoring more effectively.

3. CORRELATION BETWEEN AEROSOL OPTICAL PROPERTIES AND PM CONCENTRATION

Haze is a kind of air turbidity phenomenon that horizontal visibility is less than 10km, which is generated by plenty of fine dry particles floating in the air. The composition of haze is complex. Haze often occurs when a large number of man-made particulate matters and aerosol particles transformed by gas and particulate matter increase and accumulate.

Atmospheric aerosol is a suspension of solid and liquid particles in air whose diameter are about 0.001-100μm. Aerosol is the important component of atmosphere, which play a critical role in global climate change and atmospheric heat balance. Also aerosol is the main composition of local photochemical smog and regional air pollution, which can reduce air visibility and affect human health. Compared with other properties of aerosol, AOD has been measured by ground-based and satellite-based instruments with relatively more mature technology. AOD refers to the integral for aerosol extinction coefficient from earth's surface to top of atmosphere, representing the extinction by aerosol in atmospheric vertical column without cloud.

In recent years, the estimations of PM concentration near surface from satellite remote sensing data focus on how to establish a correlation model between AOD and PM concentration. Linear regression methods have been widely used (Wang and Christopher, 2003; Hutchison, 2003; Li, et al., 2005). The influences of vertical distribution of aerosol, relative humidity (RH) and other climate factors have been introduced to correct the model so that the correlation ratio can be improved (Donlelaar, et al., 2006; Wang, et al., 2010).

The promising correlation between AOD and PM pollution has been found base on the measurements during the serious dust haze period in January 2013 in Beijing (Zhang and Li, 2013). The data of aerosol optical and microphysics properties from AERONET (Aerosol Robotic Network) at Beijing site are used. Fine-mode aerosol optical depth is estimated by aerosol fine-mode fraction using the data measured by sun photometer (CE318). Meanwhile, the correlation between AOD and PM2.5 is established:

$$PM_{2.5} = 244.48AOD \cdot \eta + 10.6, \ R^2 = 0.77 \quad (1)$$

Height of planetary boundary layer (PBL), the mean of RH profile at the bottom of the troposphere, near-surface RH, near-surface temperature and other meteorological factors are introduced into the simple PM2.5-AOD correlation model to establish multiple regression linear and nonlinear models for near-surface PM concentration monitoring over Beijing and its surrounding area (Jia, et al., 2014). The relation between AOD through the atmosphere and PM_x is:

$$AOD = PM_x \cdot H \cdot f(RH) \cdot \frac{3\langle Q_{ext} \rangle}{4\alpha \cdot \rho \cdot r_{eff}} \quad (2)$$

where PM_x = near-surface PM concentration whose aerodynamic equivalent diameter is less than or equal to 'x' micron

ρ = near-surface PM concentration
H = height of PBL
f(RH) = function of RH profile in PBL
$\langle Q_{ext} \rangle$ = normalized PM extinction efficiency
r_{eff} = PM effective radius
α = ratio of AOD of PBL to AOD of whole atmosphere.

Furthermore, shape and size of aerosol particle, refractive index, single scattering albedo and other properties retrieved from remote sensing data provide plenty of information for investigating the composition, distribution and variation trend of pollution aerosol particles.

4. REQUIREMENTS OF PM MONITORING INSTRUMENT

Measuring aerosol optical properties accurately is the basis of improving the accuracy of particulate matters monitoring. Based on the Beijing-Tianjin-Hebei regional PM distribution feature and the empirical relations between AOD and PM concentration, the specifications including bands, viewing angles and polarization measuring requirements of the on-orbit PM monitoring instrument are discussed in this section.

Traditional optical remote sensing is only to measure the spectrum, radiance and geometrical information, and not to identify the chemical composition in aerosol and to determine the quantity and concentration of aerosol particles. Currently, most sensors used to measure aerosol are not designed for this mission specially. Aerosol retrieval need the capacity for detecting the multispectral, multi-angle and polarization information. Considering less amount of data obtained by polar orbit satellite generally with a single daily overpass, a imager with required swath and spatial resolution will be better for the regional monitoring of pollutants. The major characteristics of some aerosol measuring instruments are listed in Table 1.

Sensors	Multispectral	Multi-angle	Polarization	Swath	IFOV
POLDER(PARASOL)	√	√	√	2400km	6km×6km
MISR	√	√		360km	275m, 1.1km
AATSR	√	√		512km	1km×1km
MODIS	√			2330km	0.25km, 0.5km, 1km
MERIS	√			1150km	0.3km×0.3km
OMI	√			2600km	24km×13km
APS	√	√	√	5.6km	5.6km

Table 1. Major characteristics of aerosol measuring instruments

4.1 Bands

Based on light scattering theory, the bands used for aerosol retrieval are in visible and near infrared range of 0.4-0.9μm. Different bands are chose respectively for observation of surface, ocean, cloud, water vapour and other objects according to their wavelength sensitivity. Typical bands for aerosol polarized reflectance measurements are with the central wavelength of 443nm, 555nm, 670nm and 865nm. In addition, the bands of 1610nm and 2250nm are used to determine the surface polarized reflectance as short wave infrared (SWIR) band is insensitive to atmosphere. Thus the contribution of surface reflectance to aerosol measurement can be removed on basis of the SWIR data, and the accuracy of aerosol retrieval can be improved.

Only the clean data in visible band with no cloud can be used in AOD retrieval. For aerosol observation by visible band instrument, the data should be identified whether it is with cloud, and the cloud contamination data should be deleted. If the spatial resolution of the instrument is not enough to identify the cloud data, high resolution cameras could be coupled with it for cloud detection. For example, Aerosol Polarimetry Sensor (APS) has only one single pixel which is carried on Glory spacecraft with two Cloud Cameras. The difference between the reflectance of cloud and earth surface is usually used to identify the cloud data. This method is applicable to the region that the surface albedo is quite different from cloud, but is not working for snow or other surface with high reflectance. Strong scattered solar radiance by cirrus can be determined in the water vapour strong absorbing band around 1380nm. And upper cirrus can be determined by setting respective threshold according to the reflectance of the cirrus over different surface.

During fog-haze days which are always with lots of cloud and water vapour in the air, AOD is hard to determined by visible band instrument. UV aerosol index (UVAI) is sensitive to the content of absorptive aerosol such as mineral aerosols and carbonaceous aerosols, and not sensitive to cloud and water vapour. So the haze which usually contains 50% carbonic matter can be monitored effectively by using UVAI. AOD, absorbing aerosol optical path (AAOD) and aerosol index have been retrieved by applying OMAERO algorithm using spectral reflectance in UV and visible bands from OMI (Ozone Monitoring Instrument) (Torres, et al., 2007). AAOD which is very sensitive to absorptive aerosol is applicable for aerosol measurements over bright surface or cloud. And comparing to other instrument not covering UV band, the instrument with near UV bands obtain more information which also can be used to identify the aerosol type of weak absorbing aerosol or strong absorbing aerosol.

In conclusion, the required bands for pollution particulate matter monitoring instrument need to cover from UV to SWIR. The typical bands are 380nm, 443nm, 555nm, 670nm, 865nm, 1380nm, 1610nm, 2250nm and so on. Data of different bands are used for different aerosol type, surface models and cloud types using dedicated retrieval algorithm.

4.2 Viewing Angles

Aerosol optical properties can be retrieved by single viewing angle methods and multiple viewing angles methods according to the imaging manner of the instrument. The retrieval methods for single angle observation data include histogram matching method, dark object method, lower contrast method and type matching method. Based on the phenomenon that the reflectance of dense vegetation pixels in 2.13μm is strongly correlative to the reflectance of blue and red bands, dark object method is to calculate the reflectance of blue and red bands using band 2.13μm which is less affected by aerosols. This method is a typical and simple aerosol retrieval method with no polarization to estimate AOD by measuring surface reflectance of dark object, whose accuracy isn't much high, and not applicable to desert.

Multiple views imagers such as MISR and POLDER have been applied in remote sensing. Unlike traditional remote sensing which suppose earth surface to be Lambert, multiple views imaging can determine bidirectional reflectance accurately. The advantages of this imaging manner are:

1. Polarized reflectance of aerosol increases with the decreasing of scattering angle. The information which can be retrieved from polarized reflectance data are much less or even can be ignored when the scattering angle is up to larger than 150°. The reduction of polarized retrieval accuracy due to the large scattering angle can be solved by applying multiple viewing angles observation.

2. AOD retrieval by multiple viewing angles method only need to measure the ratio of corresponding surface reflectances of multiple views, which don't need the accurate surface reflectance itself. Furthermore, the requirement for calibration accuracy of instrument can be reduced.

The difference of the radiances through different atmospheric paths should be considered to design the view angles. This radiance difference increases along with the increasing of the zenith angle of tilt observation, thus the calculating error using known atmospheric model increases with it. Generally, zenith angle should be less than 60° to guarantee the calculation accuracy.

On basis of retrieval stability, retrieval accuracy will be higher if data of more viewing angles are brought into calculation. The instrument design will be much complex with too many angles. Selecting appropriate viewing angles can realize the required retrieval accuracy and can also adopt to the dedicated satellite platform.

4.3 Polarization

Removing the contribution of surface is very important for aerosol retrieval. Surface reflectance is much higher and with complex spatial distribution. In particular, the reflectance of bright surface changes obviously with time and different regions, which requires more higher calibration accuracy for the instrument. The change of surface reflectance of 1% can lead to the AOD error of 10%. However, the polarized reflectance of surface is much smaller and more stable than reflectance, which almost doesn't change with wavelength. And the polarization effects of aerosols scattered sunlight is much strong especially in small scattering angle. Therefore, the retrieval accuracy can be improved based on polarization determination.

Aerosol type can be identified by using polarized reflectance as polarization is sensitive to small particle. Several particular aerosol mode should be presupposed in aerosol retrieving based on look-up table. Using several typical aerosol mode can get good results because of the correlation of polarized reflectance to aerosol type. The polarization parameters such as degree of polarization, angle of polarization and polarization ellipticity are measured, which provide more dimensions of sensing information. Aerosol properties including refractive index, single scattering albedo, shape and size of aerosol particle can be retrieved from the data of multiple angles and polarization. The limitation of conventional optical measurements to determine atmospheric aerosol scattering phase function, particle size and its accuracy can be solved by remote sensing with multiple view angle and polarization.

According to polarization properties of aerosol, the determination of linear polarization characteristics is enough, while the circular polarization characteristics have very small influence to the measurements. The system of equations can be established with the values of three directions of 0°, 60° and 120° to derive the stokes parameters I, Q and U. The polarization measuring accuracy of instrument is corresponding to the extinction ratio of polarizing element, error of polarization axis, alignment error and polarization error of optical system. On ground and on orbit polarized radiometric calibration is critical for the polarization measuring accuracy so as to guarantee the retrieval accuracy of aerosol.

5. CONCLUSION

For the problems of the serious fog-haze pollution in Beijing-Tianjin-Hebei region and global particulate matter pollution, 3-D monitoring capability of ground-based measurement is limited to study the generating, developing and disappearing of fog and haze in air. Satellite remote sensing can provide regional coverage which is complementary to ground-based measurements. The remote sensing of global aerosols especially of fine PM need multispectral, multi-angle and polarization measurement in UV, visible, NIR and SWIR bands by using one instrument with all these functions or several instruments to cooperate. Optical and microphysical properties of atmospheric aerosol including AOD, shape and size of particle, refractive index and single scattering albedo can be retrieved from the remote sensing data. Furthermore, the spatial distribution and variation trend of PM2.5 can be used for the research on atmospheric environmental monitoring, climate change and weather forecast.

REFERENCES

van Donkelaar, A., R.V. Martin, and R.J. Park, 2006. Estimating ground-level PM2.5 using aerosol optical depth determined from satellite remote sensing. *Journal of Geophysical Research*, 111(d21), pp. D21201-1.

van Donkelaar, A., R.V. Martin, M. Brauer, R. Kahn, R. Levy, C. Verduzco, and P.J. Villeneuve, 2010. Global estimates of ambient fine particulate matter concentrations from satellite-based aerosol optical depth: development and application. *Environmental Health Perspectives*, 118(6), pp. 847-855.

He, K., H. Huo, and Q. Zhang, 2002. Urban air pollution in China: current status, characteristics and progress. *Annual Review of Energy and the environment*, 27, pp. 397-431.

Hutchison, K.D., 2003. Applications of MODIS satellite data and products for monitoring air quality in the state of Texas. *Atmos. Environ.*, 37, pp. 2403-2412.

Jia, S., L. Su, J. Tao, Z. Wang, L. Chen, and H. Shang, 2014. A study of multiple regression method of estimating concentration of fine particulate matter using satellite remote sensing. *China Environment Science*, 34(3), pp. 565-573.

Kacenelenbogen, M., J.-F. Léon, I. Chiapello, and D. Tanré, 2006. Characterization of aerosol pollution events in France using ground-based and POLDER-2 satellite data. *Atmos. Chem. Phys.*, 6, pp. 4843-4849.

Streets, D.G., J.H.S. Fu, C.J. Jang, J.M. Hao, K.B. He, X.Y. Tang, Y.H. Zhang, Z.F. Wang, Z.P. Li, and Q. Zhang, 2007. Air quality during the 2008 Beijing Olympics Games. *Atmos. Environ.*, 41(26), pp. 480-492.

Sui, K., Z. Wang, J. Yang, F. Xie, and Y. Zhao, 2007. Beijing persistent PM10 pollution and its relationship with general meteorological features. *Research of Environmental Sciences*, 20(6), pp. 77-82.

Torres, O., A. Tanskanen, B. Veihelmann, C. Ahn, R. Braak, P.K. Bhartia, J.P. Veefkind, and P.F. Levelt, 2007. Aerosols and surface UV products from OMI observations: an overview. *J. Geophs. Res.*, 112, pp. D24S47.

Wang, J., and S.A. Christopher, 2003. Intercomparison between satellite-derived aerosol optical thickness and PM2.5 mass: implications for air quality studies. *Geophys. Res. Lett.*, 30(21), pp. 2095.

Wang, X., and J. Li, 2003. A numerical simulation study of PM10 pollution in Beijing during summer time. *Acta Scientiarum Naturalium Universitatis Pekinensis*, 39(3), pp. 419-427.

Wang, Z., L. Chen, J. Tao, Y. Zhang, and L. Su, 2010. Satellite-based estimation of regional particulate matter (PM) in Beijing using vertical-and-RH correcting method. *Remote Sensing of Environment*, 114(1), pp. 50-63.

Xu, H., Y. Zhang, W. Hou, Y. Lv, L. Li, X. Gu, and Z. Li, 2013. Monitoring of haze in Beijing region from HJ-1 data. *Journal of Remote Sensing*, 17(2), pp. 476-477.

Zhang, Q., C. Zhao, X. Tie, Q. Wei, M. Huang, G. Li, Z. Ying, and C. Li, 2006. Characterizations of aerosols over the Beijing region: a case study of aircraft measurements. *Atmos. Environ.*, 40(24), pp. 4513-4527.

Zhang, Y., and Z. Li, 2013. Estimation of PM2.5 form fine-mode aerosol optical depth. *Journal of Remote Sensing*, 17(4), pp. 929-943.

Zhao, X., W. Pu, W. Meng, Z. Ma, F. Dong, and D. He, 2013. PM2.5 pollution and aerosol optical properties in fog and haze days during autumn and winter in Beijing area. *Environment Science*, 34(2), pp. 416-423.

21

ESTIMATION OF GRASSLAND USE INTENSITIES BASED ON HIGH SPATIAL RESOLUTION LAI TIME SERIES

S. Asam [a, b, *], D. Klein [c], S. Dech [c]

[a] Institute for Applied Remote Sensing, EURAC Research, Viale Druso, 1, 39100 Bolzano, Italy – sarah.asam@eurac.edu
[b] Department of Remote Sensing, Institute of Geography, University of Wuerzburg, Oswald-Külpe-Weg 86, 97074 Wuerzburg, Germany
[c] German Aerospace Center (DLR), German Remote Sensing Data Center (DFD), Oberpfaffenhofen, 82234 Weßling, Germany – (doris.klein, stefan.dech)@dlr.de

KEY WORDS: Leaf area index (LAI), radiation transfer modeling, RapidEye, Alpine area, grassland, management intensity

ABSTRACT:

The identification and surveillance of agricultural management and the measurement of biophysical canopy parameters in grasslands is relevant for environmental protection as well as for political and economic reasons, as proper grassland management is partly subsidized. An ideal monitoring tool is remote sensing due to its area wide continuous observations. However, due to small-scaled land use patterns in many parts of central Europe, a high spatial resolution is needed. In this study, the feasibility of RapidEye data to derive leaf area index (LAI) time series and to relate them to grassland management practices is assessed. The study area is the catchment of river Ammer in southern Bavaria, where agricultural areas are mainly grasslands. While extensively managed grasslands are maintained with one to two harvests per year and no or little fertilization, intensive cultivation practices compass three to five harvests per year and turnover pasturing.

Based on a RapidEye time series from 2011 with spatial resolution of 6.5 meters, LAI is derived using the inverted radiation transfer model PROSAIL. The LAI in this area ranges from 1.5 to 7.5 over the vegetation period and is estimated with an RMSE between 0.7 and 1.1. The derived LAI maps cover 85 % of the study area's grasslands at least seven times. Using statistical metrics of the LAI time series, different grassland management types can be identified: very intensively managed meadows, intensively managed meadows, intensively managed pastures, and extensively managed meadows and moor. However, a precise identification of the mowing dates highly depends on the coincidence with satellite data acquisitions. Further analysis should focus therefor on the selection of the temporal resolution of the time series as well as on the performance of further vegetation parameters and indices compared to LAI.

1. INTRODUCTION

The identification and surveillance of agricultural practices, especially of management intensities, is relevant for a range of ecological, conservation, and political issues. In the alpine region, livestock farming is the predominant agricultural land use, and at the same time, various different pasturing and mowing intensities exist. The ecological relevance of grassland usage intensity is linked to exchange fluxes of water, energy, and gases between the land surface and the atmosphere. Knowledge on use intensities can hence improve greenhouse gas inventories (Schaller et al. 2011). In addition, the conservation status of grassland needs to be monitored to assess the ecological value of landscapes. Traditional and other extensive types of agricultural land use can maintain the biodiversity of grassland landscapes. Especially semi-natural, extensively used grasslands play an important role as habitats with a high conservation value (Öster et al. 2008; Sullivan et al. 2010). Intensification or abandonment of these grassland managements on the other hand can cause biodiversity loss (Henle et al. 2008). Furthermore, knowledge on mowing and pasturing intensities are important for political and economic reasons, as proper grassland management is partly subsidized in the EU.

The monitoring of agricultural management intensities often relies on remote sensing data, as they provide repetitive and area wide continuous observations. Due to the small-scaled land use patterns in the alpine areas of central Europe, the used remote sensing images need a high spatial resolution, which becomes increasingly feasible with relatively new sensors such as RapidEye, Landsat 8, and, soon, Sentinel-2. Nevertheless, the potential of satellite imagery for the inventory of grassland use intensity on individual fields has barely been assessed (Franke et

al. 2012). Remote sensing based research on managed grassland mainly focused on extent, status, and primary production assessment (see e.g. Seaquist et al. 2003, Mutanga et al. 2004, Boschetti et al. 2007, Vescovo and Gianelle 2008), but only rarely on grassland management. In addition, mostly large-scale biomes such as semi-arid or subtropical grasslands have been assessed (Numata et al. 2007, Kurtz et al. 2010). Another issue is the temporal resolution of grassland observations, since with high spatial resolution data the coverage of a certain study area at regular intervals is often impeded, especially in areas prone to frequent cloud cover such as the Alpine area.

In this study, the feasibility of RapidEye data to derive leaf area index (LAI) time series and to relate them to grassland management practices is assessed. The LAI is a key parameter of vegetation structure and particularly important for quantifying exchange fluxes in the biosphere, photosynthesis and biomass production. The advantage of LAI based grassland use intensity estimation is the physical meaning of the LAI parameter. The usually used vegetation index (VI) values trace only a relative abundance and health of vegetation, and often vary e.g. with soil conditions, local viewing and illumination conditions, and canopy structure. These effects that are reduced during physical LAI estimation through taking the canopy and scene geometry specifications into account. While the canopy light absorption, which is the process influencing VI levels, can be diminished e.g. by droughts or senescence, the actual biomass status can be better represented by LAI. Furthermore, descriptive statistics such as the range or the accumulated LAI are directly comparable between sites as the resulting numbers are absolute values. Therefore, statistical metrics derived from a LAI time series for the year 2011 will be used in this study to characterize some common grassland usage schemes in the Bavarian Alpine upland.

2. STUDY AREA

The study area is the catchment of the river Ammer in the Bavarian alpine upland covering 770 km^2 (see Fig.1). This catchment constitutes the TERENO Prealpine Observatory (http://teodoor.icg.kfa-juelich.de). The rural land cover of the study area is representative for the European alpine upland. The landscape is dissected by small settlements, forest patches and small-scale agricultural patches. These agricultural areas are mainly grasslands used for pasturing and mowing. In the alpine upland, many pastures are cultivated using rotational grazing and cutting hayfield systems. With the grazing and/or vegetation cuttings followed by rapid re-growth, these grasslands undergo multiple growing cycles within a single vegetation period (Wohlfahrt and Cernusca, 2002). Apart from the intensively used grasslands, i.e. habitats consisting of few, mesophilic species, there are also extensively managed grassland types in the Ammer catchment which consist of rarer species that are more adapted to very humid, dry, cold, or nutrient poor conditions. Two European agro-environmental schemes aiming at the preservation of biodiversity are implemented in the area: the High Nature Value (HNV) farmland indicator and the habitats Natura 2000 directive. These extensively managed grasslands are semi-natural grasslands maintained with one to two harvests per year and no or little fertilization.

3. DATA AND METHODS

3.1 Remote Sensing data

All analyses conducted in this study are based on a RapidEye time series consisting of nine images acquired between April and September 2011 (see Tab. 1). The RapidEye constellation consists of five satellites located in the same sun-synchronous orbital plane. The sensors are push broom scanners with five spectral bands in the visible (blue [440 – 510 nm], green [520 – 590 nm], red [630 – 685 nm]) and infrared (red edge [690 – 730 nm] and near infrared [760 – 850 nm]) domain. The spatial resolution of RapidEye level 1B images is 6.5 m.

The preprocessing of the level 1B data consisted of a transformation into UTM (WGS 84 datum) projection by using a nearest neighbor algorithm; orthorectification using RPCs associated with the RapidEye data and a 30 m Shuttle Radar Topography Mission (SRTM) digital elevation model (DEM); precise georeferencing using ground control points; and finally a topographic and atmospheric correction using ATCOR (Richter and Schläpfer 2012).

The individual RapidEye images cover the Ammer catchment to different extents. Clouds further reduce the spatial information

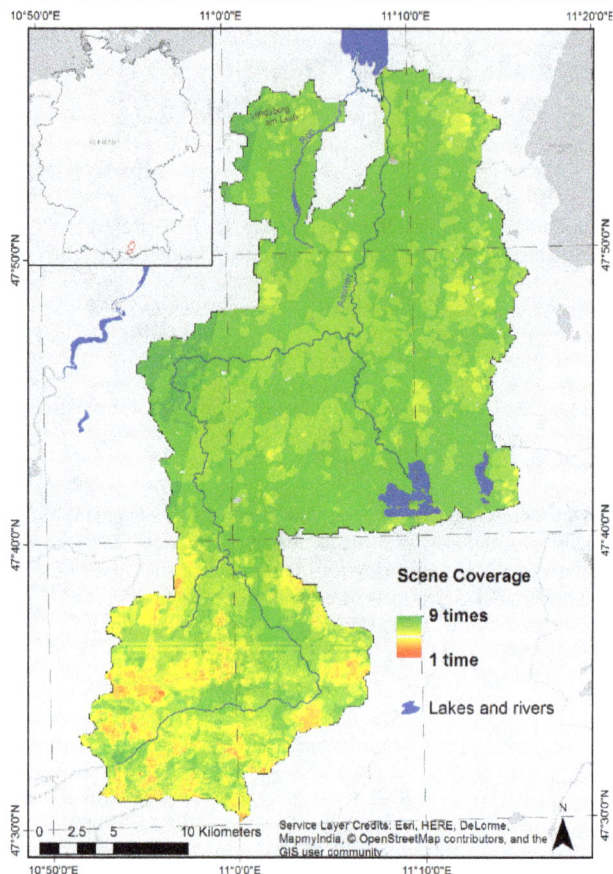

Figure 1. Location of the study area covering the Ammer catchment in the Bavarian alpine upland and number of RapidEye observations during 2011.

available individually for each scene (see Tab. 1). The resulting number of observations per pixel is illustrated in Fig. 1. A lower number of observations is given in some areas of the southern half of the study area, which is part of the Bavarian Limestone Alps, due to higher cloud occurrence close to the mountains.

3.2 Field measurements

In situ LAI measurements have been collected during four weeks contemporaneous to the RapidEye data acquisitions in May, July, and September 2011. The measurements were arranged within a two-stage nested design (Morisette et al. 2006) resulting in 20 - 33 plots per time step. LAI was measured at 20 points on two transects within each plot using a LAI-2000 Plant Canopy Analyzer (PCA) (LI-COR Biosciences, Lincoln, NE, USA), and corrected afterwards using an empirical relationship that was established between LAI-2000 measurements and destructive LAI samplings from 14 of the above mentioned locations at the respective same days. The LAI *in situ* values per plot cover a data range from 1.5 to 7.4 with a mean of 3.6. For more details on the field measurements the readers can refer to Asam et al. (2013). At the same time information on the status of management of the observation sites have been noted.

3.3 Land cover classification

All water bodies, snow covered areas, clouds, and cloud shadows were masked manually in the RapidEye images. The same water mask was applied to all images, whereas all other masks were created scene-specifically. A grassland mask was derived in a next step based on a 'random forest' land cover classification

Date	Scene Coverage [%]	Cloud Cover [%]
April 8	100	24.9
April 20	100	17.5
May 5	24.8	2.3
May 9	100	0.0
May 25	99.5	3.6
July 16	100	2.0
August 21	100	5.3
September 6	100	8.2
September 26	94.2	6.9

Table 1: Acquisition dates during the vegetation period 2011, percentage of the study area covered by the image, and percentage of pixels within the study area masked due to cloud cover of the RapidEye images used in this study.

(Breiman 2001). For the classification, three RapidEye images with a high scene coverage and only little cloud cover were used (May 9, July 16, and September 6). A multi-temporal classification approach was chosen because some of the land cover classes, for example "winter wheat" and "grasslands", show similar spectral signatures in advanced development stages, but distinctly different phenologies. The scenes were stacked into one data frame together with three VIs derived for each scene respectively, namely the Normalized Difference Vegetation Index (NDVI), the Soil Adjusted Vegetation Index (SAVI), and an adjusted NDVI with the red edge band substituting the red band, resulting in a 24 layer feature space. The training and validation data, in total about 83000 pixels, were collected *in situ* during the field campaigns and complemented by visual interpretation within the RapidEye scenes using Google Earth imagery in order to increase the number and diversity of land cover units used for classifier training. The number of trees within the random forest was set to 500 in order to achieve convergence. The overall classification error of the areas which are covered by all three images (86.1 % of the study area) is 3.4 %. No distinction between grassland types could be made at this stage, due to the relatively low number of field campaigns conducted (see above) and the very high heterogeneity of the grassland fields.

3.4 LAI derivation

LAI is derived from the RapidEye data using the inverted radiation transfer model (RTM) PROSAIL (Jacquemoud et al. 2009). RTMs simulate the interactions of radiation with vegetation elements and the soil while traveling through the canopy, i.e. absorbance, reflectance and transmittance are considered. Based on these processes, the radiation leaving a canopy can be related to the spectral and structural properties of the canopy. In order to derive LAI, the RTM is first run in forward mode to calculate reflectances for given specific canopies and observation configurations. Therefor different multiple canopy realizations are implemented using varying combinations of input parameter values. Based on a global sensitivity analysis, the influence of each parameter on the spectral domains covered by the RapidEye bands was identified

first, which was used to determine the sampling interval of each parameter. PROSAIL was then characterized by its leaf and canopy variables based on values collected in the field and on literature values (e.g. Weiss et al. 2000, Darvishzadeh et al. 2008, Feret et al. 2008; Table 2). Additionally, the local, i.e. topographically corrected, viewing and illumination were considered in this process. For all parameters, a uniform distribution was used. The parameterization resulted in 33516 to 198450 variable combinations for the different RapidEye time steps. This was mainly caused by the small sampling intervals and the differing value ranges of the LAI parameter, since this range is larger in summer than during spring and autumn (see Table 2). These values were stored in look up tables (LUTs) together with the respective calculated reflectances. After calculating the reflectances of these multiple canopy realizations, the model was inverted, i.e. for each pixel of the satellite image the parameter set (including LAI) which produced the reflectances most similar to the reflectances measured by the remote sensing sensor in all bands was selected. Additionally to the original spectral bands, two VIs, namely the ratio vegetation index (RVI, Jordan, 1969) and the *Curvature* index (Conrad et al., 2012), have been used as input features to invert the RTM. For the inversion a cost function based on the normalized root mean square error (RMSE) was applied and the median of a multiple solution sample (0.5 % of all LUT entries) was extracted as solution.

3.5 Assessment of Management Intensities

In a next step the LAI time series is analyzed regarding their suitability to discriminate four different classes of grassland use intensities:

- 'Very intensively managed meadows' that undergo four or more harvests per year,
- 'Intensively managed meadows' that undergo two to three harvests per year,
- 'Intensively managed pastures' that are alternately grazed and cut, and
- 'Extensively managed meadows and moor' that are cut at most once a year due to the framework regulation of

Parameter		Unit	Min	Max	Interval
PROSPECT					
N	Structure coefficient	-	1.3	1.9	0.3
C$_{ab}$	Chlorophyll a + b	µg*cm^{-2}	10 - 20	40 - 80	10
C$_{ar}$	Carotenoid	µg*cm^{-2}	4 - 12	4 - 12	-
C$_w$	Equivalent water thickness	cm	0.02	0.02	-
C$_m$	Dry matter	g*cm^{-2}	0.004	0.012	0.004
C$_{bp}$	Brown pigments	-	0.4	0.4	-
SAIL					
LAI	Leaf area index	m^2*m^{-2}	0.2	3.8 – 7.0	0.2
angl	Average leaf angle	°	36	78	6
hs	Hot spot parameter	m*m^{-1}	0.10 – 0.14	0.10 – 0.14	-
ρsoil	Soil reflectance coefficient	-	0.0 – 0.1	0.7 - 1	0.1
skyl	Diffuse/total incident radiation	-	0.1 – 0.18	0.1 – 0.18	-
θ$_s$	Solar zenith angle	°	*	*	*
θ$_o$	Observer zenith angle	°	*	*	*
φ$_{rel}$	Relative azimuth angle	°	*	*	*

Table 2: PROSAIL parameter settings for the 2011 RapidEye scenes. For each parameter, the minimum, maximum, and sampling interval is given. For the parameters whose upper and lower boundaries varied throughout the year, the respective highest and lowest minimum and maximum boundary values used are given. If a parameter was fixed to a certain value, no interval is indicated. The sun and sensor zenith and azimuth angles (*) were calculated for each pixel based on the scene specific sun and sensor angles as well as on an SRTM DEM, grouped into classes for the reflectance modelling, and fixed during inversion.

the Natura 2000 directive (Bundesamt für Naturschutz, 2013).

To prevent usage intensity underestimation due to a low number of RapidEye observations, the further analysis are only conducted on areas for which at least seven observation were available, so that theoretically at least one observation is available every three to four weeks. Considering the grassland areas, the available RapidEye data cover 85 % of the study area's grasslands at least seven times (see green signatures in Fig. 1). For these pixels, basic statistical properties of the LAI time series were derived in order to capture the following characteristics of the differently used grassland areas.

First of all it can be observed that grassland areas on which the human impact is kept small, i.e. extensively managed areas such as moors, flood plains, and fallow lands, show a low LAI variability since neither a quick biomass accumulation nor abrupt LAI decreases take place. This low LAI variability is identified using the standard deviation. In addition to this basic distinction between extensively managed meadows and other management forms, different levels of variability indicate differing numbers of mowing events, enabling the identification of intensively and very intensively managed meadows. Hence, the standard derivation σ of each pixel's time series of the length $n = 7$ to 9 is derived in order to detect the overall variability of the LAI values:

$$\sigma = \sqrt{\frac{1}{n-1}\sum_{i=1}^{n}(LAI_i - \overline{LAI})^2} \qquad (1)$$

A measure that is calculated in a similar way is the 'Mean Absolute Spectral Dynamic (MASD)' index introduced by Franke et al. (1012). Also in this formula, n is the number of observations, but in addition the observation date t is taken into account for a direct comparison of LAI values to the respective prior value. Thus a distinction between LAI time series with continuously increasing LAI values and fluctuating LAI time series should be improved. The MASD is hence a detailed measure for LAI level changes with respect to the prior condition, i.e. the number of harvests and the strength of a change, and should therefore be useful to distinguish intensively managed pastures from intensively managed meadows which are assumed to be characterized by more abrupt changes. Since MASD was adapted to the usage with LAI values instead of spectral reflectances in this study, it becomes the 'Mean Absolute LAI Dynamic (MALD)':

$$MALD = \frac{1}{n-1}\sum_{t=1}^{n-1}\left|LAI_i^t - LAI_i^{t+1}\right| \qquad (2)$$

In a last step each pixel's accumulated productivity should be measured to distinguish meadows and pastures. Since no continuous (e.g. daily) LAI time series exists due to the high spatial resolution, instead of accumulating the single values, the area under curve (AUC) was calculated using a linear trapezoidal method over the n time steps t_i:

$$AUC = \sum_{i=1}^{n}\frac{1}{2}(LAI_i + LAI_{i+1})(t_{i+1} - t_i) \qquad (3)$$

The statistical measures were then used in the rule set of a decision tree in order to classify the grassland use intensities (see Fig. 2). The structure as well as the thresholds applied in the

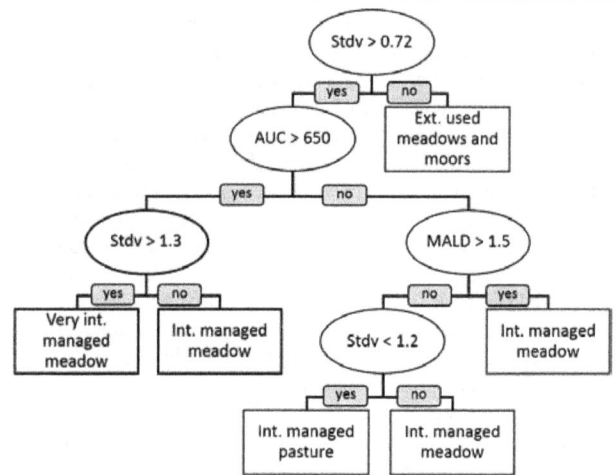

Figure 2. Decision tree rule set used for identifying the different grassland management intensities.

decision tree have been adapted empirically based on the grassland management types observed at the 20 repeatedly visited field measurement plots (see section 3.2). One advantage of a decision tree is that one class which includes different occurrences due to different evironmental factors or varying harvesting times, such as the 'intensively managed meadows', can be identified based on different rule sets.

4. RESULTS AND DISCUSSION

4.1 LAI time series

Based on the above described model set-up of the inverted PROSAIL model a high spatial resolution LAI time series with a maximum of nine LAI measurements (or less in case of cloud occurrence) was generated for the Ammer catchment. The LAI was validated for the four images for which contemporaneous field measurements exist resulting in root mean square errors (RMSEs) ranging from 0.73 and 1.14 and a relative RMSE in the range of 20 - 30 %. A similar error rate is expected for the other five LAI maps since the filed measurements cover the complete growing season.

Figure 3 displays the LAI map derived from the September 6, 2011 scene and gives an impression of the landscape structure and spatial variability of LAI. In the mountainous south-western part of the catchment, only few grassland areas exist. They can be distinguished into valley bottom areas with a high LAI, and mountain pastures with overall lower LAI, which are not intensively used due to the topography and climatic conditions. Following the further course of the Ammer River, conventionally managed grasslands with LAI values around 5 can be distinguished from dry and calcareous Natura 2000 habitats such as the 'Ammertaler Wiesmahdhänge' or moor areas 'Moore im oberen Ammertal' with a comparably lower LAI (Fig.3, upper left zoom). In the alpine foreland to the north, significantly more areas are covered by grassland. These meadows and pastures have an overall higher LAI, as they profit from more nutrient rich soils and higher temperatures. The map also displays the partly strong spatial differences in between fields resulting from different managements.

4.2 Grassland Management Intensities

Since continuous field observations with regard to management intensities were only available at few selected points, and since this information has been used to build the decision tree, only a qualitative accuracy assessment could be conducted in this study

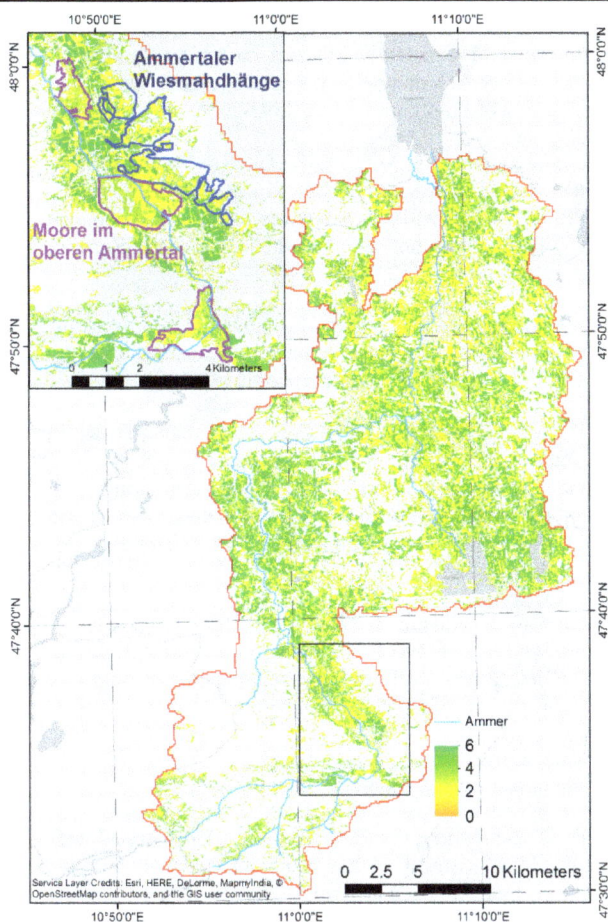

Figure 3. Grassland LAI in the Ammer catchment on September 6, 2011 including a zoom on the grassland sites in the Ammer valley with the Natura 2000 habitats hay meadows (orange outline) and moors (purple outline). While the intensively used grasslands at the valley bottom have LAI values around four to five, the Natura 2000 sites, which are subject to more extensive management, show distinctively lower LAI values.

Figure 4: Map of the derived grassland management types including a zoom on the grassland sites in the Ammer valley with the Natura 2000 habitats hay meadows (orange outline) and moors (purple outline). Pixels for which the LAI time series was shorter than 7 points in time are left blank.

(see Fig. 4). However, from the *in situ* observations as well as from the spatial patterns of the resulting classification it can be concluded that the different management intensities could be reasonably distinguished. For example Fig. 4 shows that fewer very intensively managed meadows (red signature) are located in the more alpine parts of the study area but in the northern half instead. Also, extensively managed meadows and moors (green signatures) are mapped where they would be expected, i.e. at Natura 2000 sites, at fallow lands e.g. of abandoned gravel pits, and close to lakes and rivers.

Although some redundant information is carried in the standard deviation, AUC, and MALD (the correlation coefficients between the layers range between 0.33 and 0.78), they show different sensitivities for differently managements. Especially the AUC and the standard deviation highlight specific conditions such as an accumulation of biomass on intensively used but not too often cut meadows, or the distinction between fields purely used for fodder generation and those also used for pasturing, which show a lower variability.

Nevertheless, the performance of the presented methodology relies on a precise capture of the moving dates, which in turn highly depends on the coincidence with the acquisition time of the satellite data. Figure 5 illustrates exemplarily the timing of

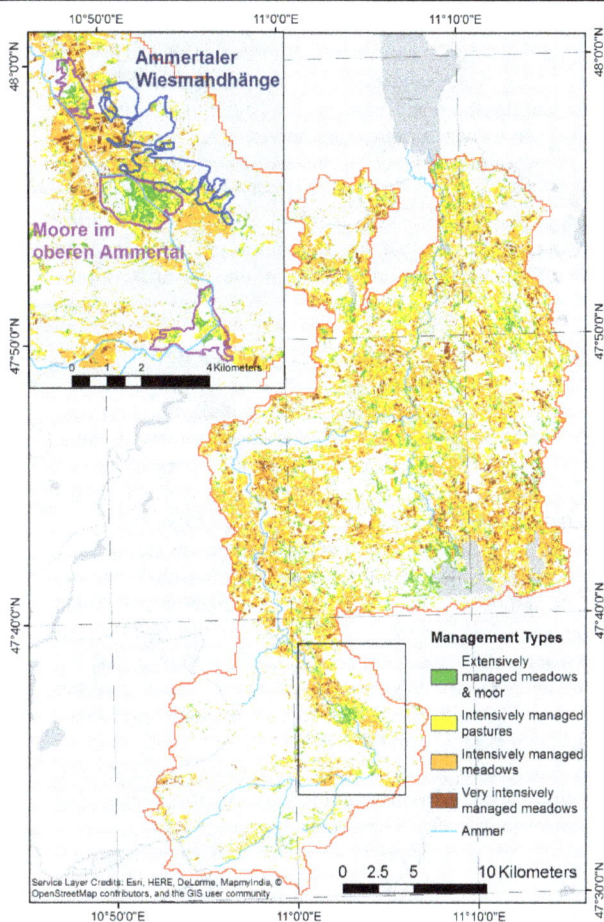

cutting events with respect to the LAI time series derived from the RapidEye images for the *in situ* measurement site near Fendt (47°49'58.48"N, 11° 3'38"E). At this site, the dates of the mowing events are derived from a hemispherical camera installed on the site. While the first, third, and fourth mowing are well reproduced, the second and the last harvest remain unrecognized. It can be assumed that the insufficient acquisition

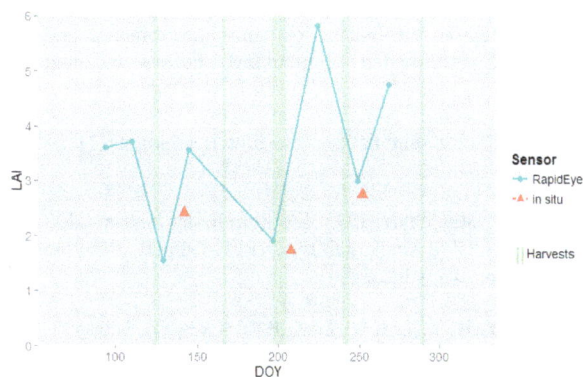

Figure 5: LAI time series with eight cloud free satellite observations derived from the RapidEye imagery of a field measurement plot near Fendt compared to field measurement values and the mowing events observed on this meadow.

frequency observed on this plot is a limiting factor also on other sites within the Ammer catchment. With two to three harvests normally taking place on very intensively managed meadows in the Bavarian alpine upland between the beginning of June and end of August, as well as one more harvest often scheduled for October, some of these events are very likely unnoticed given the above described time series. This suggests that an acquisition frequency of two to three weeks as well as a time series covering also the month of October are a requirement for a precise mapping of management intensities in the Bavarian alpine upland.

5. CONCLUSION AND OUTLOOK

In this study, the feasibility of very high spatial resolution LAI time series for the assessment of agricultural grassland usage intensities was analyzed. With the LAI providing comparable and physically meaningful measures of vegetation growth and sudden LAI reductions due to harvests, different management types could be distinguished by using statistical time series parameters. The high spatial resolution, allowing for the delineation of every single field and therewith a high pixel purity, enabled an adapted establishment of a decision tree rule set for the small-scaled heterogeneous landscape. However, it was also shown that the used time series has too large gaps during June/July and October to cover every harvest on very intensively managed meadows. While a higher acquisition frequency is preferable, further analysis should also focus on the optimal selection of remote sensing time steps. Also the performance of other vegetation parameters or indices could be tested.

The most important next step however is the establishment of a field observations data base against which the derived categorization can be quantitatively assessed.

ACKNOWLEDGEMENTS

We thank the German Aerospace Agency (DLR) for providing data from the RapidEye Science Archive (RESA) under the RapidEye project 468. This work was supported by the MICMoR research school of the Helmholtz association (http://www.micmor.kit.edu/).

REFERENCES

Asam, S., Fabritius, H., Klein, D., Conrad, C., Dech, S., 2013. Derivation of leaf area index for grassland within alpine upland using multi-temporal RapidEye data, International Journal of Remote Sensing, 34 (23), 8628-8652.

Boschetti, M., Bocchi, S., & Brivio, P. A., 2007. Assessment of pasture production in the ItalianAlps using spectrometric and remote sensing information. Agriculture, Ecosystems and Environment, 118, 267–272.

Breiman, L. 2001. Random Forests, Machine Learning 45(1), 5–32.

Bundesamt für Naturschutz, 2013. Steckbriefe der Natura 2000 Gebiete. 8431-371 Ammergebirge (FFH-Gebiet). Online available at: http://www.bfn.de/.

Conrad, C., Fritsch, S., Lex, S., Löw, F., Rücker, G., Schorcht, G., Sultanov, M., Lamers, J., 2012. Potenziale des 'Red Edge' Kanals von RapidEye zur Unterscheidung und zum Monitoring landwirtschaftlicher Anbaufrüchte am Beispiel des usbekischen Bewässerungssystems Khorezm. In E. Borg, H. Daedelow, R. Johnson (Eds.), RapidEye science archive (RESA). Vom Algorithmus zum Produkt (pp. 203–217). GITO.

Darvishzadeh, R., Skidmore, A., Schlerf, M. and Atzberger, C., 2008. Inversion of a radiative transfer model for estimating vegetation LAI and chlorophyll in a heterogeneous grassland, Remote Sensing of Environment 112, 2592-2604.

Feret, J.-B., François, C., Asner, G.P., Gitelson, A.A., Martin, R.E., Bidel, L.P., Ustin, S.L., Le Maire, G. and Jacquemoud, S., 2008. PROSPECT-4 and 5: Advances in the leaf optical properties model separating photosynthetic pigments, Remote Sensing of Environment 112, 3030–3043.

Franke, J., Keucka, V., Siegert, F., 2012. Assessment of grassland use intensity by remote sensing to support conservation schemes. Journal for Nature Conservation 20 (3), 125 – 134.

Henle, K., Alard, D., Clitherow, J., Cobb, P., Firbank, L., Kull, T., McCracken, D., Moritz, R.F.A., Niemelä, J., Rebane, M., Wascherk, D., Watt, A., Young, J., 2008. Identifying and managing the conflicts between agriculture and biodiversity conservation in Europe – A review. Agriculture, Ecosystems and Environment, 124, 60–71.

Jacquemoud, S., Verhoef, W., Baret, F., Bacour, C., Zarco-Tejada, P., Asner, G., François, C., Ustin, S., 2009. PROSPECT + SAIL models: a review of use for vegetation characterization, Remote Sensing of Environment, 113, 56-66.

Jordan, C.F., 1969. Derivation of leaf-area index from quality of light on the forest floor. Ecology 50 (4), 663–666.

Kurtz, D. B., Schellberg, J., & Braun, M., 2010. Ground and satellite based assessment of rangeland management in sub-tropical Argentina. Applied Geography, 30, 210–220.

Morisette, J., Baret, F., Privette, J., Myneni, R., Nickeson, J., Garrigues, S., Shabanov, N., Weiss, M., Fernandes, R., et al., 2006. Validation of global moderate-resolution LAI products: a framework proposed within the CEOS land product validation subgroup, IEEE Transactions on Geoscience and Remote Sensing 44, 1804-1817.

Mutanga, O., Skidmore, A., Prins, H., 2004. Predicting in situ pasture quality in the Kruger National Park, South Africa, using continuum-removed absorption features. Remote Sensing of Environment 89 (3), 393–408.

Numata, I., Roberts, D. A., Chadwick, O. A., Schimel, J., Sampaio, F. R., Leonidas, F. C., 2007. Characterization of pasture biophysical properties and the impact of grazing intensity using remotely sensed data. Remote Sensing of Environment, 109, 314–327.

Öster, M., Persson, K., & Eriksson, O., 2008. Validation of plant diversity indicators in semi-natural grasslands. Agriculture, Ecosystems & Environment, 125, 65–72.

Richter, R., and D. Schläpfer, 2012. Atmospheric / Topographic Correction for Satellite Imagery: ATCOR-2/3 User Guide", DLR-IB 565-01/12.

Schaller, L., Kantelhardt, J., Droesler, M., 2011. Cultivating the climate: Socio-economic prospects and consequences of climate-friendly peat land management in Germany. Hydrobiologia, 674, 91–104.

Seaquist, J. W., Olsson, L., and Ardö, J., 2003. A remote sensing-based primary production model for grassland biomes. Ecological Modelling, 169, 131–155.

Sullivan, C. A., Skeffington, M. S., Gormally, M. J., & Finn, J. A., 2010. The ecological status of grasslands on lowland farmlands in western Ireland and implications for grassland classification and nature value assessment. Biological Conservation, 143, 1529–1539.

Verrelst, J., Rivera, J.P., Leonenko, G., Alonso, L., & Moreno, J., 2014. Optimizing LUT-based RTM inversion for semiautomatic mapping of crop biophysical parameters from Sentinel-2 and -3 data: Role of cost functions. IEEE Transactions on Geoscience and Remote Sensing 52 (1), 257–269.

Vescovo, L., and Gianelle, D., 2008. Using the MIR bands in vegetation indices for the estimation of grassland biophysical parameters from satellite remote sensing in the Alps region of Trentino (Italy). Advances in Space Research 41 (11), 1764–1772.

Weiss, M., Baret, F., Myneni, R.B., Pragnère, A. and Knyazikhin, Y., 2000. Investigation of a model inversion technique to estimate canopy biophysical variables from spectral and directional reflectance data, Agronomie 20, 3-22.

Wohlfahrt, G., and Cernusca, A., 2002. Momentum transfer by a mountain meadow canopy: A simulation analysis based on Massman's (1997) model. Boundary-Layer Meteorology 103 (3), 391–407.

22

TIME-SERIES ANALYSIS OF SATELLITE-MEASURED VEGETATION PHENOLOGY AND AEROSOL OPTICAL THICKNESS OVER THE KOREAN PENINSULA

S. Park [a]

[a] Geography Education, Pusan National University, Busan, Korea - spark@pusan.ac.kr

ATMC-4

KEY WORDS: Aerosol, MODIS, Vegetation Index, Phenology, AOT

ABSTRACT:

The spatiotemporal influences of climatic factors and atmospheric aerosol on vegetative phenological cycles of the Korean Peninsula was analysed based on four major forest types. High temporal-resolution satellite data can overcome limitations of ground-based phenological studies with reasonable spatial resolution. Moderate Resolution Imaging Spectroradiometer (MODIS) vegetation index (VI) (MOD13Q1 and MYD13Q1) and aerosol (MOD04_D3) data were downloaded from the USGS Earth Observation and Science (EROS) Data Center and NASA Goddard Space Flight Center. Harmonic analysis was used to describe and compare the periodic phenomena of the vegetative phenology and atmospheric aerosol optical thickness (AOT). The method transforms complex time-series to a sum of various sinusoidal functions, or harmonics. Each harmonic curve, or term (or Fourier series), from time-series data us defined by a unique amplitude and a phase, indicating the half of the height and the peak time of a curve. Therefore, the mean, phase, and amplitude of harmonic terms of the data provided the temporal relationships between AOT and VI time series. The phenological characteristics of evergreen forest, deciduous forest, and grassland were similar to each other, but the inter-annual VI amplitude of mixed forest was differentiated from the other forest types. Overall, forests with high VI amplitude reached their maximum greenness earlier, and the phase of VI, or the peak time of greenness, was significantly influenced by air temperature. AOT time-series showed strong seasonal and inter-annual variations. Generally, aerosol concentrations were peaked during late spring and early summer. However, inter-annual AOT variations did not have significant relationships with those of VI. Weak relationships between inter-annual AOT and VI variations indicate that the impacts of aerosols on vegetation growth may be limited for the temporal scale investigated in the region.

1. INTRODUCTION

Surface greenness dynamics are associated with the patterns of primary productivity and atmospheric carbon exchange rates on a broad scale (Field et al. 1995; Running et al. 2004; Potter et al. 2007). Vegetation responses to climate change appear strong in Northeast Asia, where diverse forest types and tree species in the region are considered as important carbon sinks. It is reported that Northeastern China and the Korean Peninsula experience significant changes of species composition in forested areas (Jiang et al. 1999). Since atmospheric aerosols play an important role in the interactions between solar radiation and the atmosphere, they are considered one of the key variables in climate modeling (Remer et al. 2005; Lee et al. 2009). The objective of this study is to determine the spatiotemporal influences of climatic factors and atmospheric aerosol on phenological cycles of the Korea Peninsular on a regional scale.

Systematic, continuous, broad-scaled vegetation monitoring requires satellite-based, remote sensing technologies. High temporal-resolution satellite data can overcome limitations of ground-based phenological studies with reasonable spatial resolution. One set of biophysical variables formulated from remote sensing data is vegetation indices (VI). These indices are defined as dimensionless, radiometric measures that have strong correlations with biophysical parameters of green vegetation, including leaf area index (LAI). The Moderate Resolution Imaging Spectroradiometer (MODIS), a hyperspectral earth-observing satellite sensor operated by the National Aeronautics and Space Administration (NASA), provides daily satellite coverage, and its VI data products are designed to provide

consistent vegetation conditions with spatial resolution ranging from 250 m to 1 km (Justice et al., 1998). Atmospheric aerosol information has been also collected by the MODIS system. Advances in aerosol remote sensing allowed researchers to analyze satellite-measured aerosol datasets for scientific uses at various scales (Lee et al., 2009). Aerosol detection algorithms were significantly improved, and aerosol properties, such as aerosol optical thickness (AOT) and aerosol types, are retrieved by the system, providing good agreement with ground-based measurements (Holben et al., 1998; Remer et al., 2002). AOT describes atmospheric attenuation of sunlight by a column of aerosol, and it assesses vertical distribution of aerosols (Kaufman et al., 2002). This study was conducted to determine the spatiotemporal influences of climatic factors and atmospheric aerosols on vegetation phonological cycles in the Korea Peninsula on a regional scale using high-temporal satellite data.

2. MATERIALS AND METHODS

2.1 Study Area

The Korean Peninsula is the eastmost part of Northeast Asia and located west of the Pacific Ocean (Figure 1). The mean annual temperature ranges from 2.1°C to 15.9°C, and it varies substantially depending on proximity to the ocean, elevation, and latitude. The East Asian monsoon strongly contributes to the climate of the region. Temperatures in January are below 0°C for most areas, and winter precipitation, on average, is less than 10% of the annual total, which is about 1,200mm. Primary forest types of the region include boreal, temperate, and subtropical forests.

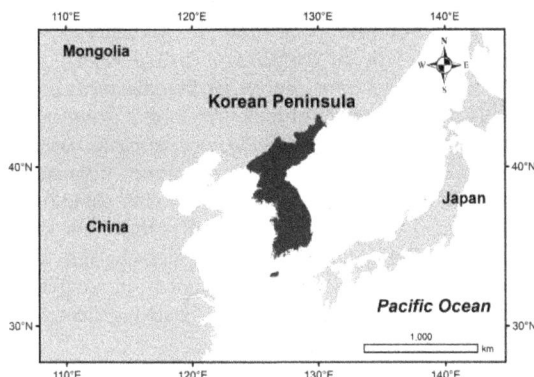

Figure 1. Study area.

2.2 Data Processing

Since MODIS provides both morning (Terra) and afternoon (Aqua) vegetation index products with high compatibility, the relatively new system is a valuable data source for environmental studies, including seasonal patterns of photosynthesis over large geographic areas (Running et al., 2004; Yang et al., 2005). MODIS 16-day composites VI data, MOD13Q1, were downloaded from the EROS data center for the 2001-2010 period (https://wist.echo.nasa.gov). Three image tiles (h27v04, h27v06, and h28v05) were mosaicked together to cover the entire Korea Peninsula with the spatial resolution of 250 meters. In total, 690 composite images were processed for cloud screening and only clear-sky pixels were used for further analyses.

Biophysical parameters, such as the fraction of photosynthetically active radiation absorbed by chloroplasts (fPAR) and leaf area index (LAI), have strong correlations with vegetation indices. In MODIS-based modeling of primary production, the Enhanced Difference Vegetation Index (EVI) has been common used. The algorithm of EVI was developed to improve its sensitivity to densely vegetated areas (Huete et al., 2002). This index incorporates the blue band (ρ_{blue}) into its formula for atmospheric correction, and has been applied to vegetation phenology studies:

$$EVI = 2.5 \times (\rho_{nir} - \rho_{red})/(\rho_{nir} + (6 \times \rho_{red} - 7.5 \times \rho_{blue}) + 1)$$

In addition to MODIS VI data, aerosol products, MOD04, were analyzed to extract AOT over the study area. Land aerosol observations are based on near-infrared (2.1μm, 3.8μm) channels, which are relatively free from atmospheric interference, and the data are provided with the spatial resolution of 10km. With improved algorithms, their accuracy is known ±0.05±0.20AOT (Remer et al., 2005; Levy et al., 2007). Distributions of atmospheric aerosols are determined from a light extinction coefficient, which is the fraction of light that is scattered or absorbed by aerosol particles as a function of altitude (Wong et al., 2009). AOT is the sum of light extinction coefficients integrated over a vertical column, and it estimates the amount of aerosols for the whole atmospheric column (Kaufman et al., 1997; Remer et al., 2005).

Although greenness signals in forest ecosystems are complex, frequently acquired VI records allow researchers to extract and identify the fluctuations of canopy greenness (Huete et al. 2006; Xiao et al. 2005). Spectral analysis, or harmonic analysis, was

used to extract major periodic fluctuation signals, or wave forms from the temporal domain of the input data (Azzali and Menenti 2000; Jakubauskas et al. 2001; Moody and Johnson 2001; Park 2009). Any complex raw data collected on a regular basis, such as VI time-series, can be reduced to a series of harmonic terms that have a unique set of amplitude, wavelength, and phase angle (Figure 2). Annual and inter-annual variations and peak points of greenness and AOT signals were analyzed by the technique. Four major forest types, mixed forest, evergreen forest, deciduous forest, and grassland, were selected from ESA GlobCover landuse data (Sophie et al. 2010), and their primary phonological cycles were extracted from the VI time-series. Monthly minimum, maximum, and mean temperatures at 90 weather stations, where no missing data were found during the study period, were collected from NOAA and Korea Meteorological Agency. Spectral analysis was also applied to these data to determine their annual variations.

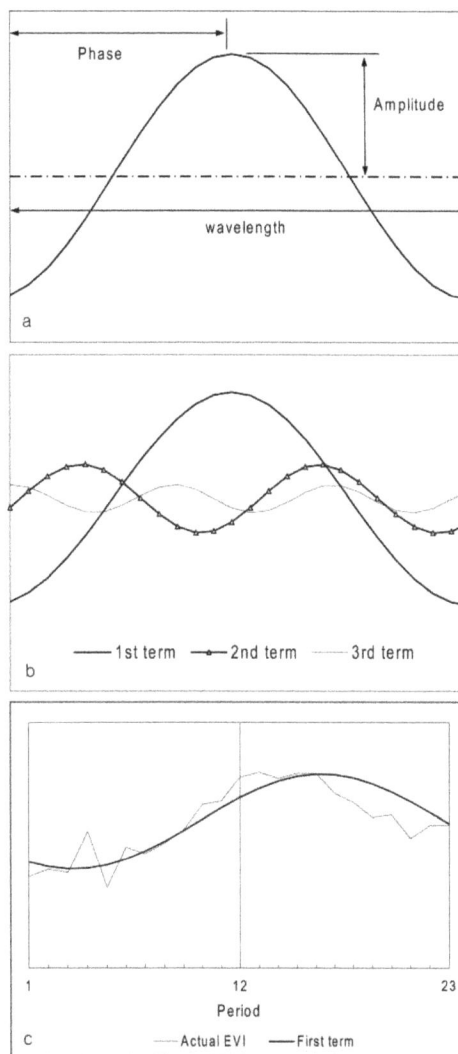

Figure 2. Main components of harmonic analysis (a), and a schematic illustration of the first three harmonic terms, 1st, 2nd, and 3rd terms (b). Principal seasonality of mid-latitude regions is mostly well represented by the first term.

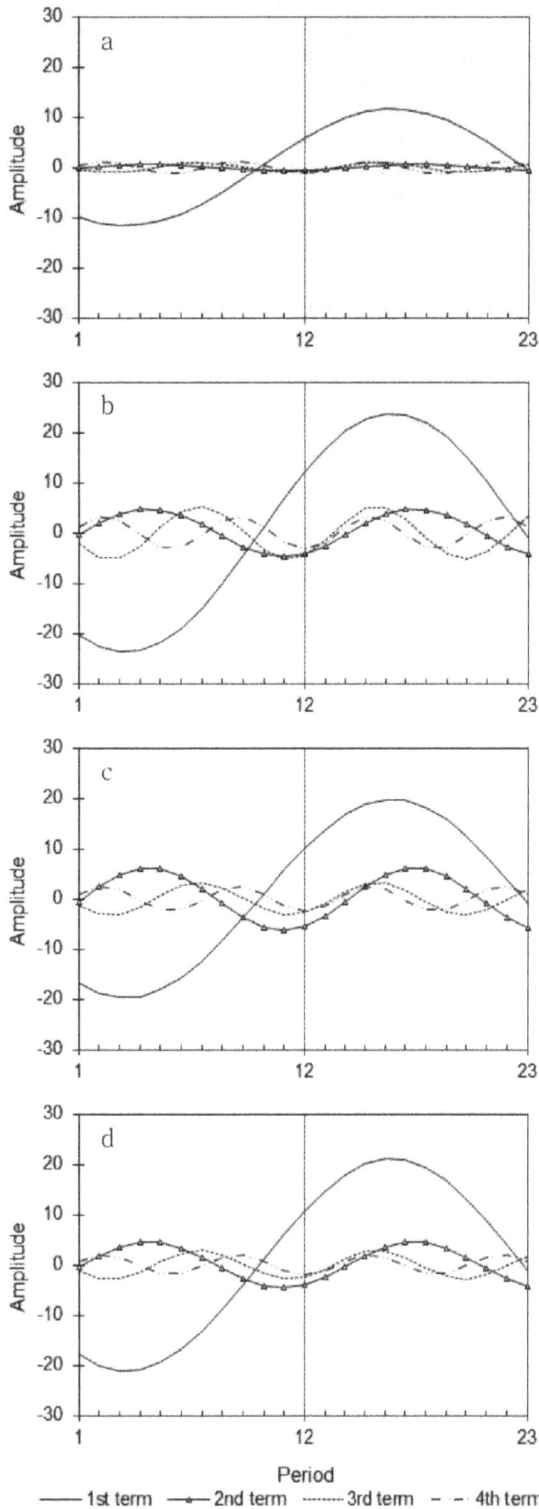

Figure 3. Harmonic patterns for the 4 land cover types (a-mixed forest, b-deciduous forest, c-evergreen forest, d-grassland).

3. SUMMARY OF RESULTS

Study results showed that phenological characteristics were similar among evergreen forest, deciduous forest, and grassland, while the inter-annual VI amplitude of mixed forest was differentiated from the other forest types (Figure 3). The VI amplitude indicates the degree of greenness fluctuations over

the year, and it was greatest for deciduous forest followed by grassland, evergreen forest, and mixed forest. Vegetation types with high VI amplitude reached their maximum VI values, but this relationship was not observed within the same forest type.

The phase of VI, or the peak time of greenness, was significantly influenced by air temperature (r=0.61~0.79). The earlier the peak air temperature, the earlier the greenness peak time. It is likely that the growth of mountain forests is typically limited by temperature throughout the Korean peninsula, where deciduous and evergreen forests dominate. However, influences of mountain elevation and aspect on forest growth should be further investigated. The pace of forest growth is important with respect to the regional climate. Recent studies showed that earlier vegetation growth had increased evapotranspiration and its cooling effect had taken place in the spring season (Jeong et al., 2009a; Jeong et al., 2009b).

AOT time-series showed strong seasonal and inter-annual variations. Generally, aerosol concentrations were highest from late spring to early summer. However, inter-annual AOT variations did not have significant relationships with those of VIs. Aerosol concentrations in the peninsula are strongly associated with industrialization in the Eastern China as well as domestic wild fires and man-made aerosol emission (Li et al., 2007). Annual variations of AOT, or amplitude, had only weak correlations with surface greenness. The phase of AOT had weak negative correlations with surface greenness (r=-0.43~-0.67). VI peak values were observed later over the year as AOT concentrations reached earlier in spring, Pixel-based, high resolution research is still required to determine the influences of the characteristics of aerosols (sources, spatial distributions, etc.) on the vegetation phenology in the region.

4. CONCLUDING REMARKS

It is apparent that AOT time-series has a strong seasonality and geographically uneven distributions in the Korean Peninsula (Kim et al., 2007; Park, 2014). Unfortunately, long-term monitoring of the natural and anthropogenic aerosols is still technically limited due to the lack of continuous atmospheric data over the country. Since intense urbanization and industrialization in the eastern China and Korean Peninsula had direct and sensitive impacts on the atmospheric quality of the region from ecological, public health, and socioeconomic perspectives, broad-scaled, satellite-based atmospheric monitoring should be systematically practiced to overcome limitations of ground-based observations of atmospheric aerosols. AOT observations, a quantitative measure of the atmospheric quality, had significant geographical variations in the region. Particularly, the maximums of monthly AOT means showed 1.9-fold to 3.2-fold variations across the peninsula. AOT measurements were apparently higher in western regions than eastern regions, indicating the strong effects of atmospheric effluents from China. Therefore, in addition to forest types, geographical characteristics should be considered as an important factor when it comes to the impacts of AOT on the vegetation phenology.

ACKNOWLEDGEMENTS

This research was financially supported by the Basic Science Research Programme through the National Research Foundation (NRF) of Korea funded by the Ministry of Education, Science and Technology (grant number NRF-2010-0024819).

REFERENCES

Azzali, S. and Menenti, M., 2000. Mapping vegetation-soil-climate complexes in southern Africa using temporal Fourier analysis using NOAA-AVHRR-NDVI data. *International Journal of Remote Sensing*, 21, pp. 973-996.

Field, C.B., Behrenfeld, M.J., Randerson, J.T., and Falkowski, P., 1998. Primary production of the biosphere: integrating terrestrial and oceanic components. *Science*, 281, pp. 237-240.

Holben, B.N., Eck, T.F., Slutsker, I., Tanre, D., Buis, J.P., Setzer, A., Vermote, E., Reagan, J.A., Kaufman, Y.J., Nakajima, T., Lavenu, F., Jankowiak, I., and Smirnov, A., 1998. AERONET-a federated instrument network and data archive for aerosol characterization. *Remote Sensing of Environment*, 66, pp. 1-16.

Huete, A.R., Didan, K, Shimabukuro, Y.E., Ratana, P., Saleska, S.R., Hutyra, L.R., Yang, W., Nemani, R.R., and Myneni, R., 2006. Amazon rainforests green-up with sunlight in dry season. *Geophysical Research Letters* 33, L06405, DOI:10.1029/2005GL025583.

Jakubauskas, M.E., Legates, D.R., and Kastens, J.H., 2001. Harmonic analysis of time-series AVHRR NDVI data. *Photogrammetric Engineering and Remote Sensing*, 67, pp. 461-470.

Jeong, S.-J., Ho, C.-H., and Jeong, J.-H., 2009a. Increase in vegetation greenness and decrease in springtime warming over East Asia. *Geophysical Research Letters*, 36, L02710, doi:10.1029/2008GL036283.

Jeong, S.-J., Ho, C.-H., Kim, K.-Y., and Jeong, J.-H., 2009b. Reduction of spring warming over East Asia associated with vegetation feedback. *Geophysical Research Letters*, 36, L18705, doi:10.1029/2009GL039114.

Jiang, H., Apps, M.J., Zhang, Y., Peng, C. and Woodard, P., 1999. Modelling the spatial pattern of net primary productivity in Chinese forests. *Ecological Modeling*, 122, pp.275-288.

Justice, C.O., Vermote, E., Townshend, J.R.G., Defries, R., Roy, D.P., Hall, D.K., Salomonson, V.V., Privette, J.L., Riggs, G., Strahler, A., Lucht, W., Myneni, R.B., Knyazikhin, Y., Running, S.W., Nemani, R.R., Wan, Z., Huete, A.R., van Leeuwen, W., Wolfe, R.E., Giglio, L., Muller, J.P., Lewis, P., and Barnsley, M.J., 1998. The Moderate Resolution Imaging Spectroradiometer (MODIS): land remote sensing for global change research. *IEEE Transactions on Geoscience and Remote Sensing*, 36, pp.1228–1249.

Kaufman, Y.J., Tanré, D., Remer, L., Vermote, E.F., Chu, A., and Holben, B.N., 1997. Operational remote sensing of troposphere aerosol over land from EOS Moderate Resolution Imaging Spectroradiometer. *Journal of Geophysical Research*, 102, pp. 17051-17067.

Kaufman, Y., Tanre, D., and Boucher, O., 2002. A satellite view of aerosols in the climate system. *Nature*, 419, pp. 215-223.

Kim, S.W., Yoon, S.C., Kim, J., Kim, S.Y., 2007. Seasonal and monthly variations of columnar aerosol optical properties over East Asia determined from multi-year MODIS, Lidar, and AERONET sun/sky radiometer measurements. *Atmospheric Environment*, 41, pp.1634-1651.

Lee, K., Li, Z., Kim, Y., and Kokhanovsky, A., 2009. Atmospheric aerosol monitoring from satellite observations: a history of three decades. In: Kim, Y., Platt, U., Gu, M., and Iwahashi, H., 2009. *Atmospheric and Biological Environmental Monitoring* DOI 10.1007/978-1-4020-9674-7_2, pp.13-38. Springer.

Levy, R.C., Remer, L., Mattoo, S., Vermote, E., and Kaufman, Y.J., 2007. A second-generation algorithm for retrieving aerosol properties over land from MODIS spectral reflectance. *Journal of Geophysical Research*, 124, pp.2046-2070.

Li, Z., Niu, F., Lee, K. H., Xin, J., Hao, W. M., Nordgren, B., Wang, Y., and Wang, P., 2007. Validation and understanding of Moderate Resolution Imaging Spectroradiometer aerosol products (C5) using ground-based measurements from the handheld Sun photometer network in China. *Journal of Geophysical Research*, 11, D22S07, doi:10.1029/2007JD008479.

Moody, A. and Johnson, D.M., 2001. Land-surface phenologies from AVHRR using the discrete Fourier transform. *Remote Sensing of Environment*, 75, pp.305-323.

Park, S., 2009. Synchronicity between satellite-measured leaf phenology and rainfall regime in Hawaiian tropical forests. *Photogrammetric Engineering and Remote Sensing*, 75, pp.1231-1237.

Park, S., 2014. Regional analyses of aerosol optical thickness over South Korea. *The Geographical Journal of Korea*, 48, pp.523-532.

Potter, C., Klooster, S., Huete, A., and Genovese, V., 2007. Terrestrial carbon sinks for the United States predicted from MODIS satellite data and ecosystem modeling. *Earth Interactions*, 11.

Remer, L.A., Kaufman, Y.J., Tanre, D., Mattoo, S., Chu, D.A., Martins, J.V., Li, R.R., Ichoku, C., Levy, R., Kleidman, R.G., Eck, T.F., Vermote, E., and Holben, B.N., 2005. The MODIS aerosol algorithm, products and validation. *Journal of Atmospheric Sciences*, 62, pp. 947-973.

Remer, L.A., Tanre, D., Kaufman, Y.J., Ichoku, C., Mattoo, S., Levy, R., Chu, D.A., Holben, B.N., Dubovik, O., Smirnov, A., Martins, J.V., Li, R.R., and Ahmad, Z., 2002. Validation of MODIS aerosol retrieval over ocean. *Geophysical Research Letters*, 29, pp.321-324.

Running, S.W., Nemani, R.R., Heinsch, F.A., Zhao, M., Reeves, M., and Hashimoto, H., 2004. A continuous satellite-derived measure of global terrestrial primary production. *BioScience*, 54, pp.547-560.

Sophie, B., Pierre, D., and Eric, V.B., 2010. GLOBCOVER 2009, *Products Description and Validation Report*, ESA.

Wong, M.S., Nichol, J.E., and Lee, K.H., 2009. Modeling of aerosol vertical profiles using GIS and remote sensing. *Sensors*, 9, pp. 4380-4389.

Xiao, X.M., Zhang, Q.Y., Saleska, S, Hutyra, L., De Camargo, P., Wofsy, S., Frolking, S., Boles, S., Keller, M., and Moore, B.,

2005. Satellite-based modeling of gross primary production in a seasonally moist tropical evergreen forest. *Remote Sensing of Environment,* 94, pp.105-122.

Yang, W., Shabanov, N.V., Huang, D., Wang, W., Dickinson, R.E., Nemani, R.R., Knyazikhin, Y., and Myneni, R.B., 2006. Analysis of leaf area index products from combination of MODIS Terra and Aqua data. *Remote Sensing of Environment,* 104, pp.297-312.

23

MERGING AIRBORNE LIDAR DATA AND SATELLITE SAR DATA FOR BUILDING CLASSIFICATION

T. Yamamoto, M. Nakagawa

Dept. of Civil Engineering, Shibaura Institute of Technology, 3-7-5 Toyosu, Koto-ku, Tokyo, Japan - (h10082, mnaka@shibaura-it.ac.jp)

Commission IV / WG 7

KEY WORDS: Urban Sensing, Building Extraction, Building Classification, Airborne LiDAR, Satellite SAR, Data Fusion

ABSTRACT:

A frequent map revision is required in GIS applications, such as disaster prevention and urban planning. In general, airborne photogrammetry and LIDAR measurements are applied to geometrical data acquisition for automated map generation and revision. However, attribute data acquisition and classification depend on manual editing works including ground surveys. In general, airborne photogrammetry and LiDAR measurements are applied to geometrical data acquisition for automated map generation and revision. However, these approaches classify geometrical attributes. Moreover, ground survey and manual editing works are finally required in attribute data classification. On the other hand, although geometrical data extraction is difficult, SAR data have a possibility to automate the attribute data acquisition and classification. The SAR data represent microwave reflections on various surfaces of ground and buildings. There are many researches related to monitoring activities of disaster, vegetation, and urban. Moreover, we have an opportunity to acquire higher resolution data in urban areas with new sensors, such as ALOS2 PALSAR2. Therefore, in this study, we focus on an integration of airborne LIDAR data and satellite SAR data for building extraction and classification.

1. INTRODUCTION

A frequent map revision is required in GIS applications, such as disaster prevention and urban planning. In general, airborne photogrammetry and LiDAR measurements are applied to geometrical data acquisition for automated map generation and revision. In the airborne photogrammetry, a geometrical modeling and object classification can be automated using color images. Stereo matching is an essential technique to reconstruct 3D model from images. Recently, structure from motion (SfM) is proposed to generate 3D mesh model from random images (Uchiyama, 2014). Although, object classification methods are automated using height data estimated with stereo matching and SfM, it is difficult to recognize construction materials, such as woods and concrete. The construction materials are significant attribute data in building modeling and mapping. Therefore, ground survey and manual editing works are required in attribute data classification.

In the LiDAR measurements, modeling and object classification are also automated by a point cloud segmentation (Sithole, 2003). The intensity data assist the object classification (Antonarakis, 2008). Moreover, data fusion approaches are proposed using aerial images and LiDAR data. These approaches focus on improvement of modeling accuracy and processing time (Uemura, 2011). However, these approaches classify geometrical attributes.

On the other hand, although geometrical data extraction is difficult, SAR data have a possibility to automate the attribute data acquisition and classification. The SAR data represent microwave reflections on various ground surfaces and buildings. There are many researches related to monitoring activities of disaster, vegetation, and urban. Moreover, we have an opportunity to acquire higher resolution data in urban areas with new sensors, such as ALOS2 PALSAR2 (Japan Aerospace

Exploration Agency, 2014). Therefore, in this study, we focus on an integration of airborne LiDAR data and satellite SAR data for building extraction and classification

2. METHODOLOGY

Our process is shown in Figure 1.

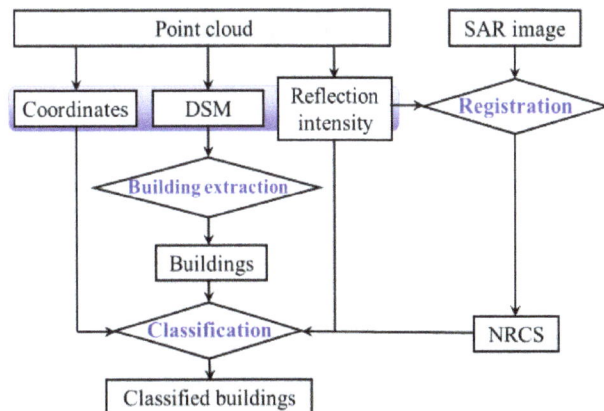

Figure 1. Our process flow

In this study, we focus on an integration of airborne LIDAR data and satellite SAR data for building extraction and classification. Firstly, we generate DSM and reflection intensity orthoimage from LiDAR point cloud data. Secondary, these data are registered using corresponded points taken from each datum. Thirdly, buildings are extracted from the DSM. Finally, buildings are classified with normalized radar cross section (NRCS) calculated using SAR data.

2.1 Building footprint extraction

Building footprints are extracted from DSM, as shown in Figure 2.

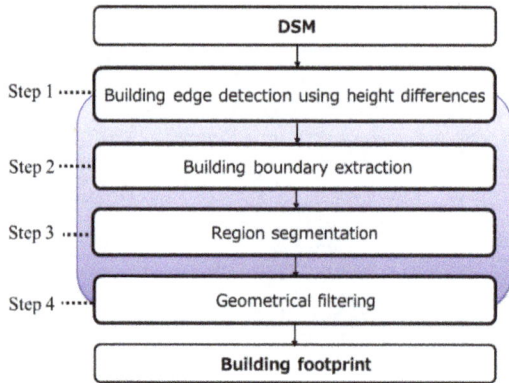

Figure 2. Process flow of building footprint extraction

DSM generated from point cloud data for building footprint extraction. Firstly, building edges are detected using height differences between building roofs and ground surfaces from DSM with a 3 × 3 operator. Although the building edges are discontinuous, approximate building features are detected in this step. Secondary, building boundaries are extracted. Discontinuous edges are connected to each other in the DSM with 8-neighborhood pixel filtering. The connected edges are defined as a building boundary. Thirdly, segmentation is applied to each region inside of the building boundary to refine building footprints. Although extracted region includes many noises, such as bridges, street trees, and automobiles, an approximate geometry of each region is extracted in this step. Finally, the region segments are filtered with their perimeter and area to extract building footprints. An example of the building extraction is shown in Figure 3.

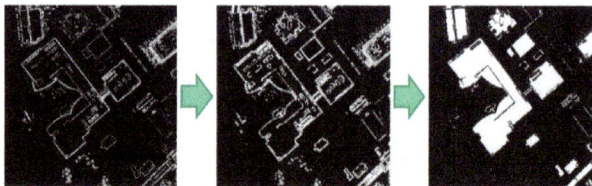

Figure 3. Example of building extraction

2.2 Building classification

Our process flow of building classification is shown in Figure 4.

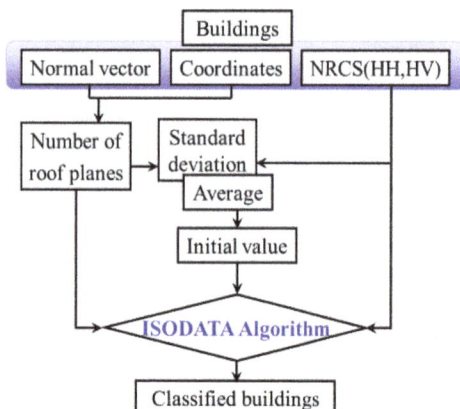

Figure 4. Process flow of building classification

In this study, buildings are classified into several groups with non-supervised classification. Two types of approaches are applied to our building classification. The first approach is a building classification based on roof materials with an average value of NRCS in each polarimetric SAR images. The second approach is a building classification based on geometrical segments taken from LiDAR data. Roof shapes have clear features in the classification. The number of roof planes is estimated for building classification with height information and normal vector of point cloud, as shown in Figure 5.

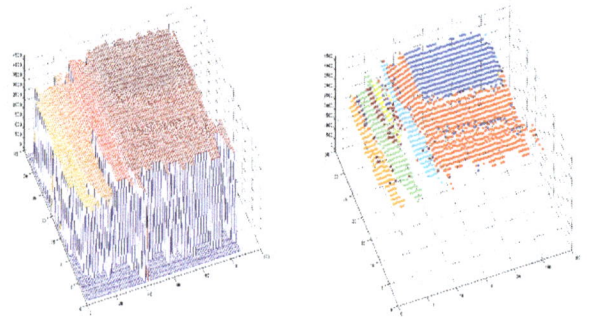

Figure 5. Building classification

3. EXPERIMENT

3.1 Study area

We selected Toyosu and Monzennakacho areas in Tokyo as our study area. These areas include various types of buildings, such as residential houses, high-rise buildings and shopping malls, as shown in Figure 6.

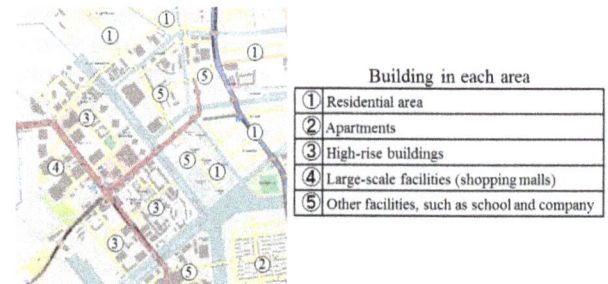

Figure 6. Study area

3.2 Data specification

We prepared point cloud data acquired with an airborne LiDAR and geocoded satellite SAR data, as shown in Table 1 and Table 2. Moreover, threshold values were used in building extraction, as shown in Table 3.

Table 1. Specification of LiDAR data

Observer	Date	Spatial resolution	number of points
Kokusai Kogyo Co., Ltd	7, March, 2011	0.5m (DSM)	4000×4000

Table 2. Specification of SAR data

Observer	Date	Spatial resolution	Geocoded	Polarized wave
JAXA	20, March, 2009	12.5m	Map North	HH, HV

Table 3. Threshold values

	Step 1	Step 2	Step 3	Step 4
Height	2 m	±0.2 m	2 m	---
Area	---	---	---	200 pixels
Perimeter	---	---	---	10000 pixels
Perimeter/area	---	---	---	0.35

3.3 Registration

In a registration between SAR and LiDAR data, corresponded points are required to be extracted from each datum. Although SAR and LiDAR data have different indices, we can recognize road intersections, rivers, and bridges as feature points in manual. An example of corresponded points between SAR and LiDAR data is shown in Figure 7.

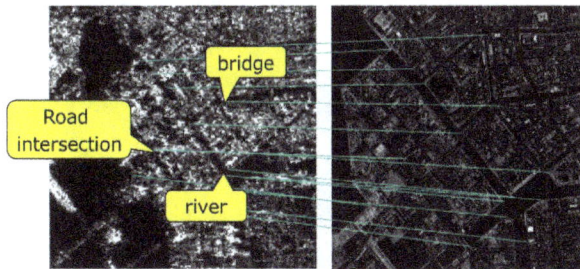

Figure 7. Example of corresponded points

Before a feature extraction procedure, two types of orthoimages were prepared as follows. Firstly, digital number (DN) of SAR image was converted into an orthoimage of NRCS. We used the following transformation formula with calibration factor (CF). We substituted -83 for the CF (ALOS User Interface Gateway, 2009).

$$NRCS(dB)=10\times\log_{10}(DN^2)+CF \qquad (1)$$

Next, the other orthoimage was generated from reflection intensity values taken from LiDAR point cloud data. In this procedure, the reflection intensity values were projected into DSM generated from LiDAR data, as shown in Figure 8 and Figure 9.

We selected several corresponding points, such as road intersections, rivers, and bridges, from each orthoimage. Moreover, the affine transformation was applied to the image registration between SAR and LiDAR data.

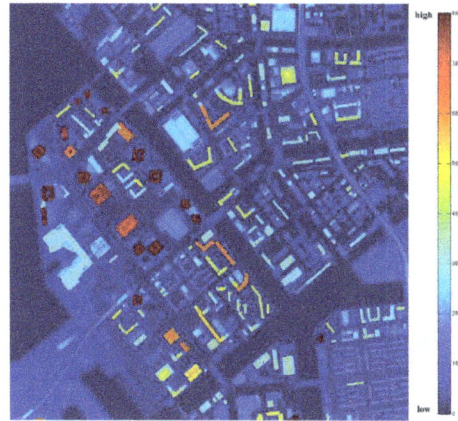

Figure 8. DSM

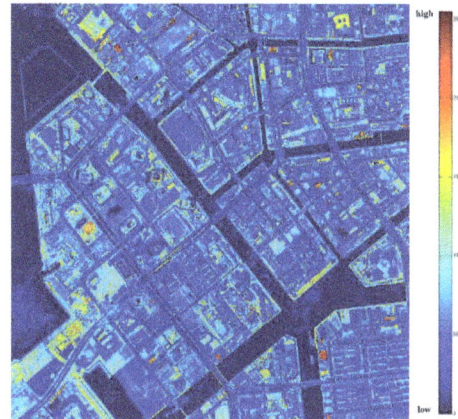

Figure 9. Reflection intensity orthoimage

4. RESULTS

4.1 Building extraction

First, Figure 10 shows a result in the step 1. White edges indicate extracted building boundaries with height differences. Next, Figure 11 shows a result in the step 2. Dilated white edges indicate refined building boundaries. Figure 12 shows a result in the step 3. White regions indicate extracted building footprints. Figure 13 shows a result in the step 4.

Figure 10. Result in the step 1

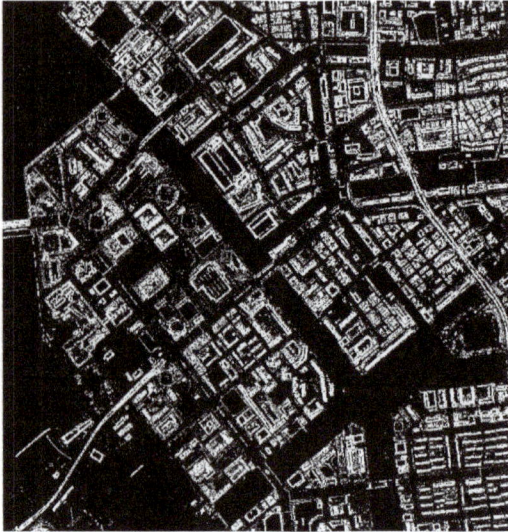

Figure 11. Result in the step 2

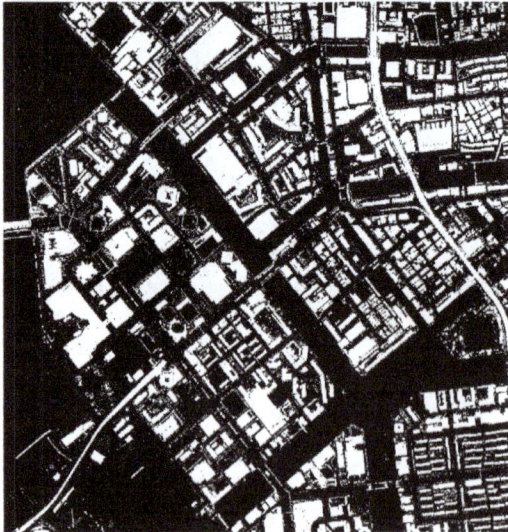

Figure 12. Result in the step 3

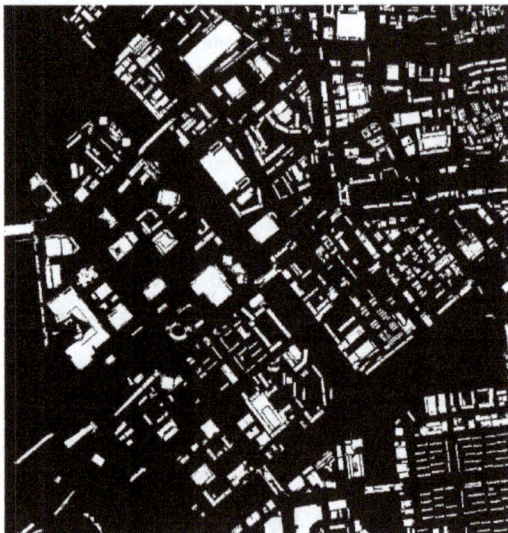

Figure 13. Result in the step 4

4.2 Building classification

We extracted 911 buildings from DSM. Our result after the building classification based on ISODATA clustering of NRCS values is shown in Figure 14. The vertical axis indicates HH NRCS values, and the horizontal axis indicates HV NRCS values. Classified buildings with NRCS values are projected into an orthoimage, as shown in Figure 15.

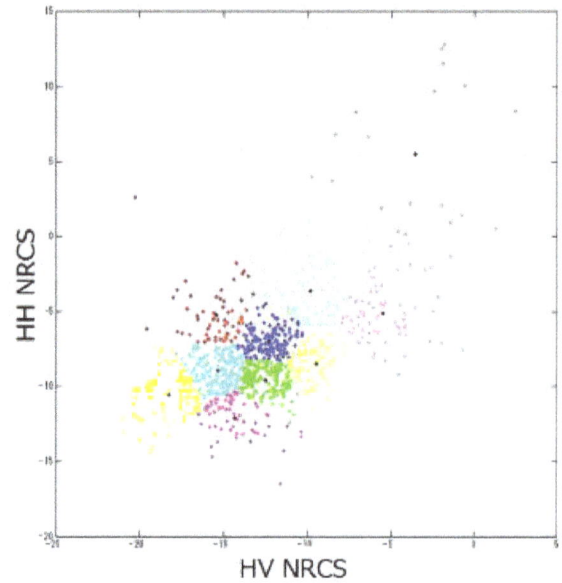

Figure 14. Building classification result using NRCS values (1)

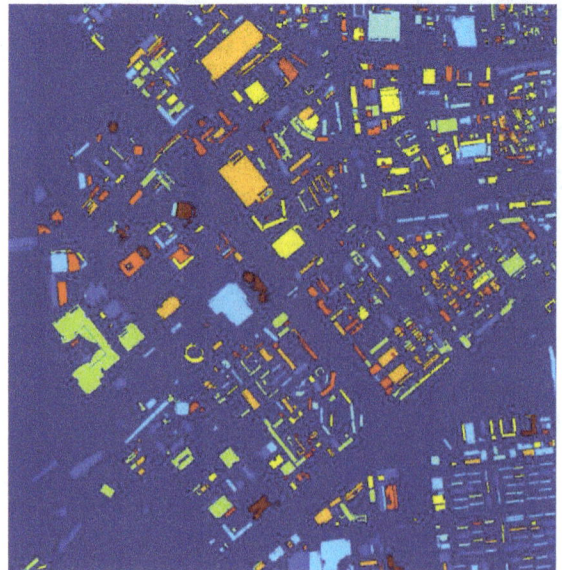

Figure 15. Building classification result using NRCS values (2)

Moreover, we classified buildings using NRCS values and the number of building roofs, as shown in Figure 16. The perpendicular axis indicates the number of roof segments, and the X and Y axes indicate HH and HV NRCS values. Classified buildings with NRCS values are also projected into an orthoimage, as shown in Figure 17.

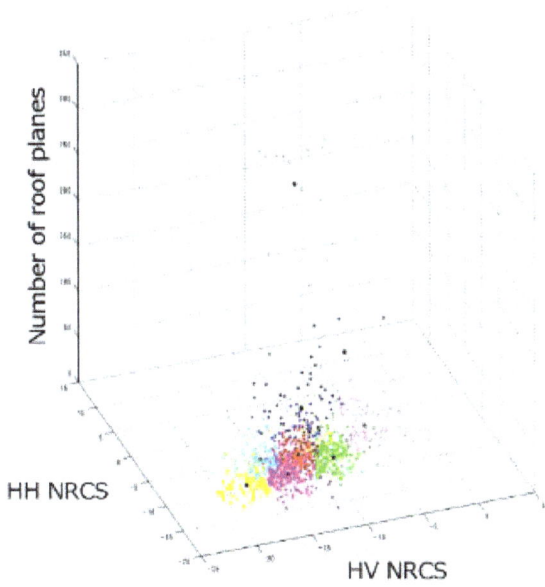

Figure 16. Building classification result using NRCS values and the number of roof segments (1)

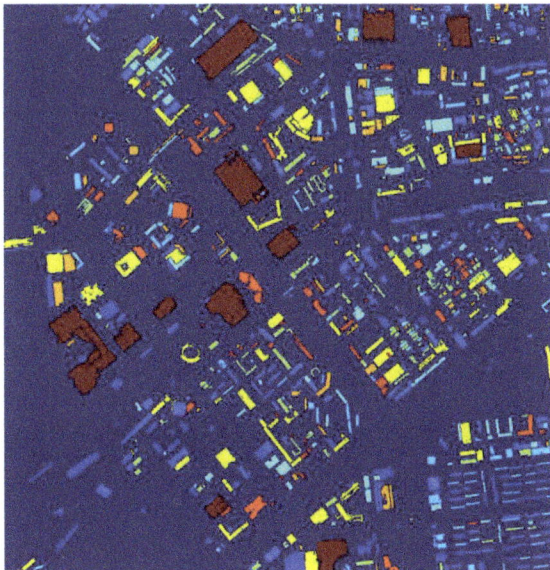

Figure 17. Building classification result using NRCS values and the number of roof segments (2)

5. DISCUSSION

Classified buildings using NRCS values in each cluster is shown in Figure 18. Classified buildings using NRCS and the number of roof segments is shown in Figure 19. In Figure 18 and 19, vertical axis indicates the number of roof segments, and horizontal axis indicates cluster numbers.

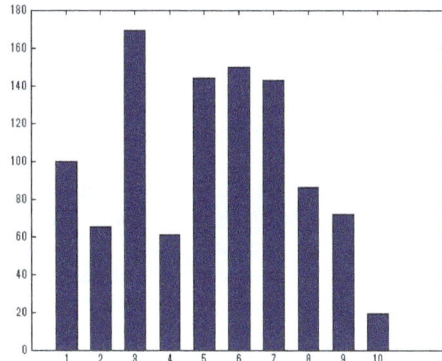

Figure 18. Classification with NRCS values in each cluster

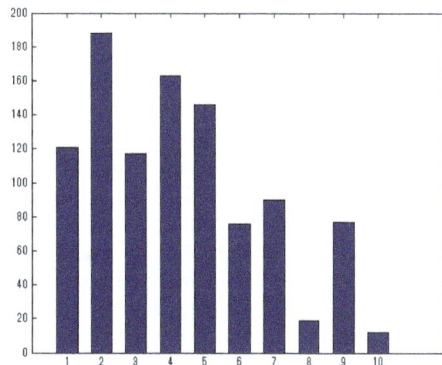

Figure 19. Classification with NRCS and the number of roof segments in each cluster

In the classification with NRCS values, we have confirmed that spatial resolution was too low to recognize small residential buildings and complex roofs of large buildings. In our experiment, the supervised classification in the building extraction was affected by speckle noises. Therefore, we would propose a speckle noise filtering before the classification. In the classification with NRCS and the number of roof segments, large buildings with complex roofs were extracted. Although we focused on building roofs, we can focus on an opportunity to acquire more detailed building features from aerial LiDAR data. We would improve our classification with a wall surface estimation and smaller object recognition, as shown in Figure 20.

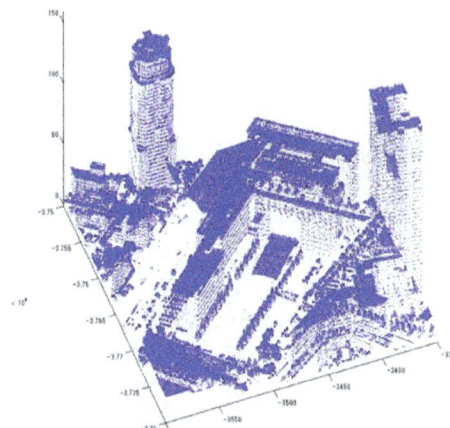

Figure 20. Visualized building with point cloud data

In our experiment, although visual checks were required to determine the best threshold values, buildings were extracted from DSM in our object extraction procedure. Moreover, several small noises such as automobiles were left as unknown objects in the DSM. We can focus on a semantic approach using road connections to improve our feature extraction accuracy. Additionally, although shadow detection is required, we can focus on a combination of LiDAR data with aerial images.

In general, SAR has several problems, such as a layover, radar shadow, and foreshortening. They are caused by undulating grounds. In this study, SAR data were strongly affected by their problem. However, when we use dense point cloud data, we can recognize ground surface in detail. Therefore, we have a possibility to avoid these problems. Moreover, cardinal effect can be analysed using 3D geometrical data generated from point cloud data.

6. CONCLUSION

In this paper, we have focused on an integration of LiDAR with SAR data to achieve the frequent map update with attribute data acquisition. Firstly, we generated DSM from point cloud acquired with airborne LiDAR. Secondary, the DSM was registered the SAR data to overlay with NRCS calculated from the SAR data. Thirdly, buildings are extracted from the DSM. Finally, we classified buildings in the DSM into several clusters. In our experiment, we prepared point cloud data acquired with an airborne LiDAR and satellite SAR data acquired with ALOS PALSAR in Tokyo. Next, we extracted 911 buildings from DSM. Although our result included noises such as bridges and automobiles, we classified buildings into 10 clusters with average NRCS values. In this study, we clarified that a combination of airborne LiDAR data and satellite SAR data can extract and classify buildings in urban area. In our future works, we will apply the supervised clustering with a semantic approach to improve our classification accuracy.

REFERENCES

Antonarakis, A, S., Richards, K, S., Brasington, J., 2008. Object-based land cover classification using airborne LiDAR, Remote Sensing of Environment, Volume 112, Issue 6, pp. 2988-2998.

ALOS User Interface Gateway 2009 PALSAR Calibration Factor Updated http://www.eorc.jaxa.jp/en/about/distribution/info/alos/2009010 9en_3.html (21 Jan 2014)

Haala, N., Kada, M., 2010. An update on automatic 3D building reconstruction, ISPRS Journal of Photogrammetry and Remote Sensing Volume 65, Issue 6, pp. 570-580.

Japan Aerospace Exploration Agency, 2014. Advanced Observing Satellite-2 "DAICHI-2" (ALOS-2). http://global.jaxa.jp/projects/sat/alos2/ (27 Jul 2013)

Sithole, G., Vosselman, G., 2003. Automatic structure detection in a point-cloud of an urban landscape, Remote Sensing and Data Fusion over Urban Areas. 2nd GRSS/ISPRS Joint Workshop on, pp. 67-71.

Tupin, F., Roux, M., 2003. Detection of building outlines based on the fusion of SAR and optical features, ISPRS Journal of Photogrammetry and Remote Sensing Volume 58, Issue 1–2, pp. 71–82.

Uchiyama, S., Inoue, H., Suzuki, H., 2014. Approaches for Reconstructing a Three-dimensional Model by SfM to Utilize and Apply this Model for Research on Natural Disasters, Volume 81, pp. 37-60.

Uemura, T., Uchimura, K., koutaki, G., 2011. Road Extraction in Urban Areas using Boundary Code segmentation for DSM and Aerial RGB images, Journal of The Institute of Image Electronics Engineers of Japan, Volume 40, No.1, pp. 74-85.

Zhang, K., Yan, J., Chen, S., 2006. Automatic construction of building footprints from airborne LIDAR data, IEEE Transactions on Geoscience and Remote Sensing, Volume 44, No. 9, pp. 2523-2533.

ACKNOWLEDGMENT

This work is supported by Japan Aerospace Exploration Agency. Moreover, our experiments are supported by Kokusai Kogyo Co. Ltd.

24

SATELLITE-BASED DROUGHT MONITORING IN KENYA IN AN OPERATIONAL SETTING

A. Klisch [a, *], C. Atzberger [a], L. Luminari [b]

[a] University of Natural Resources and Applied Life Sciences (BOKU), Institute of Surveying, Remote Sensing and Land Information (IVFL), Peter-Jordan-Straße 82, 1190 Wien, Austria - (anja.klisch, clement.atzberger)@boku.ac.at
[b] National Drought Management Authority (NDMA), Lonrho House -Standard Street , Nairobi, Kenya - luigi.luminari@dmikenya.or.ke

Commission VI, WG VI/4

KEY WORDS: Drought monitoring, Kenya, Whittaker smoother, MODIS, NDMA, Drought Contingency Funds

ABSTRACT:

The University of Natural Resources and Life Sciences (BOKU) in Vienna (Austria) in cooperation with the National Drought Management Authority (NDMA) in Nairobi (Kenya) has setup an operational processing chain for mapping drought occurrence and strength for the territory of Kenya using the Moderate Resolution Imaging Spectroradiometer (MODIS) NDVI at 250 m ground resolution from 2000 onwards. The processing chain employs a modified Whittaker smoother providing consistent NDVI "Monday-images" in near real-time (NRT) at a 7-daily updating interval. The approach constrains temporally extrapolated NDVI values based on reasonable temporal NDVI paths. Contrary to other competing approaches, the processing chain provides a modelled uncertainty range for each pixel and time step. The uncertainties are calculated by a hindcast analysis of the NRT products against an "optimum" filtering. To detect droughts, the vegetation condition index (VCI) is calculated at pixel level and is spatially aggregated to administrative units. Starting from weekly temporal resolution, the indicator is also aggregated for 1- and 3-monthly intervals considering available uncertainty information. Analysts at NDMA use the spatially/temporally aggregated VCI and basic image products for their monthly bulletins. Based on the provided bio-physical indicators as well as a number of socio-economic indicators, contingency funds are released by NDMA to sustain counties in drought conditions. The paper shows the successful application of the products within NDMA by providing a retrospective analysis applied to droughts in 2006, 2009 and 2011. Some comparisons with alternative products (e.g. FEWS NET, the Famine Early Warning Systems Network) highlight main differences.

1. INTRODUCTION

Drought is a recurrent natural phenomenon in many arid and semi-arid regions of the world. The resulting stress depends primarily on the strength, duration, timing and spatial extent of the dry spell. At the same time, different communities and economic sectors may show varying vulnerabilities and resiliencies to drought events, as available coping strategies and previous (environmental) conditions differ.

For drought-prone countries, it is important to monitor droughts and affected communities to prevent disastrous results. For this purpose, Kenya established in 2011 a National Drought Management Authority (NDMA) which mandate is to exercise general supervision and coordination over all matters relating to drought management in Kenya. In 2014, the NDMA received some Drought Contingency Funds (DCFs) from the European Union to facilitate early response to drought threads. DCFs are disbursed by the NDMA to drought-affected counties in order to implement response activities that can help mitigating the worst impacts of droughts. MODIS satellite images are used to determine the drought status of a county in an objective and reproducible way. For near real-time processing of the data, BOKU University developed and implemented an advanced filtering method for NDVI images. The processing yields reliable drought indicators at county and sub-county levels and for various aggregation times and livelihood zones. Image analysis is complemented at NDMA by field-based (socio-economic) indicators. The innovative DCF disbursement mechanisms of NDMA ensure a timely support of drought-affected counties and communities.

The present paper describes the MODIS processing chain implemented at BOKU. Through comparison with the well-established FEWS NET data, we highlight and quantify main differences between the two datasets.

2. STUDY AREA

The study covers an area of 10° x 11° centred over Kenya. We focus on the arid and semi-arid land (ASAL) mainly located in the northern and eastern parts of the country (see Figure 1). These areas are characterised by high temperatures (except elevated areas), low rainfall amounts and therefore often relatively low biomass/NDVI. This low biomass is seen in Figure 1 as average annual NDVI ≤0.4 (brownish colour).

3. METHODOLOGY

3.1 Data Processing at BOKU

The University of Natural Resources and Life Sciences (BOKU) in cooperation with the National Drought Management Authority (NDMA) has setup an operational processing of MODIS images with the aim of providing consistent NDVI and anomaly "Monday-images" in near real-time (NRT) with a 7-day update interval. The main processing stages are depicted in Figure 2.

Note that tasks shown on the left side are only run once ("offline"), whereas the remaining processes are repeated every week. To ensure a temporally consistent NDVI time series, the weekly processing steps were initiated with the start of the time series.

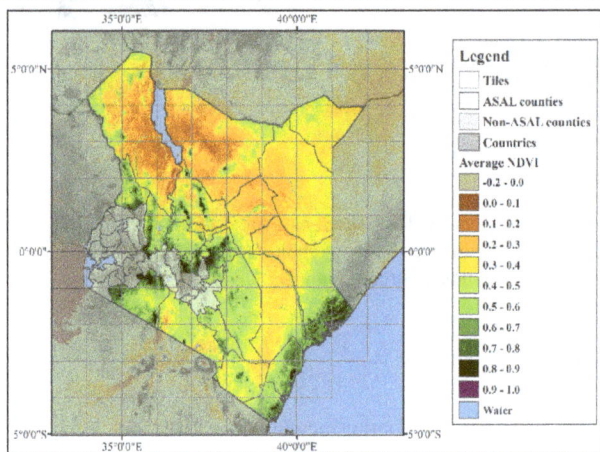

Figure 1: Average annual NDVI for ASAL counties of Kenya. Non-ASAL counties are shown in grey. The map also shows the 1° x 1° tiling system in which MODIS data is processed

3.1.1   MODIS Data: The NDMA drought indicators are derived from MOD13Q1 and MYD13Q1 NDVI collection 5 products of the MODIS Terra and Aqua satellites from LP DAAC (from 2000 onwards). These products are gridded level-3 data in approximately 250m spatial resolution in Sinusoidal projection with a (combined) temporal resolution of 8 days. The level-3 data are calculated from the level-2G daily surface reflectance gridded data (MOD09 and MYD09 series) using the constrained view angle – maximum value composite (CV-MVC) compositing method (Solano et al., 2010).

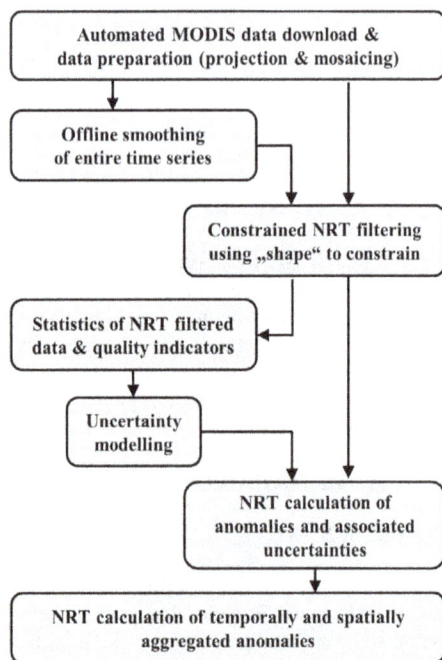

Figure 2: Processing chain of BOKU's near real-time (NRT) filtering of MODIS NDVI time series. The steps on the left side are done only once. The processes shown on the right side are repeated every week. Together the processing leads to filtered NDVI images with associated uncertainties. Based on this primary information anomaly indicators are derived and aggregated over time and for different administrative units

3.1.2   Data Acquisition and Preparation: The MODIS data are downloaded, mosaicked and re-projected to geographic coordinates (datum WGS84) with a spatial resolution of approximately 0.002232° (ca. 250 m) using nearest neighbour resampling. The images are cropped to a dedicated tile system (see Figure 1). These steps are performed on a daily basis using the R MODIS package (Mattiuzzi et al., 2012).

3.1.3   NRT Filtering: To minimize the possible impact of undetected clouds and poor atmospheric conditions, a standardized procedure temporally filters the NDVI time series based on two distinct steps: offline smoothing (only once) and near real-time filtering (every week).

The offline smoothing step uses the Whittaker smoother (Eilers, 2003), (Atzberger and Eilers, 2011a) and (Atzberger and Eilers, 2011b). It smoothes and interpolates the data in the historical archive (2000 to 2012) to daily NDVI values. The smoothing takes into account the quality of the data and the compositing day for each pixel and time step based on the MODIS VI quality assessment science dataset (Solano et al., 2010). For a detailed description of the filtering procedure and settings, see (Atzberger et al., 2014). Only every 7th image corresponding to "Mondays" is stored. The 7-day interval reduces the storage load of the archive but permits at the same time an easy restoration of daily data whenever needed. From the smoothed data, weekly statistics are calculated describing the typical NDVI paths for a given location and time. This information serves for "constraining" the Whittaker smoother during the NRT filtering.

The near real-time (NRT) filtering step also uses the Whittaker smoother. However, the filtering is executed every weekend and only uses available observations of the past 175 days. Filtered NDVI images of the successive Monday are stored but also for the past four Mondays, representing different consolidation phases of the filtered NDVI (see Figure 3 "output 0" to "output 4"). Obviously, "output 4" is more reliable (e.g. better constrained through available data) compared to the "output 0" which is always extrapolated as (reliable) MODIS observations become available only after some days.

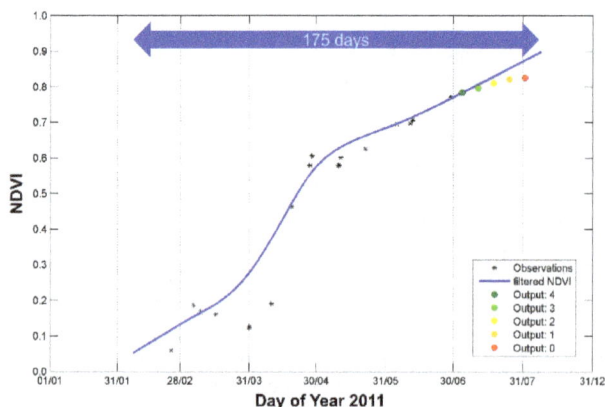

Figure 3: Principle of BOKU's constrained near real-time (NRT) filtering. Black asterisks are the observed (raw) MODIS values from 16-day MVC (both Terra and Aqua). The blue line is the fitted curve of the (unconstrained) Whittaker smoother. The five coloured dots are the final (constrained) "Monday" images, representing different consolidation phases. These "Monday" images are stored every weekend. To constrain these outputs, previously calculated statistics are used (e.g. from "offline-smoothing", not shown in the graph)

Note that the missing constraints may lead to arbitrary high or low values, particularly, at times of the year, where rapid NDVI changes take place. Thus, we apply a pixel specific constraining procedure that limits the NDVI change between consecutive "Mondays" according to weekly statistics of the offline-smoothed data. In Figure 3, the effect of the constraining can be seen as the difference between the blue line (the unconstrained Whittaker) and the coloured dots (e.g. the final output that is stored and used for drought mapping).

3.1.4 Calculation of Statistics of NRT Data and Uncertainty Modelling:
Saving every weekend the five output NDVI images plus quality information (e.g. number and quality of observations within the 175 days temporal window), allows us to keep a consistent archive of the different consolidation phases. Before starting the operational production of NDMA data, the archived NRT data are compared to a filtered "reference" time series where all observations were available (e.g. central point of the blue curve in Figure 3). The difference between "reference" time series and NRT estimates gives the error of the NRT filtering. We model this error using the stored quality information. In our operational setting, the uncertainty of a pixel filtered in NRT is estimated based on those previously established models. This is done for each output product, pixel and time step in NRT.

3.1.5 Anomalies:
From the filtered NDVI data, a weekly vegetation condition index (VCI) is calculated at pixel level (Kogan et al., 2003):

$$VCI_i = \frac{NDVI_i - NDVI_{min,w}}{NDVI_{max,w} - NDVI_{min,w}} \times 100 \qquad (1)$$

where VCI_i = vegetation condition index at time step i
   $NDVI_i$ = normalized difference vegetation index observed at time step i
   $NDVI_{min,w}$, $NDVI_{max,w}$ = lowest / highest 7-day values observed from 2003 to 2012 at week w

Conceptionally, the VCI enhances the inter-annual variations of a vegetation index (e.g. NDVI) in response to weather fluctuations while reducing the impact of ecosystem specific response (e.g. driven by climate, soils, vegetation type and topography). Other anomaly indicators are also calculated (e.g. Z-score) but will not be presented here as NDMA restricts its analysis to VCI.

To get a more concise picture of the vegetation development in the ongoing season and to identify drought-affected areas, we temporally and spatially aggregate the weekly VCI maps. Temporal aggregation includes 1-monthly and 3-monthly weighted VCI averages using the VCI images of the recent 4 and 12 weeks of the according month, respectively. During the temporal aggregation, the modelled uncertainty is taken into account down-weighting the impact of less reliable observations. Spatial aggregation averages the VCI at pixel level according to administrative units (e.g. counties, constituencies of Kenya) and/or livelihood zones. All data are imported at NDMA into SPIRITS (Eerens et al., 2014) for production of seasonal graphs, etc. Additional web-tools were developed by BOKU for educational purposes (BOKU, 2015).

3.2 Comparison with FEWS NET Data

We used for comparison the eMODIS NDVI data provided by the FEWS NET (USGS, 2013). The downloaded FEWS NET data are pentadal NDVI at 0.002413° spatial resolution (Datum WGS84) covering the area of East Africa.

The eMODIS dataset is generated by the U.S. Geological Survey (USGS) Earth Resources Observation and Science (EROS) Center from the Level 1B MODIS products of Terra (MOD09, MOD03, MOD35_L2) (USGS, 2011). The output includes near real-time and historical NDVI products that are composited in 10-day intervals every 5 days at about 250m spatial resolution. This results in 72 composite periods per year (pentades).

The historical NDVI dataset (2001-2010) is temporally smoothed by USGS with a "modified" weighted least squares linear regression approach (Swets et al., 1999). As current-year composites become available, they are added to the time series and smoothed, resulting in a smoothed composite for a given 10-day period (updated every 5 days). The eMODIS data available for download are updated during six composite periods, only after which the images become definitive (USGS, 2013). Hence, the most recent five images are produced using climatological information. For our study, only the consolidated FEWS NET data were used covering the period of 2001 to 2014.

To compare BOKU and FEWS NET datasets, temporally and spatially aggregated VCI anomalies are calculated from the FEWS NET NDVI data. First, NDVI images were cropped and resampled to the BOKU grid. Pentadal statistics (minimum, maximum) of the NDVI were derived for each pixel and the period of 2003 to 2012 similar to the BOKU dataset (see section 3.1.4). Next, pentadal VCI images were calculated using the derived statistics (Equation 1). Temporally aggregated VCI images (1 and 3 month) were obtained by averaging 6 and 18 pentades, respectively. Finally the spatial aggregation was conducted in the same way as for the BOKU data. It has to be noted that the FEWS NET indicators derived in this way, are not available in near real-time, but only after six pentades (e.g. one month). This contrasts with the NDMA data, which are derived in NRT.

4. RESULTS AND DISCUSSION

In this section, we focus on 3-monthly VCI data (VCI3M) aggregated at county level and provided on a monthly basis. The VCI3M anomalies are compared with FEWS NET anomalies. Both datasets are evaluated against food security assessment reports.

The linear regression between 3-monthly VCI datasets from BOKU and FEWS NET shows generally a good agreement between both datasets with a coefficient of determination (R^2) of 0.89 (see Figure 4). The VCI observations regularly scatter around the 1-to-1 line with a slope close to one and a (slight) positive intercept. As expected, the majority of the points (highest density, dark red points) are found in the range of 30% and 55% corresponding to near "normal" conditions. The majority of observations (88%) fall well within a range of ± 10%.

Despite the generally good agreement between the two VCI datasets, larger differences appear if the analysis is repeated month-by-month. The resulting intra-annual coefficient of determination (R^2) varies between 0.77 and 0.94 (see Figure 5 – green line). Local minima of R^2 (and maxima of RMSE) are visible in April and November. This coincides very well with Kenya's long and short rains that normally occur in March–June and October–December.

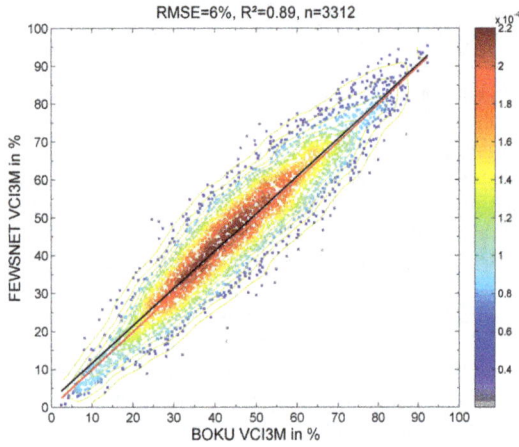

Figure 4: Scatterplot of 3-monthly aggregated VCI (VCI3M) derived from FEWS NET and BOKU datasets of ASAL counties across all months between 2003 and 2014. 1-to-1 line (red) and regression line (black)

The two seasons are captured by the average monthly NDVI profile derived from the weekly NDVI values between 2003 and 2012 for all ASAL pixels (orange line in Figure 5). Obviously, the largest differences in the VCI3M anomalies occur in parallel to significant NDVI changes - these are periods of prime interest for image interpreters and NDMA.

The inter-annual agreement/disagreement between the two datasets is depicted in Figure 6. Results show some variability from year to year. Interestingly, the RMSE slightly drops down to local minima for the years of 2005, 2009 and 2011. According to assessment reports of the Kenya Food Security Steering Group (KFSSG, 2005), (KFSSG, 2006), (KFSSG, 2011a) and (KFSSG, 2011b), the years 2005 and 2011 coincide well with major droughts. In 2009, a poor performance of the long rains was reported (KFSSG, 2009b). Consequently, our current assumption is (to be validated) that the two datasets show a better agreement in years with (extreme) droughts.

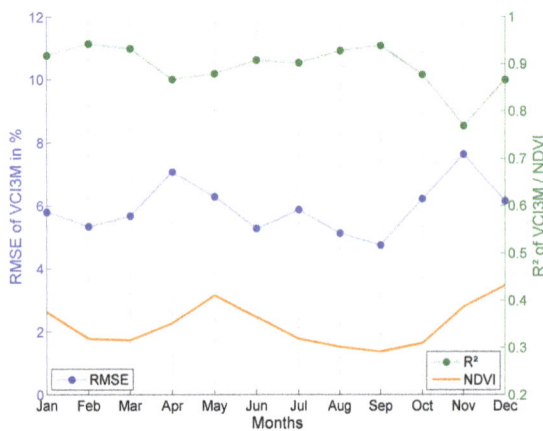

Figure 5: Intra-annual agreement/disagreement between monthly updated FEWS NET and BOKU VCI3M values. (green) coefficient of determination (R^2), (blue) root mean square error (RMSE) between the two data sets. In (orange) time course of the average NDVI of ASAL counties. Lines are only shown for reader's convenience

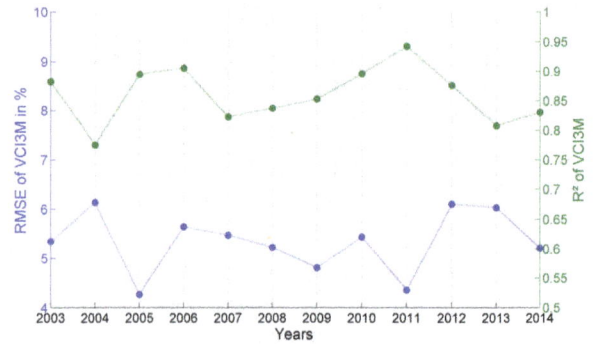

Figure 6: Inter-annual agreement/disagreement between monthly FEWS NET and BOKU VCI3M values. (green) coefficient of determination (R^2), (blue) root mean square error (RMSE). Lines are only shown for reader's convenience

The spatial variation of RMSE for the ASAL counties is shown in Figure 7. The RMSE values were calculated across all months and years of the two datasets. The resulting RMSE ranges between 4% and 9% and show some spatial coherence. Large arid counties (e.g. Turkana, Marsabit, Wajir) as well as southwestern semi-arid counties (e.g. Narok, Kajiado) show a relatively good agreement between the FEWS NET and BOKU anomalies (see Figure 7, green and dark-green).

Largest variations occur in the centre (e.g. Kitui) as well as in Mandera. Although the counties seem to build spatial groups, no obvious relation to aridity can be seen.

To further reveal county-specific differences, we prepared a detailed analysis for the counties of Laikipia, Mandera and Kitui. Laikipia represents a semi-arid county with a very good agreement (low RMSE), whereas Mandera belongs to the arid counties exhibiting medium RMSE values. Kitui is again semi-arid, but shows larger RMSE values of slightly more than 7%. All counties experienced major droughts during the period of 2003 and 2014. We employ matrix plots to display the monthly

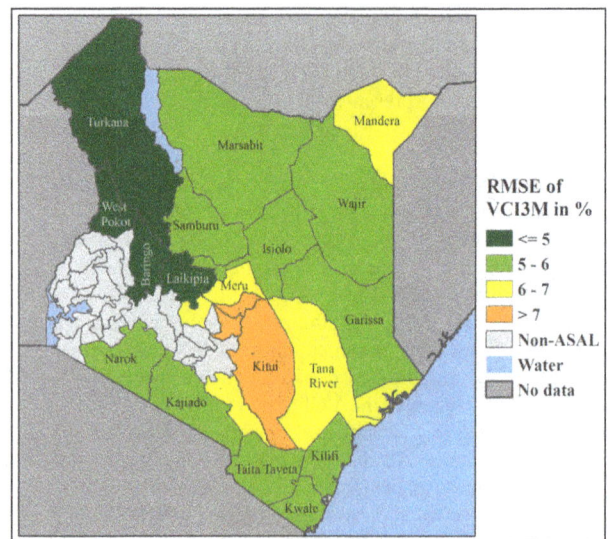

Figure 7: Variations of root mean square error (RMSE) between monthly FEWS NET and BOKU VCI3M values for ASAL counties. Counties outside the ASAL area are shown in grey. The observed minimum RMSE was 4.3 and the maximum 9.7

Colour	VCI3M in %	Drought category
	≥ 50	Wet
	35 to 50	No Drought
	21 to 34	Moderate Drought
	10 to 20	Severe Drought
	< 10	Extreme Drought

Table 1: Thresholds for monthly updated VCI3M and related drought categories

VCI3M anomalies for all years per county and dataset. Analysts at NDMA operationally use these matrix plots for their monthly bulletins distinguishing five drought categories (Table 1).

The results for **Laikipia** are displayed in Figure 8 (top) for FEWS NET and in Figure 8 (bottom) for BOKU anomalies. The low RMSE of Laikipia is confirmed by the very similar seasonal pattern throughout all years between both datasets. Major droughts can be observed for the years of 2006, 2009 and 2011 classified in both datasets as severe and extreme droughts. In 2005 and 2010 short rains failed in the region (KFSSG, 2006) and (KFSSG, 2011a), which led to the detected low VCI in February and March.

A particular situation is captured by the VCI3M anomaly in 2009. In 2009 (KFSSG, 2009b) reported poor long rains but still classified the situation in Laikipia as not exceptionally bad. BOKU data, as well as FEWS NET, reveal on the contrary an extreme drought in 2009, which is for example also confirmed by (Zwaagstra et al., 2010) employing NOAA AVHRR data.

For Laikipia, the only visible difference between BOKU and FEWS NET data relates to the evolution of the drought in 2009; the BOKU data show a gradual degradation of the situation (from dark green to red), whereas FEWS NET saw "wet" conditions in January 2009 (dark green) immediately followed by "moderate" drought in February (yellow). Despite this, the overall agreement between the two datasets is very good for Laikipia.

The drought detected for **Mandera** is displayed in Figure 9. One can observe slight differences between the drought categories of both datasets, but the differences never exceed more than one drought category. The overall pattern is still quite similar.

Mandera experienced unfavourable long rains in 2005 as well as late, poorly distributed and early ending short rains in 2005 (KFSSG, 2005) and (KFSSG, 2006). As a result, Mandera was affected by a major drought at the end of 2005 and the beginning of 2006 as depicted in Figure 9.

A complete season failure of the short rains 2010 and the long rains 2011 (partially less than 10% of normal rains) was reported for the central and northern part of Kenya including Mandera (KFSSG, 2011a) and (KFSSG, 2011b). Again, this is well reflected by both datasets.

The matrix plots of **Kitui** are displayed in Figure 10. We added the differences between the FEWS NET and BOKU datasets at the bottom of Figure 10.

For 2005/2006 a moderate to severe drought is captured in both datsets (see Figure 10 top and centre) for the same reasons as in Mandera (KFSSG, 2005) and (KFSSG, 2006).

The 2008 short rains in the southeast including Kitui were exceptionally poor, delayed by 20-40 days and lasted less than three weeks (KFSSG, 2009a). Parts of Kitui received on average only 10-20% of normal long rains in 2009 (KFSSG, 2009b). The situation is depicted in 2009 by both datasets. However, the BOKU anomalies show a clear offset reaching a VCI difference of more than 10% in October 2009 (see Figure 10 bottom), which might be explained by the NRT filtering.

The drought of 2011 was again caused by unfavourable rains both in the short and long season of 2010/2011 but to a lesser extent than for e.g. Mandera (KFSSG, 2011a) and (KFSSG, 2011b). In particular, Kitui hardly experienced rainfall onsets

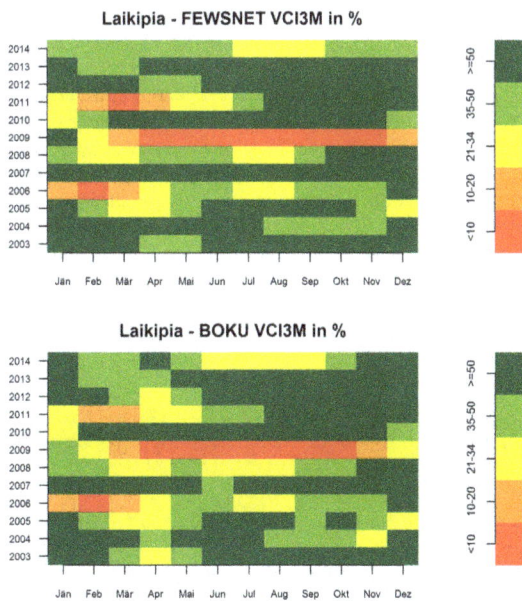

Figure 8: Seasonal matrix plot of categorised monthly VCI3M anomalies for the county of Laikipia (2003-2014) derived from (top) FEWS NET dataset and (bottom) BOKU dataset

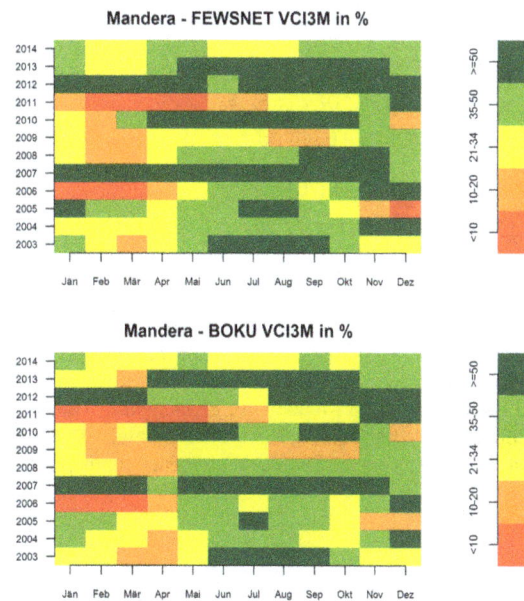

Figure 9: Seasonal matrix plot of categorised monthly VCI3M anomalies for the county of Mandera (2003-2014) derived from (top) FEWS NET dataset and (bottom) BOKU dataset

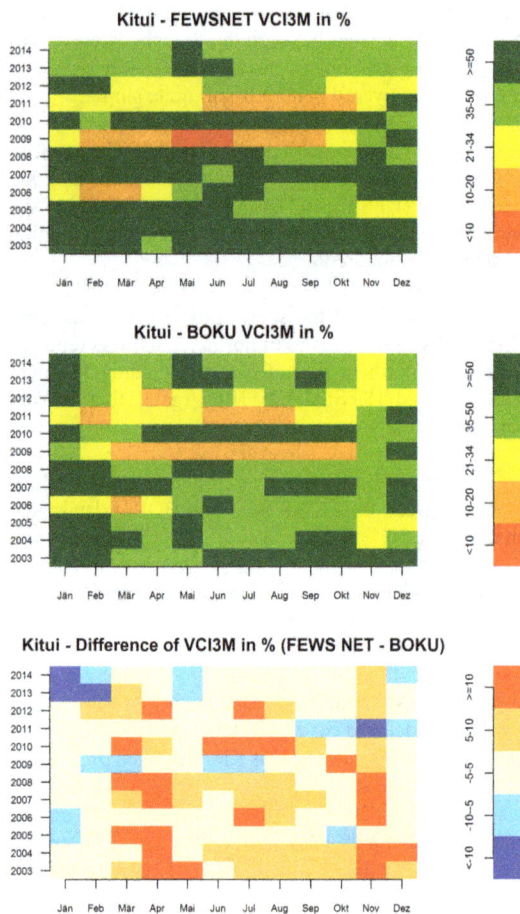

Figure 10: Seasonal matrix plot of categorised monthly VCI3M anomalies for the county of Kitui (2003-2014) derived from (top) FEWS NET dataset, (centre) BOKU dataset and (bottom) difference between both datasets (FEWS NET minus BOKU)

but had very short episodes of 10-20 day rainfall. The rains ceased unusually early in late April. Consequently, both datasets show severe droughts starting from June 2011. BOKU anomalies recover a little early than the one derived from FEWS NET as indicated by the negative VCI differences in September to December 2011 in Figure 10 (bottom).

5. CONCLUSIONS

FEWS NET provides relevant and well established data for drought monitoring based on satellite observations. With our research, we aimed to studying to which extent we can re-produce the drought indicators provided by FEWS NET in near real-time that is without waiting for the end of the consolidation period. The BOKU data analysed in this paper are provided within 2-3 days after the last Monday in a given month. The data are operationally used by NDMA for their monthly drought bulletins and for triggering the disaster contingency funds (DCF) of Kenya.

In summary, our results clearly show an overall good correspondence between the two chosen datasets. An RMSE in the order of 6 was found for the more closely investigated 3-monthly VCI products. Some larger differences were observed at the onset of vegetation growth that is before the short and the long rains in Kenya. Generally, the driest years were modelled best. Interestingly too, the spatial pattern of the differences between FEWS NET and BOKU-derived VCI was non-random.

Together, these findings indicate some potential systematic differences between the two datasets, which deserve more research. In our future work, we will also focus to quantify how much the FEWS NET data quality degrades if delivered in near real-time.

ACKNOWLEDGEMENTS

The MODIS MOD13Q1, MYD13Q1 and MCD12Q1 data processed in BOKU's NRT processing chain were obtained through the online Data Pool at the NASA Land Processes Distributed Active Archive Center (LP DAAC), USGS/Earth Resources Observation and Science (EROS) Center, Sioux Falls, South Dakota (https://lpdaac.usgs.gov/get_data). The eMODIS NDVI data were obtained through the Famine Early Warning Systems Network (FEWS NET) data portal provided by the USGS FEWS NET Project, part of the Early Warning and Environmental Monitoring Program at the USGS Earth Resources Observation and Science (EROS) Center. This is acknowledged.

We further acknowledge the support of Matteo Mattiuzzi for R code development and Francesco Vuolo and Martin Siklar (all from BOKU) for setting up the web-tool for data display (http://ivfl-geomap.boku.ac.at/html/demo_WG/kenya/).

Some of the computations of BOKU have been done using the Vienna Scientific Cluster (VSC). This is acknowledged.

REFERENCES

Atzberger, C., Eilers, P. H. C., 2011a. Evaluating the effectiveness of smoothing algorithms in the absence of ground reference measurements. International Journal of Remote Sensing 32, pp. 3689–3709.

Atzberger, C., Eilers, P. H. C., 2011b. A time series for monitoring vegetation activity and phenology at 10-daily time steps covering large parts of South America. International Journal of Digital Earth 4, pp. 365–386.

Atzberger, C., Klisch, A., Mattiuzzi, M., Vuolo, F., 2014. Phenological metrics derived over the European continent from NDVI3g data and MODIS time series. Remote Sensing 6, pp. 257–284.

BOKU, 2015. Web-tools for displaying drought indicators in Kenya, BOKU //ivfl-geomap.boku.ac.at/html/demo_WG/kenya/ (24 Mar 2015).

Eerens, H., Haesen, D., Rembold, F., Urbano, F., Tote, C., Bydekerke, L., 2014. Image time series processing for agriculture monitoring. Environmental Modelling & Software 53, pp. 154-162.

Eilers, P. H. C., 2003. A perfect smoother. Analytical Chemistry 75, pp. 3631–3636.

KFSSG, 2005. Kenya long rains assessment report 2005. Tech. Rep., Kenya Food Security Steering Group (KFSSG) http://documents.wfp.org/stellent/groups/public/documents/ena/wfp083955.pdf (24 Mar 2015).

KFSSG, 2006. Kenya short rains assessment report 2005. Tech. Rep., Kenya Food Security Steering Group (KFSSG) http://www.disasterriskreduction.net/fileadmin/user_upload/drought/docs/Kenya_2006_SRA_Report.pdf (24 Mar 2015).

KFSSG, 2009a. The 2008/'09 short-rains season assessment report. Tech. Rep., Kenya Food and Security Steering and Group (KFSSG) http://reliefweb.int/report/kenya/200809-short-rains-season-assessment-report-kenya-food-security-steering-group-kfssg (24 March 2015).

KFSSG, 2009b. Kenya long rains assessment report 2009. Tech. Rep., Kenya Food Security Steering Group (KFSSG) http://documents.wfp.org/stellent/groups/public/documents/ena/wfp208056.pdf (24 Mar 2015).

KFSSG, 2011a. Kenya short rains assessment report 2010. Tech. Rep., Kenya Food Security Steering Group (KFSSG) http://home.wfp.org/stellent/groups/public/documents/ena/wfp241326.pdf (24 Mar 2015).

KFSSG, 2011b. Kenya long rains assessment report 2011. Tech. Rep., Kenya Food Security Steering Group (KFSSG) http://documents.wfp.org/stellent/groups/public/documents/ena/wfp240180.pdf (24 Mar 2015).

Kogan, F., Gitelson, A., Zakarin, E., Spivak, L., Lebed, L., 2003. AVHRR-based spectral vegetation index for quantitative assessment of vegetation state and productivity: calibration and validation. Photogrammetric Engineering & Remote Sensing 69, pp. 809–906.

Mattiuzzi, M., Verbesselt, J., Hengl, T., Klisch, A., Evans, B., Lobo, A., 2012. Modis: Modis download and processing package. processing functionalities for (multi-temporal) MODIS grid data. First International Workshop on "Temporal Analysis of Satellite Images" Mykonos Island, Greece, May 23-25, 2012 (Poster).

Solano, R., Didan, K., Jacobson, A., Huete, A., 2010. MODIS vegetation index user's guide (MOD13 Series), version 2.00, May 2010 (collection 5). Vegetation Index and Phenology Lab, The University of Arizona. http://vip.arizona.edu/documents/-MODIS/MODIS_VI_UsersGuide_01_2012.pdf (24 Mar 2015).

Swets, D. L., Reed, B. C., Rowland, J. D., Marko, S. E., 1999. A weighted least-squares approach to temporal NDVI smoothing. In: Proceedings of the 1999 ASPRS Annual Conference from Image to Information Portland Oregon.

USGS, 2011. eMODIS Africa product guide version 1.0. USGS EROS Data Center: Sioux Falls, SD, USA.

USGS, 2013. eMODIS TERRA Normalized Difference Vegetation Index (NDVI). http://earlywarning.usgs.gov/fews/-africa/web/readme.php?symbol=zd (24 Mar 2015).

Zwaagstra, L., Sharif, Z., Wambile, A, de Leeuw, J., Said, M.Y., Johnson, N., Njuki, J., Ericksen, P., Herrero, M., 2010. An assessment of the response to the 2008-2009 drought in Kenya. A report to the European Union Delegation to the Republic of Kenya. Tech. Rep., ILRI (International Livestock Research Institute), Nairobi, Kenya.

25

LAND COVER CHANGE ANALYSIS IN MEXICO USING 30M LANDSAT AND 250M MODIS DATA

R. R. Colditz [a, *], R. M. Llamas [a], R. A. Ressl [a]

[a] National Commission for the Knowledge and Use of Biodiversity (CONABIO), Av. Liga Periférico-Insurgentes Sur 4903, Parques del Pedregal, Tlalpan, CP 14010, Mexico City, DF, Mexico - (rene.colditz, ricardo.llamas, rainer.ressl)@conabio.gob.mx

36th International Symposium on Remote Sensing of Environment (ISRSE)

KEY WORDS: Land cover time series, Change detection, Spatial-temporal analysis, MODIS, Mexico

ABSTRACT:

Change detection is one of the most important and widely requested applications of terrestrial remote sensing. Despite a wealth of techniques and successful studies, there is still a need for research in remote sensing science. This paper addresses two important issues: the temporal and spatial scales of change maps. Temporal scales relate to the time interval between observations for successful change detection. We compare annual change detection maps accumulated over five years against direct change detection over that period. Spatial scales relate to the spatial resolution of remote sensing products. We compare fractions from 30m Landsat change maps to 250m grid cells that match MODIS change products. Results suggest that change detection at annual scales better detect abrupt changes, in particular those that do not persist over a longer period. The analysis across spatial scales strongly recommends the use of an appropriate analysis technique, such as change fractions from fine spatial resolution data for comparison with coarse spatial resolution maps. Plotting those results in bi-dimensional error space and analyzing various criteria, the "lowest cost", according to a user defined (here hyperbolic) cost function, was found most useful. In general, we found a poor match between Landsat and MODIS-based change maps which, besides obvious differences in the capabilities to detect change, is likely related to change detection errors in both data sets.

1. INTRODUCTION

Change detection of land cover and land use is one of the foremost applications of remote sensing data. Even though well studied over the past five decades, there is still on-going research in many fields such as method development (see summaries in Lu et al. (2004) and Coppin et al (2004)), combining spatial scales and multiple data sets (Colditz et al. 2012a), application-specific developments, e.g. for urban planning (Tapiador and Casanova 2003), or robust regional to continental change detection with automated methods (Pouliot et al. 2014). For successful change studies one needs to consider several factors, such as available resources, image availability, accessibility to ground observations and ancillary data, availability and experience with change detection algorithms, area of expertise, intended use of the product, etc. (Kennedy et al. 2009).

This study addresses two important issues: temporal and spatial scales of change detection. The temporal scale of change is important and one should select the appropriate data sets carefully. For instance, abrupt change may only persist for a short period of time while subtle change processes may not be detectable at short temporal intervals. For temporal scales, the study analyzes annual change products over five years which were accumulated and compared to direct change detection between the initial and final year. Differences in spatial scales are studied using data of different spatial resolution; in this study comparing change maps obtained from 30m Landsat data to 250m MODIS products. Landsat change maps were generated for minimum mapping units (MMUs) of 1ha, 5ha,

and 10ha. For adequate map comparison across spatial resolutions an algorithm calculated the fraction of change from fine spatial resolution data for each coarse cell, and several criteria for defining the appropriate change fraction were tested.

2. DATA AND STUDY AREA

The Canada Centre for Remote Sensing (CCRS) of Natural Resources Canada (NRCan) processed MODIS calibrated radiances for the entire North American continent and provided monthly image composites to an international research network. Specifically, the following processing steps were carried out: projection to Lambert Azimuthal Equal Area (LAEA), downscaling to obtain images of 250m spatial resolution for all 7 reflective bands, and compositing to monthly data (for overview see Latifovic et al. 2012). In this study we employed monthly composites of MODIS data from 2005 to 2010.

Table 1 lists all Landsat images of path 046 – row 020 that were analyzed in this study. All corresponding Landsat 7ETM+ images were composited to reduce data gaps in the primary image by filling with valid pixels from a secondary image. In 2008, no cloud-free data were available for the period March to May. Therefore a set, marked as 2008A, was used for change detection with the composite of 2007 and 2008B with data from 2009; all other composites were employed for both pairs. All images were preprocessed using LEDAPS (Masek et al. 2006) for obtaining surface reflectance and FMASK (Zhu et al. 2012) for detecting, clouds, shadow and invalid data due to the failure of the scan-line corrector since May 2003.

---

* Corresponding author

Year	Sensor	Primary	Secondary
2005	L7 ETM$^+$	April 10th	April 26th
2006	L7 ETM$^+$	April 29th	April 13th
2007	L7 ETM$^+$	March 15th	April 16th
2008A	L7 ETM$^+$	January 29th	January 13th
2008B	L7 ETM$^+$	August 24th	November 28th
2009	L7 ETM$^+$	April 5th	May 7th
2010	L5 TM	January 26th	NA

Table 1. Sensor and dates of Landsat images per year.

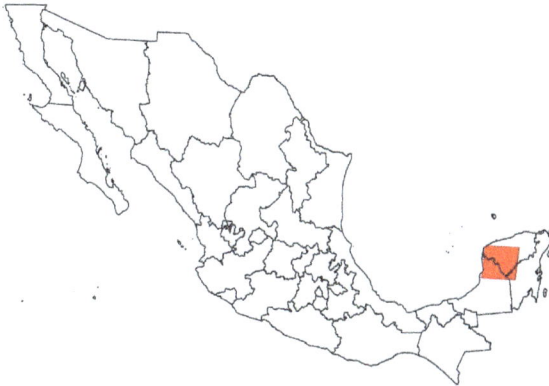

Figure 1. Study areas: the national terrestrial territory of Mexico with its states and location of Landsat path-row 046-020 on the Yucatan peninsula, marked in red.

In a first test, the national terrestrial territory of Mexico (1.972 Mio km^2) will be analyzed using MODIS data. Next, the area of path-row 046-020 will be studied in detail (Figure 1) using Landsat and MODIS-based change detection products. This region, located in the north-western portion of the Yucatan peninsula and south of the city of Merida, depicts a transition from deciduous to evergreen broadleaf tropical forests towards the East. Slash-and-burn agriculture, such as frequently applied to *Milpas*, is characteristic to fertilize poor karstic soils for a short period of 3-5 years, planting maize, squash, and beans.

3. METHODS

3.1 Change detection

3.1.1 MODIS data: A base-line land cover classification with 19 classes for North America (15 for Mexico) using a standardized, LCCS-compatible legend was generated for the year 2005 (Latifovic et al. 2012). For Mexico, supervised decision tree (C5.0, Quinlan 1993) ensemble classifiers were trained with a 121,000 sample points from field-based analysis and on-screen digitalization (Colditz et al, 2012b). A first version of this map was recently improved in selected areas mainly for classes "urban and built-up" and "water" (Colditz et al. 2014a).

An algorithm was developed to obtain potential areas of land cover change using normalized bi-annual difference images for all month and bands as well as additional data such as the NDVI and texture filters for improved edge detection. An algorithm was trained with Landsat images for a large area in northern Mexico for the period 2005 to 2010. Optimization resulted in the following thresholds: 1 and 99 percentile of each difference image and a frequency of 25% from all features (Colditz et al. 2014b).

Next, a map updating strategy was used to assign new land cover classes to pixels flagged as potential change (Colditz et al. 2014a). Therefore there are two change products: (1) potential change by biannual differences and (2) actual change for areas with a change in the class label. In this study we employ maps without minimum mapping unit, thus the smallest change object is 6.25ha, the area of a 250m MODIS pixels.

3.1.2 Landsat data: Change from Landsat images was detected by visual interpretation using two analysts: one detected changes between 2005 and 2010 directly, the other consecutive annual changes between all years. The analysts digitized polygons at high spatial detail and also identified areas which they could not map, either due to clouds or too large data gaps between scan lines. To ensure consistency, a third analyst verified and, if necessary, revised polygons. For standardized products, MMUs of 1ha, 5ha, and 10ha were applied.

3.2 Change analysis

From each sensor two data change sets were generated: (1) direct land cover change between 2005 and 2010 without analyzing the years in between and (2) annual consecutive land cover change between 2005 and 2010 (2005-2006, 2006-2007, 2007-2008, 2008-2009, and 2009-2010). The latter was aggregated to accumulated annual change between 2005 and 2010 and compared to direct land cover change for the same period. For MODIS there are potential change maps, for Landsat change products were obtained at 1ha, 5ha, and 10ha between 2005 and 2010.

3.2.1 Change area and polygons: The area of each change map, effectively the number of pixels detected as change, is calculated and expressed in area as square kilometre and percent. In addition, the number of patches of change, i.e. a pixel or group of connected pixels surrounded by pixels of no change, was calculated using the eight-neighbour rule and reported as simple count and number of patches per square kilometre. The relative numbers are useful to compare change results across different temporal and spatial scales.

3.2.2 Change comparison at the same resolution: A simple matrix (Figure 2) was used to compare two maps of no change (0) and change (1) of the same resolution of which the common change area (N_{11}) was analyzed. Specifically the error (E), expressed in percent, of change in map A against B and *vice versa* was calculated following equations (1) and (2).

		\multicolumn{3}{c}{Map A}
		0	1	Σ
Map B	0	N_{00}	N_{10}	N_{+0}
	1	N_{01}	N_{11}	N_{+1}
	Σ	N_{0+}	N_{1+}	N

Figure 2: Matrix of correspondence between change (1) in map A and B.

$$E_A = 1 - \frac{N_{11}}{N_{1+}} \qquad (1)$$

$$E_B = 1 - \frac{N_{11}}{N_{+1}} \qquad (2)$$

	Potential change				Actual change			
	km²	Area %	Patches	Patches/km²	km²	Area %	Patches	Patches/km²
2005 - 2006	6187.94	0.3156	7647	0.0039	1500.31	0.0765	4519	0.0023
2006 - 2007	5621.94	0.2867	8094	0.0041	1643.06	0.0838	5542	0.0028
2007 - 2008	5103.06	0.2602	7433	0.0038	1571.31	0.0801	5034	0.0026
2008 - 2009	6265.50	0.3195	8268	0.0042	2196.25	0.1120	6228	0.0032
2009 - 2010	6535.25	0.3333	8300	0.0042	1996.25	0.1018	5866	0.0030
2005 - 2010 accumulative	16861.56	0.8598	22237	0.0113	6572.88	0.3352	14166	0.0072
2005 - 2010 direct	7023.31	0.3582	12258	0.0063	3304.94	0.1685	9206	0.0047

Table 2. Area and patches for potential and actual change of MODIS data at the national scale.

3.2.3 Change comparison at different resolutions: Analyzing change of maps with different spatial resolution is more complicated. We employed an algorithm which calculates the area proportions (or fractions) of change from a fine spatial resolution map in a coarse spatial resolution grid (Colditz et al. 2012a). Next, we calculate the matrix of Figure 2 and errors of equations (1) and (2) for all potential fractions (0% to 100% at 1% intervals) and present this result in bi-dimensional space (E_A, E_B). There are several ways to determine the best change fraction, e.g. the 50% fraction, i.e. at least 50% of the coarse cell were mapped as change in the fine spatial resolution pixel, the fraction at which E_A and E_B are equal (if they intersect), or the lowest cost using a cost function such as the hyperbolic in equation 3. In addition, the Pareto boundary for the lowest achievable error bound (Boschetti et al. 2004) was calculated for each fine spatial resolution set.

$$Cost(x) = 1 - [(1 - E_A) \cdot (1 - E_B)] \quad (3)$$

4. RESULTS AND ANALYSIS

4.1 MODIS based change at the national scale

The MODIS-based change detection was developed for the national area of Mexico and therefore shall be analyzed first for this extent. Table 2 shows the area in square kilometres and percent and patches as absolute number and per square kilometre. The area of potential change is, on average, three times larger than actual change of class labels. Even though there are more patches of potential change than actual change, the average ratio of 1.46 is lower than for area.

Notable is also the sum of annual changes between 2005 and 2010, which is much larger than the accumulated change area, e.g. 29,713km² compared to 16,861km² for potential change. This indicates that a significant area was mapped several times as change. Figure 3A shows the percent of area that was detected one to five times as change in annual change maps. For potential change, 42% of the area was at least twice detected as change and 3.3% of the change pixels were detected in all bi-annual change maps. It is clear that actual change was less likely detected several times and this number could be further reduced be applying rules of change persistency similar to Pouliot et al. (2009).

The third notable result is that direct change between 2005 and 2010 is clearly lower than accumulative annual change. In fact, many potential annual changes are only slightly smaller than 2005-2010 direct change. For actual change the difference is higher but direct change is still only half of the area of

accumulative annual change. This indicates that longer time spans between dates may not detect several changes that occur at shorter intervals.

The change maps between 2005 and 2010 obtained by direct comparison and accumulative annual changes were compared and summarized in a matrix similar to Figure 2. Table 3 shows the respective errors for accumulative and direct change. It is clear that the error is higher for accumulative change as a 2.5 times larger area was detected, thus at best the error cannot be lower than 60%. Still, only half of the that area (21.9% of accumulative change) was also found with direct change. Errors for direct change are lower but still approximately half of the total change area. This indicates that both maps are clearly different and mark changes at distinct temporal scales.

4.2 MODIS-based change for Yucatan

Table 4 reports change areas and patches from MODIS for the study site on the Yucatan peninsula. The total area (column Yucatan) corresponds to the valid area of Landsat data (>50% of valid area proportion) and relative numbers such as area in percent and patches by area are relative to this number.

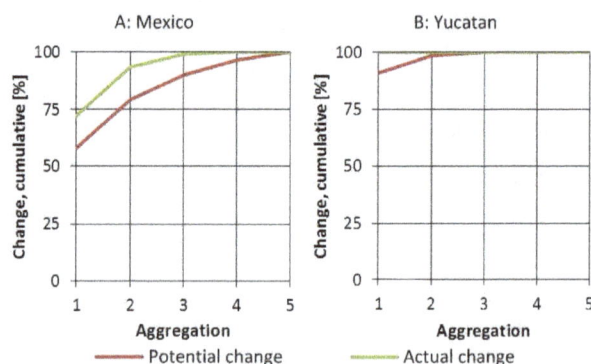

Figure 3. Proportion of change pixels detected one to five times in bi-annual change maps for Mexico (A) and Yucatan (B).

Error	Mexico		Yucatan	
	Potential	Actual	Potential	Actual
Accumulative	78.15	77.88	65.07	44.69
Direct	47.53	56.01	63.67	65.98

Table 3. Error between direct change and accumulative annual change maps between 2005 and 2010 for Mexico and Yucatan.

Change	Yucatan	Potential change				Actual change			
	km^2	km^2	Area %	Patches	Patches/km^2	km^2	Area %	Patches	Patches/km^2
2005 – 2006	14364	1.44	0.0100	48	0.0033	0.94	0.0065	25	0.0017
2006 – 2007	13807	10.75	0.0779	72	0.0052	7.06	0.0511	47	0.0034
2007 – 2008	27250	6.50	0.0239	65	0.0024	2.13	0.0078	32	0.0012
2008 – 2009	26697	32.69	0.1224	363	0.0136	9.19	0.0344	100	0.0037
2009 – 2010	32515	6.13	0.0188	65	0.0020	3.06	0.0094	40	0.0012
2005 – 2010 accumulative	33712	52.06	0.1544	255	0.0076	22.38	0.0664	106	0.0031
2005 – 2010 direct	32466	45.69	0.1407	182	0.0056	33.25	0.1024	119	0.0037

Table 4. Area and patches for potential and actual change of MODIS data for Yucatan.

Change	Yucatan	1ha				5ha				10ha			
	km^2	km^2	Area %	Patches	Patches /km^2	km^2	Area %	Patches	Patches /km^2	km^2	Area %	Patches	Patches /km^2
2005 – 2006	13845	32.48	0.23	667	0.0482	21.73	0.15	182	0.0131	17.08	0.12	106	0.0077
2006 – 2007	13300	26.93	0.20	446	0.0335	21.58	0.16	170	0.0128	17.34	0.13	98	0.0074
2007 – 2008	26418	26.01	0.09	415	0.0157	21.19	0.08	201	0.0076	16.46	0.06	118	0.0045
2008 – 2009	25782	26.21	0.10	675	0.0262	17.53	0.06	201	0.0078	11.43	0.04	84	0.0033
2009 – 2010	32056	15.39	0.04	316	0.0099	10.80	0.03	91	0.0028	7.16	0.02	35	0.0011
2005 – 2010 accumulative	33337	123.98	0.37	2264	0.0679	90.45	0.27	760	0.0228	67.52	0.20	404	0.0121
2005 – 2010 direct	31843	505.91	1.58	7709	0.2421	393.05	1.23	2889	0.0907	285.51	0.89	1150	0.0361

Table 5. Area and patches for changes of Landsat with 1ha, 5ha and 10ha minimum mapping unit for Yucatan.

In comparison to area percentages at the national scale, MODIS detected clearly less changes in the Yucatan site. However, the number of patches per area did not reduce notably, which indicates that the area of change patches is much smaller in Yucatan than at the national scale.

The percentage of annual changes detected several times is lower. In fact, Figure 3B depicts only some notable double and almost no triple detection for potential changes while all actual changes occurred only once.

The correspondence between direct and accumulative change maps between 2005 and 2010 in Table 3 show an reversal in the magnitude of errors with a the higher area of accumulative *versus* direct change for the potential change maps and *vice versa* for actual change. Overall, however, the errors are still quite high at the local level for Yucatan.

4.3 Landsat-based change

Annual Landsat-based change areas for the Yucatan site (Table 5) depict area proportions similar to the national scale of MODIS but clearly higher than those of the local level. An average decrease of 25% in change area is indicated between 1 and 5ha MMU and 45% between 1 and 10ha. Across all minimum mapping units, however, there is the four times higher area for direct changes between 2005 and 2010 as compared to accumulative annual changes for the same period, a contradictory result on comparison to the findings obtained from MODIS.

The numbers of patches reduce, on average, by 65% between 1 and 5ha and 82% between 1 and 10ha. This marked decrease

indicates a high number of small polygons. In comparison to MODIS at the national scale the number of change patches per square kilometre is approximately a magnitude higher for 1ha, roughly the same for 5ha which is similar to the 6.25ha area of a 250m MODIS pixel, and a magnitude lower for 10ha. This indicates the importance of spatial resolution for detecting change.

The comparison of Landsat change maps of accumulative annual changes and direct changes between 2005 and 2010 (Table 6) shows the expected pattern: a high error for direct changes due to the much larger area detected as changes. The numbers vary only slightly among minimum mapping units. There was no noteworthy multiple detection of Landsat pixels in annual change maps.

4.4 Comparison of Landsat and MODIS-based change

Comparison between 30m Landsat and 250m MODIS products requires a technique which relates the fine spatial resolution data to coarse spatial resolution cells. In this study we calculate the fraction of change from Landsat products with different MMUs for each 250m cell and compare this result to potential change from MODIS.

Error	1ha	5ha	10ha
Accumulative	43.97	41.90	40.88
Direct	86.86	87.25	86.68

Table 6. Error between direct change and accumulative annual change maps between 2005 and 2010 for 1ha, 5ha and 10ha minimum mapping units.

A: Pareto boundaries

B: MODIS-Landsat comparison

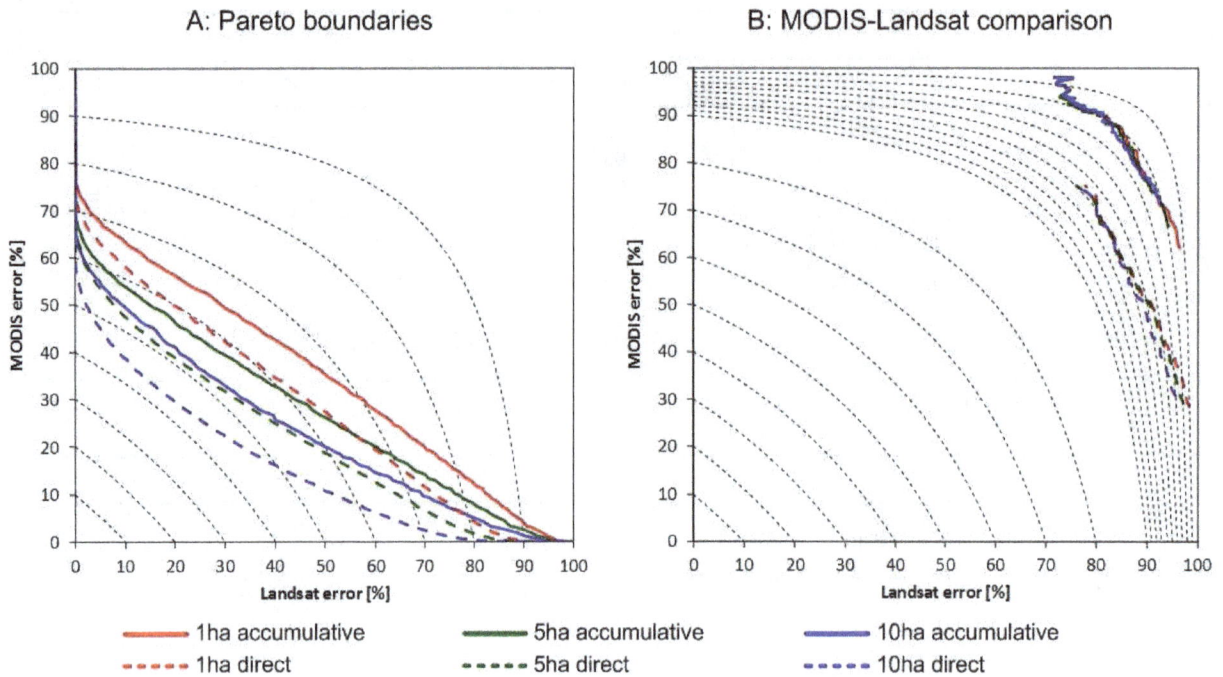

- ———— 1ha accumulative   ———— 5ha accumulative   ———— 10ha accumulative
- - - - - 1ha direct   - - - - 5ha direct   - - - - 10ha direct

Figure 4. MODIS-Landsat error plots for Pareto boundaries (A) as best achievable result and Landsat-MODIS data comparison (B) using direct and aggregated annual change between 2005 and 2010 from MODIS potential changes and Landsat change with 1ha, 5ha and 10ha MMU.

There are multiple criteria to define the most appropriate fraction and corresponding errors which we will explore in the following. The simplest is to only consider coarse cells with a fraction of at least 50% of change from Landsat. Another option is to select the intersection between both errors for a set of fractions. Third, one may choose the minimum cost according to a user-defined (here hyperbolic) cost function for the bi-dimensional error space. In the following we will first assess the results for the Pareto boundary and then for the actual MODIS-Landsat change data.

4.4.1   Error for Pareto boundary: The Pareto boundary forms a line of optimal classifications for a set of different fractions. The asymmetric curves in Figure 4A indicate a skewed error distribution and Table 7 shows best result for above-mentioned criteria. Assuming "change fraction >= 50%" the error for Landsat is almost twice as high as for MODIS. Employing the criterion of an equal error between Landsat and MODIS resulted in lower costs than the 50% threshold. All selected change fractions were smaller than 50% which indicates that cells with less than 50% change proportion in Landsat are deemed sufficient for defining change in MODIS. The lowest cost criterion decreases the cost slightly as compared to the intersection between both errors, which in all cases also resulted in an even lower change fraction. This indicates that using the 50% fraction is clearly inappropriate in this analysis, i.e. change fractions of approximately 30% should be employed. Also, the error for Landsat dropped further and is often just half as much as for MODIS. This result is meaningful as change detection with Landsat should be more accurate than with MODIS.

The direct change detection between 2005 and 2010 shows generally lower costs and comparatively lower errors than accumulative annual change detection between those years. It becomes also clear that Landsat products with larger MMUs,

which reduce the noise in change detection maps, resulted in lower costs and thus smaller errors.

4.4.2   MODIS-Landsat change data: In comparison to the Pareto boundary, change comparison between MODIS and Landsat shows high errors (Figure 4B, Table 8). For accumulative changes, assuming the 50% threshold, errors for MODIS and Landsat were almost equal, thus these results are very similar to criterion of error intersection. For direct comparison between 2005 and 2010 Landsat, errors were very high and MODIS moderate. However, the errors curves of both data sets never intersected with increasing change fractions and thus the equal error measure cannot be used for assessment. The lowest cost resulted in fractions between 30 and 40% for accumulative change and above 90% for direct change. Notable is the higher error for Landsat than MODIS, which could indicate issues in the Landsat data set.

	Accumulative			Direct		
	1ha	5ha	10ha	1ha	5ha	10ha
Change fraction >= 50%						
Cost	70.78	62.40	55.89	64.46	55.19	46.06
Landsat error	59.25	48.31	40.81	52.23	40.41	31.22
MODIS error	28.29	27.26	25.48	25.61	24.81	21.58
Equal error between Landsat and MODIS						
Fraction	35	39	40	37	42	44
Cost	65.98	58.57	53.50	60.55	52.57	44.58
Error	41.36	35.56	31.66	36.50	30.40	25.93
Lowest cost						
Fraction	26	27	33	29	29	33
Cost	64.62	56.93	52.45	59.40	51.05	43.62
Landsat error	29.60	21.21	24.86	27.89	18.55	16.66
MODIS error	49.74	45.34	36.72	43.70	39.91	32.36

Table 7. Cost, fraction and error statistics (all in percent) for the Pareto boundary.

	Accumulative			Direct		
	1ha	5ha	10ha	1ha	5ha	10ha
Change fraction = 50%						
Cost	97.76	97.72	97.63	95.86	95.69	94.76
Landsat error	87.15	86.42	84.83	92.38	92.06	90.23
MODIS Error	82.59	83.19	86.39	45.69	45.69	46.37
Equal error between Landsat and MODIS						
Fraction	60	58	52	NA	NA	NA
Cost	97.88	97.88	97.64	NA	NA	NA
Error	85.59	85.59	84.75	NA	NA	NA
Lowest cost						
Fraction	42	35	32	93	93	93
Cost	97.57	97.46	97.43	93.67	93.67	93.65
Landsat error	88.35	88.70	87.20	80.71	80.71	80.66
MODIS error	79.11	77.35	79.95	67.17	67.17	67.17

Table 8. Cost, fraction and error statistics (all in percent) for MODIS *versus* Landsat change data.

Similar to Pareto boundary assessment, the errors were lower for direct change detection than accumulative change and slightly diminished with larger MMUs, however, this effect was less notable.

5. DISCUSSION AND CONCLUSIONS

A major presumption for all successful change detection studies using spatial data sets is near-to-perfect spatial co-registration (Lu et al. 2004, Boschetti et al. 2004, Colditz et al. 2014b). In this study we only performed visual comparisons among different data sets. We could not find noteworthy spatial displacements, which confirms the generally good spatial registration of MODIS and Landsat data (Wolfe et al. 2002, Masek et al. 2006) and thus good co-registration between both sensors. An approach for quantitative analysis of spatial co-registration between data sets of different spatial resolution was shown in Colditz et al. (2012a) and applied to change detection in Colditz et al. (2014b).

5.1 Annual *versus* five year intervals

There are notable differences in temporal scales that differed by sensor products. For MODIS, accumulative annual change showed larger change area than direct change between 2005 and 2010. Given that the data sets (MODIS monthly composites) and change detection method (bi-annual normalized difference images) were the same, the reason is related to the time intervals. This pattern was expected as many changes occur abruptly and may persist only for a short interval. For instance, plots for the above-mentioned slash-and-burn agriculture only exist a few years before the land is abandoned and secondary, mostly shrubby vegetation regrows before tropical forest takes over in several successional stages. Some changes may not be observable even at annual scales, e.g. burnt areas in pastures and low shrubby vegetation (Ressl et al. 2009, Colditz et al. 2014).

Even though annual accumulative changes detect larger total area than direct changes over several years, the locality of changes differs. The main reason is that change detection at annual intervals can hardly reveal subtle change processes at a slower pace over several years. Despite the availability of trend-based and time series analysis techniques for detecting change processes (Kennedy et al. 2010, Latifovic and Pouliot 2014,

Verbesselt et al. 2010), bi-annual difference images over longer time scales may be an alternative.

Landsat change results show a contrary pattern with more changes detected for direct comparison between 2005 and 2010 than accumulative annual changes. In this case the reason is likely related to differences in visual change detection. Even though we intended to harmonize visual digitalization among different analysts, the issue could not be fully resolved. The analyst that detected changes between 2005 and 2010 directly worked approximately three weeks on this data set while the other, responsible for annual change detection between 2005 and 2010, spent less than two weeks for all five data sets (two days per bi-annual data set). Post-processing and applying minimum mapping units could not fully resolve the differences in the level of detail between both data sets.

5.2 Spatial resolution

The number of patches depends on the spatial resolution and minimum mapping unit. The reduction of small patches was illustrated for Landsat which highly affected the number of patches but only moderately the area. Notable is also the similarity between the patches per square kilometre for Landsat with 5ha MMU and MODIS with a cell size of 6.25ha at the national level.

Relating spatial scales is still an emerging topic with only a few studies. For the dichotomous case as in this study (change / no change) we adapted algorithms from file monitoring (Boschetti et al. 2004, Ressl et al. 2009, Csiszar et al. 2006, Morisette et al. 2005). The Pareto boundary, adapted to spatial data in the field of remote sensing by Boschetti et al. (2004), indicates the optimal line for a set of change fractions. The area below this curve cannot be reached due to differences in the spatial resolution between both data sets. The lower errors for Landsat data with larger MMU are therefore meaningful as the difference in resolution diminishes with increasing MMUs of finer spatial resolution data. Nevertheless, in absolute terms the Pareto boundary is still high, in comparison to Boschetti et al. (2004) or Colditz et al. (2014b).

The analysis of different criteria for defining an appropriate change fraction analysed three approaches. The rationale for choosing the 50% threshold is simply that of the majority rule, i.e. the class that makes up the largest proportion of area will be assigned. Although valid in some cases this assumption does not hold up to reality due to an uneven probability distribution function. For instance, a pixel in coarser resolution data may be flagged even though the area proportion that corresponds to this class is much smaller, a case frequently found in fire mapping (Ressl et al. 2009). Choosing the equal value between both errors may be desirable in some cases but many studies prefer to minimize one of the two errors (Colditz et al. 2014b). The intersection may even not exist as we have shown for direct comparisons. The lowest cost could be a viable alternative; however the actual fraction and errors highly depend on the selected cost function. This study and others (Boschetti et al. 2004, Colditz et al. 2014b) have deemed useful the hyperbolic cost function as defined in equation 3.

In general there is a poor relation between Landsat and MODIS change detection products, also noted by the high difference between Landsat-MODIS data comparison and the Pareto boundary, which is related to two major facts. First, MODIS change was extracted from a product designed for the national scale which in some areas works well but in this site relatively poor, mainly due to changes that are too small to detect at 250m spatial resolution, that cannot be discerned spectrally, that occur on different time scales, as well as remaining data issues due to frequent cloud cover during rainy season from May to September (Colditz et al. 2014b). The second reason is the approach of visual analysis of Landsat with two different analysts that worked at different levels of detail. This introduced inconsistencies to our data set which we could not fully resolve. It would have been desirable to obtain automatically detected changes with Landsat or even higher spatial resolution data, but opportunities are limited due to inconsistent image acquisition. For instance, Landsat 5TM was turned off over the study area from 2002 to 2009 and Landsat 7ETM+ images suffer from scan-line off data gaps in addition to frequent could cover and shadows. In order to obtain larger areas of valid data two images were composited. The selected image dates, which differ in 2008 and 2010 from the normal pattern of choosing images at the end of the dry season, illustrate the difficulties of finding appropriate data.

ACKNOWLEDGEMENTS

We kindly thank our analysts Cesar Feliciano Rodriguez and Armando Gandarilla Ramirez for digitizing change from Landsat image composite pairs.

REFERENCES

Boschetti, L., Flasse, S. P., & Brivio, P. A., 2004. Analysis of the conflict between omission and commission in low spatial resolution dichotomic thematic products: The Pareto Boundary. *Remote Sensing of Environment*, 91(3-4), pp. 280–292.

Colditz, R. R., Acosta-Velázquez, J., Díaz Gallegos, J. R., Vázquez Lule, A. D., Rodríguez Zúñiga, M. T., Maeda, P., Ressl, R., 2012a. Potential effects in multi-resolution post-classification change detection. *International Journal of Remote Sensing*, 33(20), pp. 6426–6445.

Colditz, R. R., López Saldaña, G., Maeda, P., Argumedo Espinoza, J., Meneses Tovar, C., Victoria Hernández, A., Ressl, R., 2012b. Generation and analysis of the 2005 land cover map for Mexico using 250m MODIS data. *Remote Sensing of Environment*, 123, pp. 541–552.

Colditz, R. R., Pouliot, D., Llamas, R. M., Homer, C., Latifovic, R., Ressl, R. A., Richardson, K., 2014a. Detection of North American Land Cover Change between 2005 and 2010 with 250m MODIS Data. *Photogrammetric Engineering & Remote Sensing*, 80(10), pp. 918–924.

Colditz, R. R., Llamas, R. M., & Ressl, R., 2014b. Detecting Change Areas in Mexico Between 2005 and 2010 Using 250 m MODIS Images. *IEEE Journal of Selected Topics in Applied Earth Observations and Remote Sensing*, 7(8), pp. 3358–3372.

Coppin, P., Jonckheere, I., Nackaerts, K., Muys, B., & Lambin, E., 2004. Review ArticleDigital change detection methods in ecosystem monitoring: a review. *International Journal of Remote Sensing*, 25(9), pp. 1565–1596.

Csiszar, I. A., Morisette, J. T., & Giglio, L., 2006. Validation of Active Fire Detection From Moderate-Resolution Satellite Sensors: The MODIS Example in Northern Eurasia. *IEEE Transactions on Geoscience and Remote Sensing*, 44(7), pp. 1757–1764.

Kennedy, R. E., Townsend, P. A., Gross, J. E., Cohen, W. B., Bolstad, P., Wang, Y. Q., & Adams, P., 2009. Remote sensing change detection tools for natural resource managers: Understanding concepts and tradeoffs in the design of landscape monitoring projects. *Remote Sensing of Environment*, 113(7), pp. 1382–1396.

Kennedy, R. E., Yang, Z., & Cohen, W. B., 2010. Detecting trends in forest disturbance and recovery using yearly Landsat time series: 1. LandTrendr - Temporal segmentation algorithms. *Remote Sensing of Environment*, 114(12), pp. 2897–2910.

Latifovic, R., Homer, C., Ressl, R., Pouliot, D., Hossain, S. N., Colditz, R. R., Victoria, A., 2012. North American Land-Change Monitoring System. In: *Remote Sensing of Land Use and Land Cover: Principles and Applications*, Boca Raton, FL., pp. 303–324.

Latifovic, R., & Pouliot, D., 2014. Monitoring Cumulative Long-Term Vegetation Changes Over the Athabasca Oil Sands Region. *IEEE Journal of Selected Topics in Applied Earth Observations and Remote Sensing*, 7(8), pp. 1–13.

Lu, D., Mausel, P., Brondízio, E., & Moran, E., 2004. Change detection techniques. *International Journal of Remote Sensing*, 25(12), pp. 2365–2401.

Masek, J. G., Vermote, E. F., Saleous, N. Z., Wolfe, R. E., Hall, F., Huemmrich, Lim, T. K., 2006. A Landsat surface reflectance dataset for North America, 1990-2000. *Geoscience and Remote Sensing Letters, IEEE*, 3(1), pp. 68 – 72.

Morisette, J. T., Giglio, L., Csiszar, I., & Justice, C. O., 2005. Validation of the MODIS active fire product over Southern Africa with ASTER data. *International Journal of Remote Sensing*, 26(19), pp. 4239–4264.

Pouliot, D., Latifovic, R., Fernandes, R., & Olthof, I., 2009. Evaluation of annual forest disturbance monitoring using a static decision tree approach and 250 m MODIS data. *Remote Sensing of Environment*, 113, pp. 1749–1759.

Pouliot, D., Latifovic, R., Zabcic, N., Guindon, L., & Olthof, I., 2014. Development and assessment of a 250m spatial resolution MODIS annual land cover time series (2000–2011) for the forest region of Canada derived from change-based updating. *Remote Sensing of Environment*, 140, pp. 731–743.

Quinlan, J. R., 1994. C4 . 5□: Programs for Machine Learning. *Machine Learning*, 16, pp. 235–240.

Ressl, R., Lopez, G., Cruz, I., Colditz, R. R., Schmidt, M., Ressl, S., & Jiménez, R., 2009. Operational active fire mapping and burnt area identification applicable to Mexican Nature Protection Areas using MODIS and NOAA-AVHRR direct readout data. *Remote Sensing of Environment*, 113(6), pp. 1113–1126.

Tapiador, F. J., & Casanova, J. L., 2003. Land use mapping methodology using remote sensing for the regional planning directives in Segovia, Spain. *Landscape and Urban Planning*, 62(2), pp. 103–115.

Verbesselt, J., Hyndman, R., Newnham, G., & Culvenor, D., 2010. Detecting trend and seasonal changes in satellite image time series. *Remote Sensing of Environment*, 114(1), pp. 106–115.

Wolfe, R. E., Nishihama, M., Fleig, A. J., Kuyper, J. a., Roy, D. P., Storey, J. C., & Patt, F. S., 2002. Achieving sub-pixel geolocation accuracy in support of MODIS land science. *Remote Sensing of Environment*, 83(1-2), pp. 31–49.

Zhu, Z., & Woodcock, C. E., 2012. Object-based cloud and cloud shadow detection in Landsat imagery. *Remote Sensing of Environment*, 118, pp. 83 – 94.

26

PERFORMANCE OF THE ENHANCED VEGETATION INDEX TO DETECT INNER-ANNUAL DRY SEASON AND DROUGHT IMPACTS ON AMAZON FOREST CANOPIES

Benjamin Brede,* Jan Verbesselt, Loïc P. Dutrieux, Martin Herold

Laboratory of Geo-Information Science and Remote Sensing, Wageningen University,
Droevendaalsesteeg 3, 6708PB, Wageningen, The Netherlands –
(Benjamin.Brede, Jan.Verbesselt, Loic.Dutrieux, Martin.Herold)@wur.nl

KEY WORDS: Amazon rainforests, Enhanced Vegetation Index, dry season leaf flush, sun-sensor geometry effects, drought impact

ABSTRACT:

The Amazon rainforests represent the largest connected forested area in the tropics and play an integral role in the global carbon cycle. In the last years the discussion about their phenology and response to drought has intensified. A recent study argued that seasonality in greenness expressed as Enhanced Vegetation Index (EVI) is an artifact of variations in sun-sensor geometry throughout the year. We aimed to reproduce these results with the Moderate-Resolution Imaging Spectroradiometer (MODIS) MCD43 product suite, which allows modeling the Bidirectional Reflectance Distribution Function (BRDF) and keeping sun-sensor geometry constant. The derived BRDF-adjusted EVI was spatially aggregated over large areas of central Amazon forests. The resulting time series of EVI spanning the 2000-2013 period contained distinct seasonal patterns with peak values at the onset of the dry season, but also followed the same pattern of sun geometry expressed as Solar Zenith Angle (SZA). Additionally, we assessed EVI's sensitivity to precipitation anomalies. For that we compared BRDF-adjusted EVI dry season anomalies to two drought indices (Maximum Cumulative Water Deficit, Standardized Precipitation Index). This analysis covered the whole of Amazonia and data from the years 2000 to 2013. The results showed no meaningful connection between EVI anomalies and drought. This is in contrast to other studies that investigate the drought impact on EVI and forest photosynthetic capacity. The results from both sub-analyses question the predictive power of EVI for large scale assessments of forest ecosystem functioning in Amazonia. Based on the presented results, we recommend a careful evaluation of the EVI for applications in tropical forests, including rigorous validation supported by ground plots.

1. INTRODUCTION

The Amazon rainforests represent the largest connected forested area in the tropics. They play an integral role in the global carbon cycle and store ~93±23 Pg C in above-ground living biomass alone (Malhi et al., 2006). In recent years the discussion about their resilience against transition to other stable ecosystem states has become more intense. Climate change and deforestation put pressure on the forests (Malhi et al., 2008) and the question appeared if they may partly, but abruptly transform into savannahs (White et al., 1999; Cox et al., 2004; Hirota et al., 2010).

The discussion of drought impacts has been vivid in the past years partially due to the severe droughts in 2005 and 2010 and its implications. In a fast response survey Phillips et al. (2009) assessed the impact of the 2005 drought on tree biomass in long-term forest census plots and found major biomass losses, which ended a long-term trend in tree carbon assimilation. Phillips et al. (2010) refined this analysis. They found evidence for elevated mortality rates 2 years after the meteorological event. Furthermore, large trees had a disproportionately high mortality during the drought. This is supported by de Toledo et al. (2013).

On the other hand, Schwalm et al. (2010) found a positive response of Net Primary Productivity (NPP) to relative drought. Actually, some studies found higher rates of carbon uptake during the dry compared to the wet season (Saleska et al., 2003; Hutyra et al., 2007). Instead of precipitation regimes, carbon uptake was controlled by phenology and light conditions (Hutyra et al., 2007). This does not contradict the aforementioned studies. The idea would be that water shortage as it occurs regularly in dry seasons does not necessarily lead to water shortage for the trees as some

have access to soil water via deep roots (Nepstad et al., 1994). However, prolonged dryness depletes the soil water storage and eventually shows effects in tree health and carbon uptake. This means that regular dry seasons may first lead to increased carbon uptake, but the effect turns when the dry season turns into drought.

Besides ground plots, satellite remote sensing delivers information about spatial and temporal patterns of drought impacts. The dry season resilience was observed with optical remote sensing data and especially Moderate-Resolution Imaging Spectroradiometer (MODIS) Enhanced Vegetation Index (EVI). An increase in Enhanced Vegetation Index (EVI) during the dry months was explained with flushing of new leafs (Myneni et al., 2007; Samanta et al., 2012). This would be a combined effect of an increase in Leaf Area Index (LAI) and a change in leaf optical properties. However, other studies raise concern about sun and viewing geometry effects in EVI (Sims et al., 2011; Galvão et al., 2011; Moura et al., 2012). Galvão et al. (2011) studied a seasonal evergreen forest in the tropical forest savannah transition zone and found the highest EVI at the end of the dry season, when Solar Zenith Angle (SZA) was smallest. Morton et al. (2014) corrected MODIS EVI for sun and viewing geometry effects with the help of a kernel based Bidirectional Reflectance Distribution Function (BRDF) model for MODIS tiles h11v09 and h12v09, which are situated over the central parts of the Amazon rainforests. Analysis of this data showed no remaining seasonality.

The objective of this study was to use a data set independent from Morton et al. (2014), but with the same study area, and check the consistency of the results. Additionally, the intra-annual effect of dry season strength expressed with drought indices was investigated for the whole of the Amazon rainforests.

---

* Corresponding author

2. METHOD

2.1 Study Area

We identified tropical forests in the Amazon basin with the MODIS MCD12Q1 Land Cover Type product V051 with 500 m spatial resolution. In order to focus on mostly undisturbed forests, we set the requirement for each pixel that at least two thirds of all land cover observations in the temporal domain have to be ever-green broadleaf forest according to the International Geosphere-Biosphere Programme (IGBP) classification scheme. Additionally, only pixels within the hydrological Amazon river basin (Mayorga et al., 2012) were considered. The resulting landcover mask can be seen in Figure 1.

2.2 Data

A major aim of this study was to correct Bidirectional Reflectance Factors (BRFs) for sun sensor geometry effects. The MCD43 product suite provides unique opportunities for this purpose. It contains results of fitting pixel- and band-wise reflectance observations of 16 days to a semi-empirical BRDF model, the RossThick-LiSparse-Reciprocal (RTLSR) model (Schaaf et al., 2002, 2011). This kernel based model describes the surface BRDF as the sum of isotropic, volumetric and geometric scattering. The volumetric kernel assumes a dense canopy of randomly oriented facets, while the geometric kernel is based on non-overlapping shadows cast by randomly distributed objects (Roujean et al., 1992; Strahler et al., 1999; Lucht et al., 2000). With the three parameters for the isotropic f_{iso}, volumetric f_{vol} and geometric f_{geo} kernels BRFs at any given SZA and View Zenith Angle (VZA) can be modeled. The statistical fitting of the model to observations requires several measurements. Therefore, both Aqua and Terra MODIS instruments deliver input to the inversion process (Strahler et al., 1999; Lucht et al., 2000). A back-up algorithm offers model inversions when only few observations are available. This algorithm makes use of the BRDF properties of previous full inversions and scales these shapes to new measurements. This back-up algorithm is termed magnitude inversion.

For this study we downloaded MCD43A1 (inversion parameters), MCD43A2 (quality and auxiliary data) and MCD43A4 (BRFs modeled at nadir and local solar noon, Nadir BRDF-Adjusted Reflectance (NBAR) product) data from the MODIS data pool (https://lpdaac.usgs.gov/) from between February 18, 2000 and November 25, 2013 and spatial tiles with horizontal numbers 10 to 12 and vertical numbers 8 to 9 (9 tiles in total), which cover

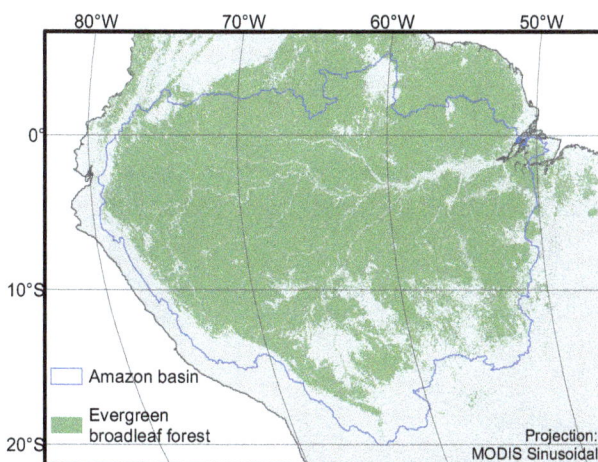

Figure 1: Study Area (land boundaries based on data set of Bjorn Sandvik www.thematicmapping.org).

northern South America. The temporal resolution is quasi 8 daily, as the production is temporally overlapping.

For spatial analysis of precipitation we used the Tropical Rainfall Measuring Mission Multisatellite Precipitation Analysis (TMPA) 3B42 V7 daily accumulated precipitation product at 0.25° spatial resolution (Huffman et al., 2010). We downloaded data with the same time span as for the MODIS BRDF products.

2.3 EVI Seasonality

Based on the MCD43 products we produced 7 time series for the whole of the study period. These time series represent spatial averages over MODIS tiles h11v09 and h12v09. Morton et al. (2014) also focused on these two tiles. All products were filtered with the landcover information and with the MCD43A2 quality flags to allow only full BRDF inversions. The time series produced were:

- NBAR Near Infrared Reflectance (NIR) reflectance, Normalized Difference Vegetation Index (NDVI) and EVI;

- BRDF-adjusted NIR reflectance, NDVI and EVI modeled at VZA = 0° and SZA = 30°;

- SZA at local solar noon.

The seasonal behavior of these time series were analyzed with a special focus on green-up phenomenon during the dry season as found by other studies (Huete et al., 2006; Xiao et al., 2006; Myneni et al., 2007; Samanta et al., 2012).

2.4 Impact of Drought on EVI

For the comparison of MODIS derived products with drought metrics based on TMPA the MODIS products were first spatially aggregated to the TMPA resolution of 0.25° (~30 km at equator). For the landcover mask the requirement for a pixel to be valid in the aggregated map was that at least 90% of the underlying MODIS pixels were evergreen broadleaf forest. The MODIS reflectance products were quality filtered (only full inversions) and averaged. On the other hand, TMPA daily precipitation was temporally averaged to the quasi 8-day periods of the MCD43 products. In this way both datasets were synchronized to 0.25° spatial and 8-day temporal resolution for the whole Amazon basin.

Based on the aggregated MODIS products we derived BRDF-adjusted EVI standardized anomalies over the temporal domain, which means standardized anomalies per pixel time series. This is a useful approach as the anomaly describes the deviation of EVI in relation to its mean for each pixel. Furthermore, anomalies were derived only for data from Septembers. This is the last month in the Amazon dry season, commonly defined as July to September (Samanta et al., 2012). September should show highest impact of precipitation related impact on greenness. The TMPA was further processed to drought indices, namely Standardized Precipitation Index (SPI) (McKee et al., 1993) and Maximum Cumulative Water Deficit (MCWD) (Aragão et al., 2007). The SPI behaves similar as seasonal anomalies with values oscillating around 0 and deviations meaning more or less precipitation than usual. The MCWD assumes an evapotranspirational demand of 100 mm month^{-1} of tropical forests and holds the available precipitation against this. Subsequent months with less precipitation than 100 mm month^{-1} let the MCWD build up, which typically happens in the dry season.

The connection between greenness anomalies and drought was analyzed with pixel-based linear regression models, where September greenness anomaly was the dependent and the respective drought index the independent variable. Per 0.25° pixel this produced slope estimates for the statistical relationship and how likely greenness was dependent on the drought index, i. e. its significance. These values were analyzed as maps. Additionally, the greenness anomaly-drought index pairs were analyzed on a MODIS tile basis. This means all greenness anomaly values were regressed against all drought index values per MODIS tile.

3. RESULTS

3.1 EVI Seasonality

Figure 2 shows spatially averaged BRFs and vegetation indices time series with 8 day temporal resolution, which emulate the results of Morton et al. (2014). From the start of the time line until circa 2002 occasional gaps appeared, when no data was available for the whole study area. Additional observations by Aqua MODIS after 2002 explain the fewer gaps. When considering the data density over time for the h11v09 and h12v09 tiles only very few observations during the wet season were available.

Another dependence was the one between SZA and NBAR NIR reflectance. As seen in Figure 2 NBAR NIR reflectance followed the SZA without a temporal lag. Even the two peaks around the turn of the year occurred in both time series. The SZA pattern marks the sun's apparent movement during the course of the year: The equinoxes occur on September 23 and March 20, and result in the lowest SZAs. A low Kendall's τ between SZA and NBAR NIR reflectance of -0.725 underlined this strong and reversed relationship. The SZA–NIR reflectance coupling was especially strong in the NBAR product, because it uses the SZA at local solar noon as an input for SZA in the RTLSR model.

On the other hand, this coupling was much weaker for the BRDF-adjusted NIR reflectance and EVI. The double peak structure of SZA did not appear in all years for the BRDF-adjusted NIR reflectance. Additionally, the magnitude of change during austral summer was much lower compared to NBAR processed products. A relatively low Kendall's τ of -0.325 between BRDF-adjusted NIR reflectance and SZA underpinned this weaker coupling. An explanation for this is that the BRDF-adjusted NIR reflectance used a fixed SZA, so that a direct impact of the SZA on NIR reflectance should not be possible. Nonetheless, correlation between

BRDF-adjusted EVI and SZA showed mostly no lags and some years showed the double peak pattern observed in the NBAR EVI.

3.2 Impact of Drought on EVI

The map representations of pixel-based linear regression models of BRDF-adjusted EVI seasonal anomalies against MCWD showed distinct spatial patterns (Figure 3). Slope parameters close to 0 dominated eastern and southern Amazonia, indicating low predictive power of MCWD for EVI anomalies. North-western regions showed a mixture of both strong positive and negative slopes. This would indicate that EVI anomalies vary with MCWD, which indicates drought impact on canopy greenness. However, in most of these areas the regression slopes were not significant. After examining scatterplots for single pixels from these areas (graphs not shown), it was clear that these steep slopes were mostly produced by single MCWD outliers with magnitudes of up to 50 mm. This is still much lower than MCWD values at the southern and eastern rainforest borders. Additionally, pairs with low MCWD values were mostly missing in these cases, possibly due to high cloud cover, which lead to missing EVI points. This combination led to randomly high or low slope values. The same was true for a strip along the foothills of the Andes at the western edge of the study area, where orographic cloud formation throughout the year hampers good quality observations.

The scatterplots for the tile-based approach showed the main characteristic of MCWD (Figure 4): Values clustered at a certain lower threshold. This is 0 for regions that did not experience notable dry seasons in terms of absolute water deficit (h11v09, h12v09). In tile h12v10 values only started at about 100 mm pointing to distinct dry seasons in this area. Independent from that, all point clouds were approximately equally distributed around the 0 EVI anomaly line. In all plots except h12v10 a point of high density towards the lower end of the MCWD axis was observed. Most regression models showed significant negative relationships of MCWD with EVI anomaly (p < 0.001). However, most models explained less than 5% of the variance in EVI. This suggests that the linear models are not meaningful in this context.

The resulting maps for SPI showed different patterns than for MCWD (Figure 5). A cluster of positive responses could be found over western Amazonia. This implies that EVI was low during relative droughts, which means greenness sensitivity to drought.

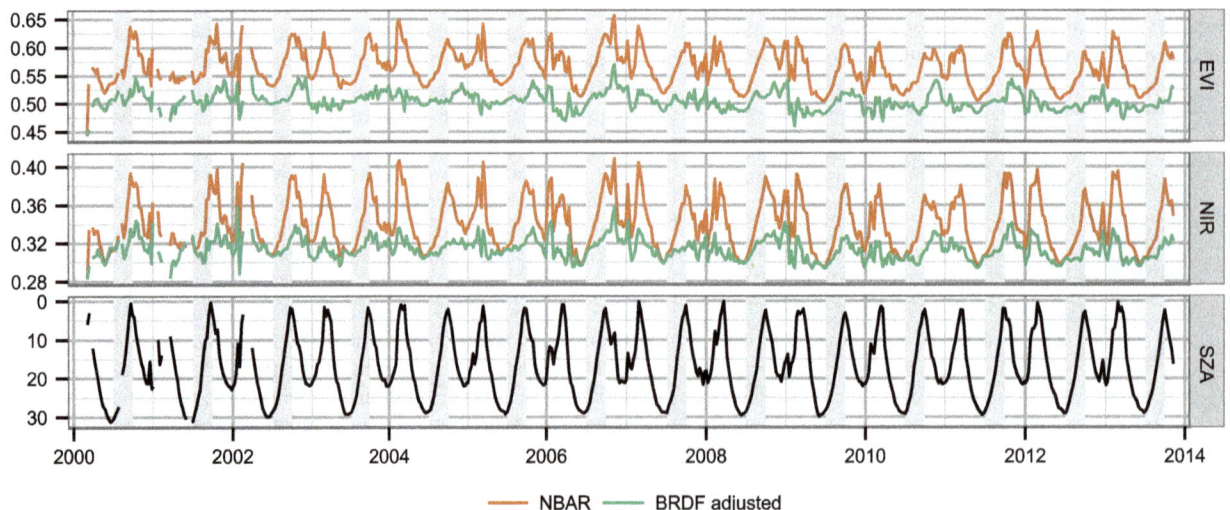

Figure 2: Spatially averaged viewing zenith angle (NBAR) and viewing and sun zenith angle adjusted (BRDF-adjusted, VZA = 0°, SZA = 30°) EVI and NIR for MODIS tiles h11v09 and h12v09, lines connect data points, solar zenith angles are reversed to facilitate comparison with vegetation index trends, 8 day temporal resolution, grey bands indicate typical dry seasons (July to September).

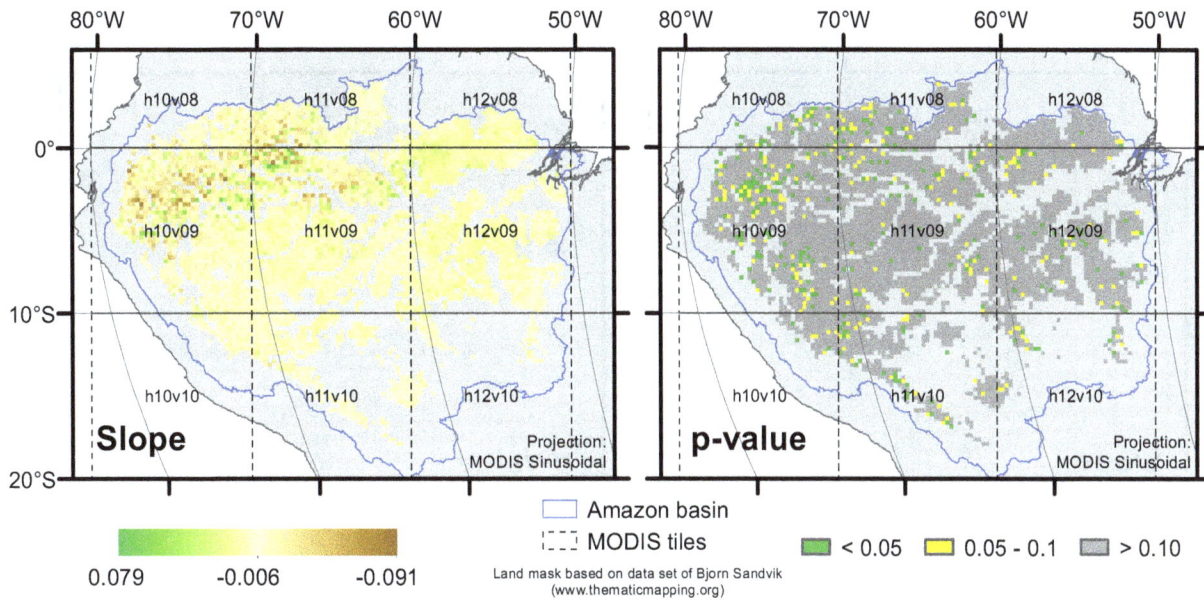

Figure 3: Pixel-based linear regressions of September BRDF-adjusted EVI seasonal anomalies against MCWD on 30 km (~0.25°) spatial resolution.

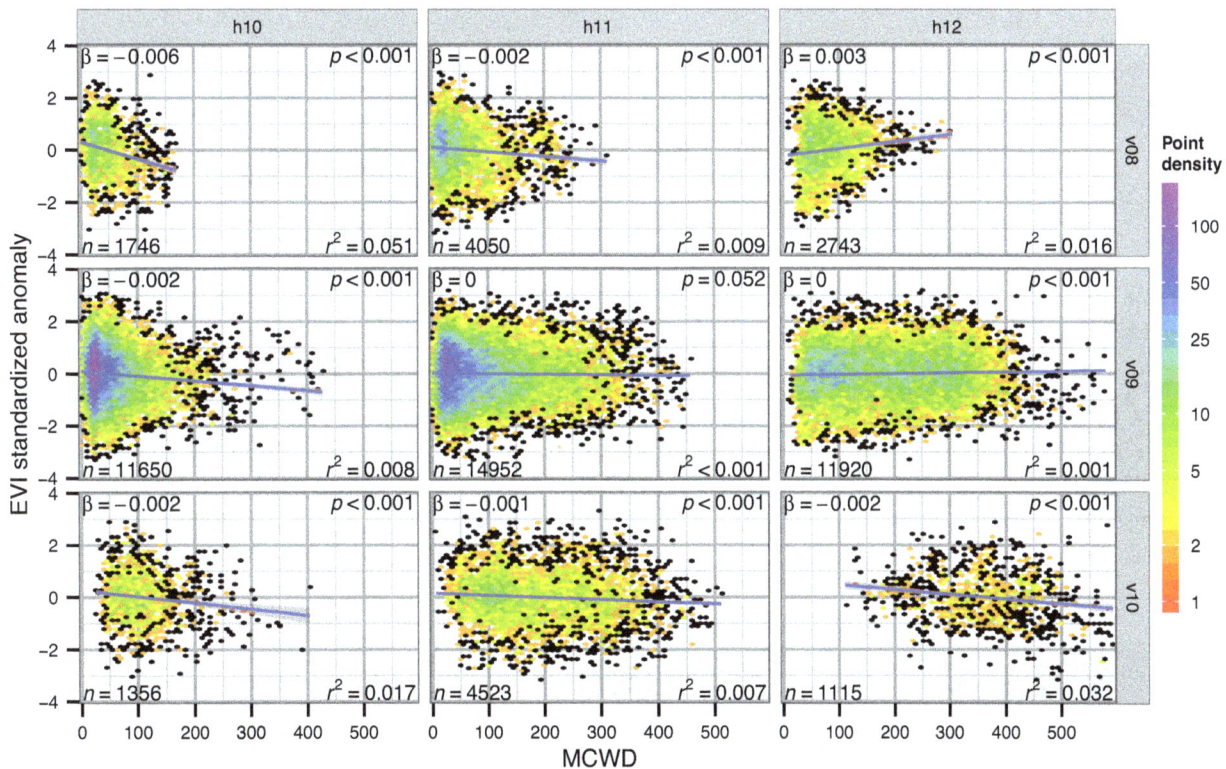

Figure 4: Tile-based scatterplots and regressions of September BRDF-adjusted EVI seasonal anomalies against MCWD, EVI anomalies dimensionless, MCWD in mm, plot grid pattern hints at spatial pattern of MODIS tiles, blue lines are linear regression models, numbers are characteristics of linear models per tile (β is the regression slope), shaded areas are 95%-confidence intervals for regression lines, density color scale is logarithmic.

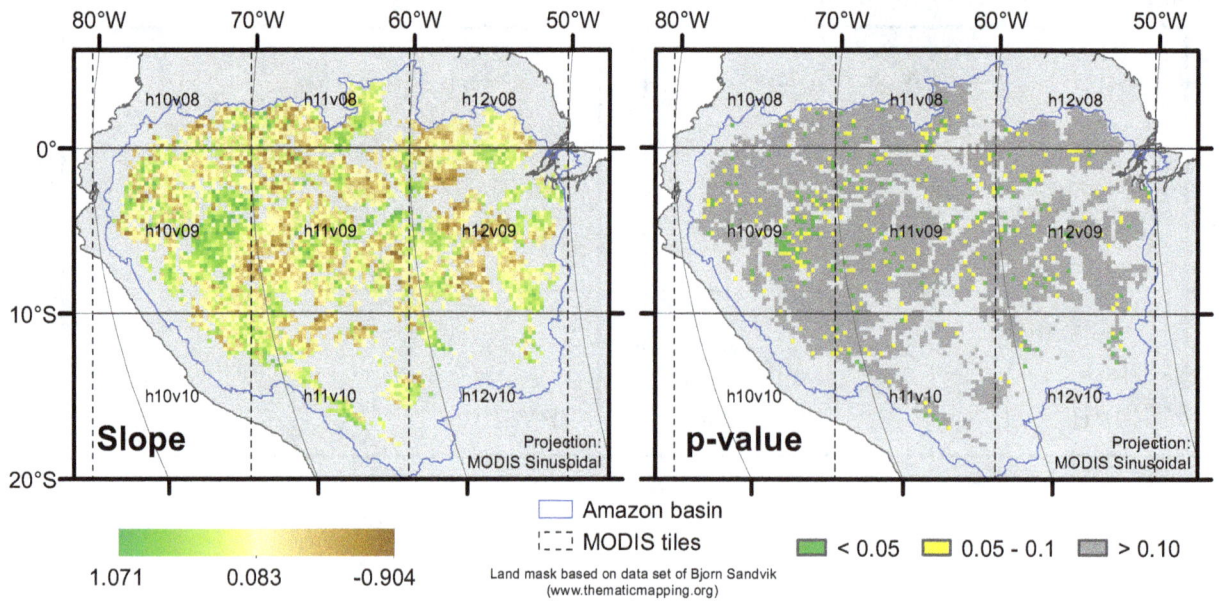

Figure 5: Same as Figure 3, but for SPI.

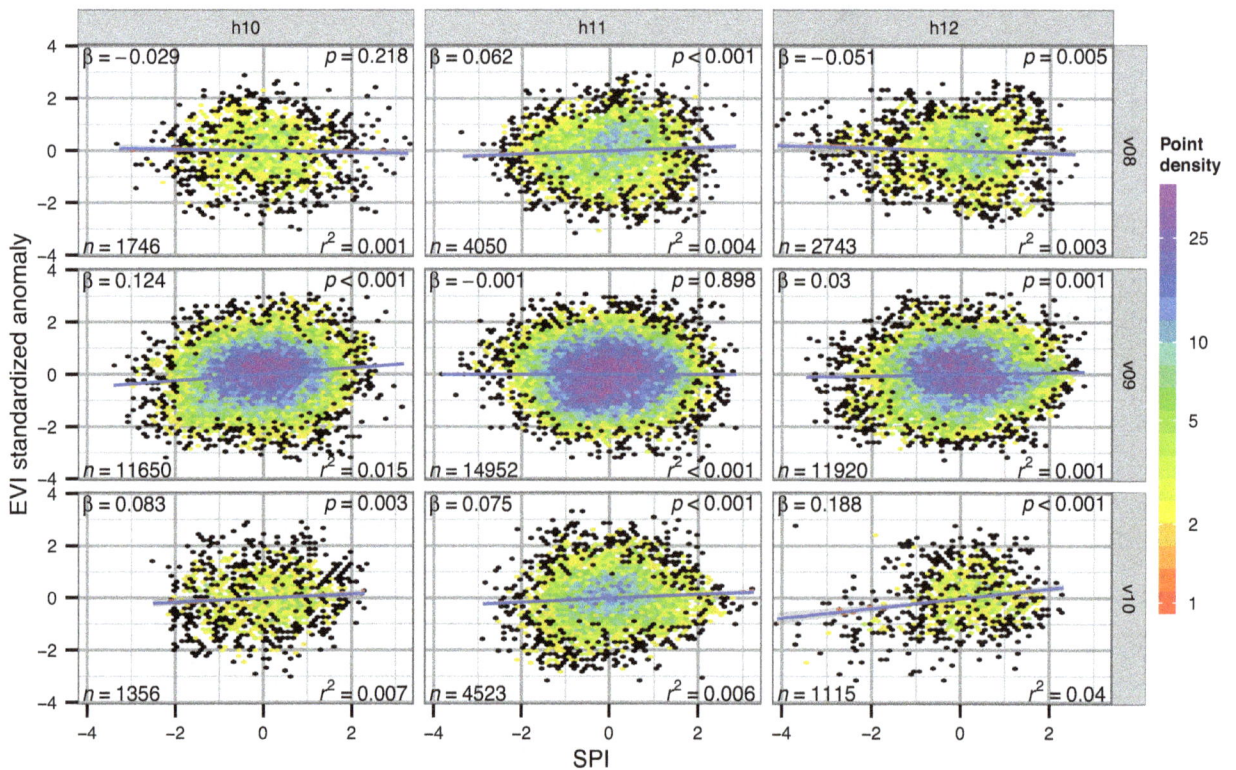

Figure 6: Same as Figure 4, but for SPI.

Some of these pixel-based regressions were statistically significant (p < 0.1). Other positive responding regions were located at the Andean foothills in the South, but no significant relationships could be found for them. This was also true for smaller patches of negative relationships in the eastern basin. Otherwise no broad scale spatial patterns could be identified. The relationship was rather patchy with high variation on short ranges. Additionally, the total number of significant cells at the 0.05 level was 4.1% of all valid pixels and the histogram of p-values for whole Amazonia showed a nearly uniform distribution. This points to a random process that produced significant slopes only by chance.

All point clouds for the EVI anomaly-SPI relationship were clustered at the point of origin for the respective scatterplots (Figure 6). This is typical for anomalies, as the observations are distributed around the overall mean. The linear models were significant (p < 0.01) for all tiles but h10v08 and h11v09. However, all models explained less than 5% of the variance in EVI anomalies expressed as r^2. Considering the point clouds it becomes clear that any model would have problems to fit the EVI anomaly-SPI relationship.

4. DISCUSSION

In this study, we investigated the performance of EVI to study inner-annual dry season and drought impacts on the canopy of undisturbed Amazon forests. We focused on recently discussed sun sensor geometry effects, which contaminate the temporal EVI signals. Correcting for these effects with the MODIS MCD43A1 product, we found distinct seasonal dynamics of EVI. However, these patterns were suspiciously correlated to SZA patterns. Additionally, EVI was not sensitive to canopy moisture stress represented by two moisture stress indices, although droughts have the potential to alter forest canopy functioning (Brando et al., 2008) and structure (Phillips et al., 2010).

From the results the question arises why BRDF-adjusted EVI derived with the MCD43 product suite and EVI derived by Morton et al. (2014) showed different temporal patterns while both products are modeled for the same viewing and sun geometries and cover roughly the same area. This might be explained by the aggregation period for the BRDF modeling approaches. While MCD43 uses 16 day, Morton et al. (2014) used monthly intervals. De Abelleyra and Verón (2014) showed that shorter aggregation intervals result in better noise reduction in NDVI time series for agricultural crops. Taking into account possible rapid changes in Amazon foliage (Doughty and Goulden, 2008), a monthly aggregation period might not take up fast changes, so that EVI time series are overly strong smoothed.

Anomalies of BRDF-adjusted EVI showed no clear response to drought expressed as MCWD and SPI. These results have particular significance, because they were derived on both coarse and medium spatial scales. Moreover, they did not only focus on drought years like 2005 or 2010, but were based on a general assessment over all years when data was available. In this way positive and negative anomalies were assessed in a broader context; chances were lower to interpret spatial or temporal noise in EVI as a response to drought. This can happen if only single drought events are considered for analysis. Furthermore, using seasonal anomalies to analyze EVI sensitivity is robust against seasonal patterns. Anomalies are more robust against seasonal patterns such as viewing and sun geometries and highlight dissimilarities between the years. However, this analysis focused on direct impacts, which occur in the same year as the drought event. Impacts on longer time scales as found by Saatchi et al. (2013) were beyond the scope of this study and cannot be excluded.

Moreover, the static evapotranspirational demand of 100 mm month[-1] in the formulation of MCWD is only a coarse approximation that excludes adaptations of trees to drier conditions and the spatial distribution of these adaptations, which is only poorly understood for the Amazon forests (Aragão et al., 2007). This static threshold adds to the explanation for spatial patterns in the NW in Figure 3. These areas lie in the inner Amazon regions and likely experience no water stress, so that the patterns are rather artifacts produced by errors in TMPA retrievals and the assumption of a static evapotranspirational demand.

The question remains why BRDF-adjusted EVI showed no overall sensitivity to the drought indices. On the one hand, this can be caused by a bad representativeness of the drought indices as discussed above. On the other hand, the overall canopy spectral response to drought might be lower than anticipated in earlier studies. For instance, NDVI shows much higher seasonal stability in the study of Samanta et al. (2012). This is usually explained with the saturation of NDVI over dense canopies (Huete et al., 2002). However, NDVI is more robust against sun and viewing angle effects than EVI (Sims et al., 2011). In this sense NDVI might point to the stability of canopy greenness, while EVI is mostly affected by angular effects. Taking into account that drought had fundamental impacts on forest functioning and structure as found in ground studies (Nepstad et al., 2002; Brando et al., 2008), greenness might not be an appropriate proxy for ecosystem functioning in tropical forests.

5. CONCLUSIONS

Morton et al. (2014) questioned changes in Amazon rainforest greenness expressed as EVI as a response to dry season and drought conditions. This conclusion was not supported by BRDF-adjusted EVI derived with the MODIS MCD43 product. However, seasonal patterns still weakly followed the course of SZA over the year. This might be caused by actual changes in canopy greenness or result from problems in the BRDF modeling process. On the other hand, BRDF-adjusted EVI was mostly insensitive to drought expressed with two drought indices. This could mean that Amazonian forest canopies are resilient against drought impacts. However, studies on forest plots suggest otherwise, so that it is rather EVI's sensitivity for canopy processes that is questionable. Based on these findings, we suggest a careful assessment of EVI as a proxy for tropical rainforest functioning.

References

Aragão, L. E. O. C., Malhi, Y., Roman-Cuesta, R. M., Saatchi, S., Anderson, L. O. and Shimabukuro, Y. E., 2007. Spatial patterns and fire response of recent Amazonian droughts. Geophysical Research Letters 34(7), pp. L07701.

Brando, P. M., Nepstad, D. C., Davidson, E. A., Trumbore, S. E., Ray, D. and Camargo, P., 2008. Drought effects on litterfall, wood production and belowground carbon cycling in an Amazon forest: results of a throughfall reduction experiment. Philosophical transactions of the Royal Society of London. Series B, Biological sciences 363(1498), pp. 1839–1848.

Cox, P. M., Betts, R. A., Collins, M., Harris, P. P., Huntingford, C. and Jones, C. D., 2004. Amazonian forest dieback under climate-carbon cycle projections for the 21st century. Theoretical and Applied Climatology 78(1-3), pp. 137–156.

de Abelleyra, D. and Verón, S. R., 2014. Comparison of different BRDF correction methods to generate daily normalized MODIS 250m time series. Remote Sensing of Environment 140, pp. 46–59.

de Toledo, J. J., Magnusson, W. E. and Castilho, C. V., 2013. Competition, exogenous disturbances and senescence shape tree size distribution in tropical forest: Evidence from tree mode of death in Central Amazonia. Journal of Vegetation Science 24(4), pp. 651–663.

Doughty, C. E. and Goulden, M. L., 2008. Seasonal patterns of tropical forest leaf area index and CO2 exchange. Journal of Geophysical Research 113, pp. G00B06.

Galvão, L. S., dos Santos, J. R., Roberts, D. A., Breunig, F. M., Toomey, M. and de Moura, Y. M., 2011. On intra-annual EVI variability in the dry season of tropical forest: A case study with MODIS and hyperspectral data. Remote Sensing of Environment 115(9), pp. 2350–2359.

Hirota, M., Nobre, C., Oyama, M. D. and Bustamante, M. M. C., 2010. The climatic sensitivity of the forest, savanna and forest-savanna transition in tropical South America. The New Phytologist 187(3), pp. 707–719.

Huete, A., Didan, K., Miura, T., Rodriguez, E., Gao, X. and Ferreira, L., 2002. Overview of the radiometric and biophysical performance of the MODIS vegetation indices. Remote Sensing of Environment 83(1-2), pp. 195–213.

Huete, A. R., Didan, K., Shimabukuro, Y. E., Ratana, P., Saleska, S. R., Hutyra, L. R., Yang, W., Nemani, R. R. and Myneni, R., 2006. Amazon rainforests green-up with sunlight in dry season. Geophysical Research Letters 33(6), pp. L06405.

Huffman, G. J., Adler, R. F., Bolvin, D. T. and Nelkin, E. J., 2010. The TRMM Multi-Satellite Precipitation Analysis (TMPA). In: M. Gebremichael and F. Hossain (eds), Satellite Rainfall Applications for Surface Hydrology, Springer Netherlands, Dordrecht, pp. 3–22.

Hutyra, L. R., Munger, J. W., Saleska, S. R., Gottlieb, E., Daube, B. C., Dunn, A. L., Amaral, D. F., de Camargo, P. B. and Wofsy, S. C., 2007. Seasonal controls on the exchange of carbon and water in an Amazonian rain forest. Journal of Geophysical Research 112(G3), pp. G03008.

Lucht, W., Schaaf, C. and Strahler, A., 2000. An algorithm for the retrieval of albedo from space using semiempirical BRDF models. IEEE Transactions on Geoscience and Remote Sensing 38(2), pp. 977–998.

Malhi, Y., Roberts, J. T., Betts, R. A., Killeen, T. J., Li, W. and Nobre, C. A., 2008. Climate change, deforestation, and the fate of the Amazon. Science 319(5860), pp. 169–172.

Malhi, Y., Wood, D., Baker, T. R., Wright, J., Phillips, O. L., Cochrane, T., Meir, P., Chave, J., Almeida, S., Arroyo, L., Higuchi, N., Killeen, T. J., Laurance, S. G., Laurance, W. F., Lewis, S. L., Monteagudo, A., Neill, D. A., Vargas, P., Nunez Pitman, N. C. A., Quesada, C., Alberto Salomao, R., Silva, J. N. M., Lezama, A. T., Terborgh, J., Martinez, R. and Vasquez Vinceti, B., 2006. The regional variation of aboveground live biomass in old-growth Amazonian forests. Global Change Biology 12(7), pp. 1107–1138.

Mayorga, E., Logsdon, M., Ballester, M. and Richey, J., 2012. BA-ECO CD-06 Amazon River Basin Land and Stream Drainage Direction Maps. http://dx.doi.org/10.3334/ORNLDAAC/1086.

McKee, T. B., Doesken, N. J. and Kleist, J., 1993. The relationship of drought frequency and duration to time scales. In: Eighth Conference on Applied Climatology, 17-22 January 1993, Anaheim, California, pp. 17–22.

Morton, D. C., Nagol, J., Carabajal, C. C., Rosette, J., Palace, M., Cook, B. D., Vermote, E. F., Harding, D. J. and North, P. R. J., 2014. Amazon forests maintain consistent canopy structure and greenness during the dry season. Nature 506, pp. 221–224.

Moura, Y. M., Galvão, L. S., dos Santos, J. R., Roberts, D. A. and Breunig, F. M., 2012. Use of MISR/Terra data to study intra- and inter-annual EVI variations in the dry season of tropical forest. Remote Sensing of Environment 127, pp. 260–270.

Myneni, R. B., Yang, W., Nemani, R. R., Huete, A. R., Dickinson, R. E., Knyazikhin, Y., Didan, K., Fu, R., Negrón Juárez, R. I., Saatchi, S. S., Hashimoto, H., Ichii, K., Shabanov, N. V., Tan, B., Ratana, P., Privette, J. L., Morisette, J. T., Vermote, E. F., Roy, D. P., Wolfe, R. E., Friedl, M. A., Running, S. W., Votava, P., El-Saleous, N., Devadiga, S., Su, Y. and Salomonson, V. V., 2007. Large seasonal swings in leaf area of Amazon rainforests. Proceedings of the National Academy of Sciences of the United States of America 104(12), pp. 4820–4823.

Nepstad, D. C., de Carvalho, C. R., Davidson, E. A., Jipp, P. H., Lefebvre, P. A., Negreiros, G. H., da Silva, E. D., Stone, T. A., Trumbore, S. E. and Vieira, S., 1994. The role of deep roots in the hydrological and carbon cycles of Amazonian forests and pastures. Nature 372(6507), pp. 666–669.

Nepstad, D. C., Moutinho, P., Dias-Filho, M. B., Davidson, E., Cardinot, G., Markewitz, D., Figueiredo, R., Vianna, N., Chambers, J., Ray, D., Guerreiros, J. B., Lefebvre, P., Sternberg, L., Moreira, M., Barros, L., Ishida, F. Y., Tohlver, I., Belk, E., Kalif, K. and Schwalbe, K., 2002. The effects of partial throughfall exclusion on canopy processes, aboveground production, and biogeochemistry of an Amazon forest. Journal of Geophysical Research 107(D20), pp. 8085.

Phillips, O. L., Aragão, L. E. O. C., Lewis, S. L., Fisher, J. B., Lloyd, J., López-González, G., Malhi, Y., Monteagudo, A., Peacock, J., Quesada, C. A., van der Heijden, G., Almeida, S., Amaral, I., Arroyo, L., Aymard, G., Baker, T. R., Bánki, O., Blanc, L., Bonal, D., Brando, P., Chave, J., de Oliveira, A. C. A., Cardozo, N. D., Czimczik, C. I., Feldpausch, T. R., Freitas, M. A., Gloor, E., Higuchi, N., Jiménez, E., Lloyd, G., Meir, P., Mendoza, C., Morel, A., Neill, D. A., Nepstad, D., Patiño, S., Peñuela, M. C., Prieto, A., Ramírez, F., Schwarz, M., Silva, J., Silveira, M., Thomas, A. S., Steege, H. T., Stropp, J., Vásquez, R., Zelazowski, P., Alvarez Dávila, E., Andelman, S., Andrade, A., Chao, K.-j., Erwin, T., Di Fiore, A., Honorio C, E., Keeling, H., Killeen, T. J., Laurance, W. F., Peña Cruz, A., Pitman, N. C. A., Núñez Vargas, P., Ramírez-Angulo, H., Rudas, A., Salamão, R., Silva, N., Terborgh, J. and Torres-Lezama, A., 2009. Drought sensitivity of the Amazon rainforest. Science 323(5919), pp. 1344–1347.

Phillips, O. L., van der Heijden, G., Lewis, S. L., López-González, G., Aragão, L. E. O. C., Lloyd, J., Malhi, Y., Monteagudo, A., Almeida, S., Dávila, E. A., Amaral, I., Andelman, S., Andrade, A., Arroyo, L., Aymard, G., Baker, T. R., Blanc, L., Bonal, D., de Oliveira, A. C. A., Chao, K.-J., Cardozo, N. D., da Costa, L., Feldpausch, T. R., Fisher, J. B., Fyllas, N. M., Freitas, M. A., Galbraith, D., Gloor, E., Higuchi, N., Honorio, E., Jiménez, E., Keeling, H., Killeen, T. J., Lovett, J. C., Meir, P., Mendoza, C., Morel, A., Vargas, P. N. n., Patiño, S., Peh, K. S.-H., Cruz, A. P. n., Prieto, A., Quesada, C. A., Ramírez, F., Ramírez, H., Rudas, A., Salamão, R., Schwarz, M., Silva, J., Silveira, M., Slik, J. W. F., Sonké, B., Thomas, A. S., Stropp, J., Taplin, J. R. D., Vásquez, R. and Vilanova, E., 2010. Drought-mortality relationships for tropical forests. The New Phytologist 187(3), pp. 631–646.

Roujean, J.-L., Leroy, M. and Deschamps, P.-Y., 1992. A bidirectional reflectance model of the Earth's surface for the correction of remote sensing data. Journal of Geophysical Research 97(D18), pp. 20455.

Saatchi, S., Asefi-Najafabady, S., Malhi, Y., Aragão, L. E. O. C., Anderson, L. O., Myneni, R. B. and Nemani, R., 2013. Persistent effects of a severe drought on Amazonian forest canopy.

Proceedings of the National Academy of Sciences of the United States of America 110(2), pp. 565–570.

Saleska, S. R., Miller, S. D., Matross, D. M., Goulden, M. L., Wofsy, S. C., da Rocha, H. R., de Camargo, P. B., Crill, P., Daube, B. C., de Freitas, H. C., Hutyra, L., Keller, M., Kirchhoff, V., Menton, M., Munger, J. W., Pyle, E. H., Rice, A. H. and Silva, H., 2003. Carbon in Amazon forests: unexpected seasonal fluxes and disturbance-induced losses. Science 302(5650), pp. 1554–1557.

Samanta, A., Knyazikhin, Y., Xu, L., Dickinson, R. E., Fu, R., Costa, M. H., Saatchi, S. S., Nemani, R. R. and Myneni, R. B., 2012. Seasonal changes in leaf area of Amazon forests from leaf flushing and abscission. Journal of Geophysical Research 117(G1), pp. G01015.

Schaaf, C. B., Gao, F., Strahler, A. H., Lucht, W., Li, X., Tsang, T., Strugnell, N. C., Zhang, X., Jin, Y., Muller, J.-P., Lewis, P., Barnsley, M., Hobson, P., Disney, M., Roberts, G., Dunderdale, M., Doll, C., D'Entremont, R. P., Hu, B., Liang, S., Privette, J. L. and Roy, D., 2002. First operational BRDF, albedo nadir reflectance products from MODIS. Remote Sensing of Environment 83(1-2), pp. 135–148.

Schaaf, C. B., Liu, J., Gao, F. and Strahler, A. H., 2011. Aqua and Terra MODIS albedo and reflectance anisotropy products. In: B. Ramachandran, C. O. Justice and M. J. Abrams (eds), Land Remote Sensing and Global Environmental Change – NASA's Earth Observing System and the Science of ASTER and MODIS, Springer, New York, Dordrecht, Heidelberg, London, chapter 23, pp. 579–602.

Schwalm, C. R., Williams, C. A., Schaefer, K., Arneth, A., Bonal, D., Buchmann, N., Chen, J., Law, B. E., Lindroth, A., Luyssaert, S., Reichstein, M. and Richardson, A. D., 2010. Assimilation exceeds respiration sensitivity to drought: A FLUXNET synthesis. Global Change Biology 16(2), pp. 657–670.

Sims, D. A., Rahman, A. F., Vermote, E. F. and Jiang, Z., 2011. Seasonal and inter-annual variation in view angle effects on MODIS vegetation indices at three forest sites. Remote Sensing of Environment 115(12), pp. 3112–3120.

Strahler, A. H., Muller, J.-P. and MODIS Science Team Members, 1999. MODIS BRDF/Albedo Product: Algorithm theoretical basis document version 5.0. http://modis.gsfc.nasa.gov/data/atbd/atbd_mod09.pdf.

White, A., Cannell, M. and Friend, A., 1999. Climate change impacts on ecosystems and the terrestrial carbon sink: A new assessment. Global Environmental Change 9, pp. S21–S30.

Xiao, X., Hagen, S., Zhang, Q., Keller, M. and Moore, B., 2006. Detecting leaf phenology of seasonally moist tropical forests in South America with multi-temporal MODIS images. Remote Sensing of Environment 103(4), pp. 465–473.

27

DETECTION AND CHARACTERIZATION OF COLOMBIAN WETLANDS: Integrating geospatial data with remote sensing derived data. USING ALOS PALSAR AND MODIS IMAGERY

L.M. Estupinan-Suarez [a], C. Florez-Ayala [a], M.J. Quinones [b], A.M. Pacheco [a, c], A.C. Santos [d]

[a] Alexander von Humboldt Institute for Research on Biological Resources (IAvH), Scientific and Applied Projects Assistance Office, The Wetlands Project, Cll. 72 No 12-65 piso 7 Edf. Skandia, Bogota, Colombia – (lestupinan, cflorez)@humboldt.org.co
[b] SarVision Application in Remote Sensing, Agro business Park 10 6708 PW Wageningen, The Netherlands –quinones@sarvision.nl
[c] IAvH and Institute of Hydrology, Meteorology and Environmental Surveys (IDEAM) partnership –anafo185@gmail.com
[d] National University of Colombia, Faculty of Engineering, Cr. 45 No 26–85 Edf. 453, Bogota, Colombia –acsantosr@unal.edu.co

KEY WORDS: wetlands, radar imagery, modis, hydrology, water bodies, flooded vegetation

ABSTRACT:

Wetlands regulate the flow of water and play a key role in risk management of extreme flooding and drought. In Colombia, wetland conservation has been a priority for the government. However, there is an information gap neither an inventory nor a national baseline map exists. In this paper, we present a method that combines a wetlands thematic map with remote sensing derived data, and hydrometeorological stations data in order to characterize the Colombian wetlands. Following the adopted definition of wetlands, available spatial data on land forms, soils and vegetation was integrated in order to characterize spatially the occurrence of wetlands. This data was then complemented with remote sensing derived data from active and passive sensors. A flood frequency map derived from dense time series analysis of the ALOS PALSAR FBD /FBS data (2007-2010) at 50m resolution was used to analyse the recurrence of flooding. In this map, flooding under the canopy and open water classes could be mapped due to the capabilities of the L-band radar. In addition, MODIS NDVI profiles (2007-2012) were used to characterize temporally water mirrors and vegetation, founding different patterns at basin levels. Moreover, the Colombian main basins were analysed and typified based on hydroperiods, highlighting different hydrological regimes within each basin. The combination of thematic maps, SAR data, optical imagery and hydrological data provided information on the spatial and temporal dynamics of wetlands at regional scales. Our results provide the first validated baseline wetland map for Colombia, this way providing valuable information for ecosystem management.

1. INTRODUCTION

Wetlands are well known to provide a large number of ecosystem services. Among many functions, they are considered as key ecosystem for their capacity of regulating floods, filtering out pollutants and improve water quality as well as been a proper ecosystem for bird migration (Finlayson et al., 2005). Since 1971 their relevance was recognized in the Ramsar convention, which was the first global instrument seeking wetland ecosystem conservation (Matthews, 1993). In Colombia, wetlands are a government priority in conservation policies and risk management since 1997 when the Ramsar convention was signed (MADS, 2002). However, good spatial information is still missing for most of the Colombian territory, there is also lack of understanding on the water pulses and dynamics. Neither an inventory nor a national wetlands baseline exists.

After the flooding events in 2010-2011 caused by La Niña phenomenon, one of the Colombian government's strategies was to identify and delineate wetlands and *Paramos* (highlands water catchment ecosystems), considering their importance to conserve and regulate water flows. Wetlands in this study were understood as "an ecosystem that due to landforms and hydrological conditions accumulates water, temporally o permanently, and all these drives to a specific type of soils (hydromorphic) and organisms adapted to these conditions" (Vilardy *et al.* 2014). In this sense, Colombian wetlands are diverse and vary across the country; there might be permanent and temporal lakes, shrub swamps, swamp forests, seasonally flooded forests, meadows, rivers, mangroves, estuarine waters, coastal freshwater lagoons, saline lagoons, and many more. This diversity of wetlands types requires the integration of different data sources and techniques to showcase ecosystem spatial and temporal dynamics when a national product is developed.

In Colombia, the wetlands' surveys have focused on the Magadalena-Cauca basin, which holds more than 80% of population (Hydrology, Meteorology and Environmental Studies Institute (IDEAM) and Cormagdalena, 2001). SarVision and The Nature Conservancy (TNC) generated a flooding frequency map using Alos Palsar imagery for this basin, within the Kyoto and Carbon initiative (K&C) and the Japan Aerospace Exploration Agency (JAXA) collaboration (Quiñones 2013, Quiñones et al 2013). After La Niña phenomena 2010-2011, a flooding report was made by IDEAM and the National Geographic Institute Agustin Codazzi (IGAC) (2011) which centred in the Magdalena and Cauca regions reporting high damage. On the other hand, less populated and some of the largest basins in the country, such as the Amazons, Orionoco and Pacific, present a lack of information.

Worldwide wetlands' surveys have had different remote sensing approaches and have been used for multiple purposes among countries. For example, Albania leaded its wetlands inventory through unsupervised classification algorithms of LandSat imagery. They calculated the Normalized Difference Vegetation Index (NDVI) and the wettest Tasseled Cap band, and used both as an input in a classification algorithm (Apostolakis 2008). Tasseled cap was described by Cris and Cicone in 1984, it is one of the first imagery transformation that highlights water and is still used. After, some other indexes have been developed for water mirror delineation using mainly LandSat images.

Some examples are de Normalized Water Difference Extraction Index NDWI (McFeeters 1996), the Modified Normalized Water Difference Extraction Index MNDWI (Xu 2006) and the Automated Water Extraction Index (Feyisa et al. 2014). All seek to improve separability between water, build-ups and shadows.

Moreover, NDVI has been demonstrated to be an indirect measure with good performance in surveys related to wetlands, taking into account that it includes information of water reflectance but also from vegetation. A methodology to detect wetlands using NDVI MODIS images was proposed by Landmann et al. (2006), it is based on the comparison of images from the dry and rainy season.

Detection of wetlands must be supplemented with hydrological data because the hydro-period brings with it an understanding of the ecological function of wetlands (van Dongen, et al 2012). Information on the ecosystem types and dynamics is required to establish conservation and management strategies.

This study aims to integrate a thematic wetlands map based on spatial geo-information (landforms, soils and Vegetation data) with information captured by active and passive sensors (Alos Palsar I MODIS respectively) and hydrological stations. Our outcomes include spatial and temporal information that can guide national and local stakeholders, ONGs and environmental agencies of the government towards a better planning of the territory.

2. METHODS

2.1 Study Area

This study was carried out on the continental territory of Colombia. It is located in the equatorial zone between -66.76° and -79.09° of longitude and -4.31° and 12.5° of latitude. Colombia is a country of high landscape diversity with mountainous and lowland ecosystems. Altitude ranges from 0 to 5.800 m.a.s.l. with the Andes as the main mountain chain, with diverse ecosystems with very dry or very high precipitation regimes. The annual mean rainfall varies from 500 to 10.000 mm and the mean temperature at 0 m.a.s.l. is 28°C and 6°C at 4.000 m.a.s.l. (IDEAM 2005). The mountain system in Colombia and their interaction with atmospheric circulation determine the regional hydrological behavior. Consequently, an altitudinal distribution pattern indicates a clear increase in precipitation and runoff with increased levels of relief. Runoff production is directly related to precipitation and evaporation, which are conditioned by elements of the landscape and the location of Colombia on the globe. Other factors such as vegetation cover, soils, land use, geology and others affect runoff through their influence on precipitation and evaporation processes (IDEAM 2010a).

Toward a better understanding of Colombian wetlands, the results were analysed at regional scales. The country was divided into eight zones that correspond to the main river basins of Colombia and were proposed by IDEAM (2010a). They are: Amazon, Caribbean, Catatumbo, Cauca, High Magdalena, Middle Magdalena, Orinoco and Pacific (Figure 1).

2.2 Wetlands thematic map

The wetland's thematic map has incorporated data from geophysical variables. Firstly, a conceptual approach was

designed with a literature review and expert's workshops identifying the criteria that allow the spatial identification of wetlands (Cortes-Duque and Rodríguez-Ortiz, 2014). The selected criteria were landforms, hydromorphic soils, wet vegetation and the river networks including water bodies. Each criteria has its own spatial layer (IDEAM 2010, IGAC 2014a, IGAC 2014b) which was assessed by thematic experts. Each expert calculated an association level of the variables related to wetlands ecosystems. Later, a cartographic edition process was performed manually to integrate scales at 1:100.000. Final integration was based on weighted overlay analysis performed in ArcGIS 10.1. following the criteria and weight defined by each thematic expert. The accuracy of the wetland map is subject of an ongoing study.

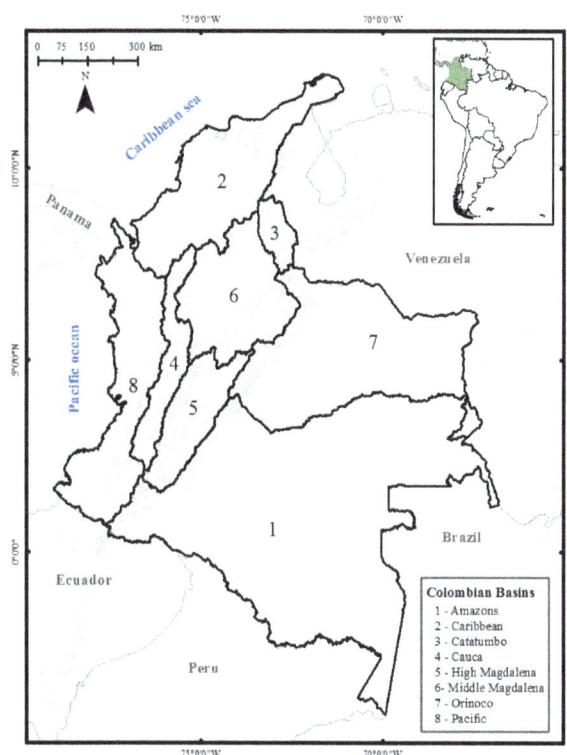

Figure 1. Colombian basins based on IDEAM (2010a).

2.3 Frequency flooding map

2.3.1   Image processing: A dense time series of the Alos PALSAR FBS and FBD-HH radar (50 m) was used to generate a flood frequency map of the whole Colombian continental territory. This map presented the number of times each pixel was detected to be flooded during the study period (2007-2010). Two main classes are shown in the map: the open water class and the flooded under the canopy class.

In total 168 strips of the Alos FBD and FBS were processed covering the Colombian territory. A total of 7 coverages over the whole territory were completed for the period between 2007 and 2010. Single look complex (SLC) Strips were processed using Gamma software following the processing steps showed in Figure 2.   Detail processing can be found in Quinones (2013).

Five Alos PALSAR –HH mosaics at 50 m resolution were created for the whole country. Each of this HH mosaics was filtered and classified using ENVI classification techniques.

Each classified map showed two classes, corresponding to open water and flooding under canopy. A flood frequency map was created integrating the information for these maps.

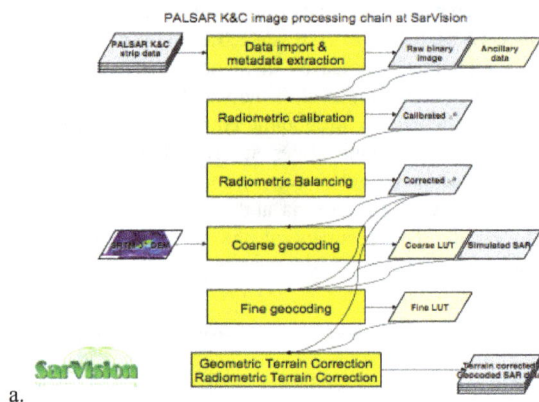

Figure 2. Processing process carried out to each FBD- FBS Alos PalSAR SLC strips.

2.3.2   Validation process: IDEAM carried out the validation process of each of these classified mosaics. Although the flooding map was generated for the all the Colombian continental territory, the validation process was made for 19 representative windows in strategic areas of the country.

Six main steps were followed during the validation process: (i) at the initial stage, the information was tested by Colombian landscape experts from IDEAM, IAvH and SarVision. They visually validated different well-known areas of the country, (ii) a stratified random sample was designed for the accuracy assessment following the Olofsson *et al* (2013) and (2014) methodology, (iii) samples were revised and compared with both high-resolution radar and optical images, (iv) matrix error evaluation, which is a cross-tabulation of the remote sensed data against the verified reference, (v) estimating area and uncertainty and (vi) confidence intervals.

2.3.3   Correlation analysis between the thematic map and the flooding map: A spatial correspondence analysis was made between the identified wetlands of the thematic map with the flooding areas detected by radar. All flooding frequency was reclassified to a value of 1. Then, the comparison was done without taking into account the type and frequency of flooding. The   processing was performed in ArcGIS 10.1.

2.4 NDVI MODIS map

2.4.1   NDVI MODIS acquisition and pre-processing: The products MOD13Q1 from MODIS Terra Sensor were downloaded   from   the   NASA   website http://reverb.echo.nasa.gov. To cover the territory of Colombia six tiles were required (h10v07-v08-v09; h11v07-v08-v09). In total 138 images for each tile were acquired from 2007 to 2012. The NDVI band of all images were imported in ERDAS IMAGINE 2010. Afterwards a layer stack was built with all NDVI images guaranteeing the time sequence. In this sense, the first image in the stack (band 1) is from January 2007 and the last one (band 138) is from December 2012. The process was performed for all the tiles to finally create a mosaic. The same procedure was done for the quality band.

2.4.2   TIMESAT processing: The NDVI image stack for Colombia was converted to binary data format with the same quality stack. The TIMESAT 3.1 (Eklundh and Jönsson 2012) program parameters were adjusted on pilot windows for the monomodal and bimodal regions of Colombia (IDEAM 2005). The selected mathematical function to adjust the NDVI profiles was Savitzky-Golay. Pixel with clouds or without processing were excluded from the analysis.

2.4.3   Imagery classification and multitemporal analysis: A TIMESAT output file was imported to ERDAS IMAGINE and its values rescaled to digital numbers (0-255). The ISODATA clustering unsupervised classification ran from 30 to 120 classes. Based on divergence statistics criteria, the most adequate number of classes was chosen (de Bie et al. 2011, Ali et al. 2013). The classes associated with wetlands were identified and assessed among the main hydrological Colombian basins.

2.4.4   Wetlands characterization using MODIS NDVI information: The NDVI annual profile of each class was assessed to assign to which land cover correspond. Next, it was calculated the NDVI classes distribution on wetlands from the thematic map, and finally a comparison within basins was done.

2.5 Hydroperiod of Colombian basins

A characterization of the main watersheds in Colombia was created based on hydroperiods. The hydroperiod is the seasonal pattern or the level of a wetland and is the wetland's hydrologic signature   (Mitsch   and   Gosselink,   2000).   From hydrometeorological gauging stations distributed along the country with recording for the period between 1974-2012, was possible to identify the season and variability of the discharge and precipitation. This analysis articulated hydrographic zoning (IDEAM, 2013) and relief allowing the characterization of regions with similar hydroperiods.

3. RESULTS

3.1 Wetlands thematic map

The map shows two classes: (1) identified wetlands which refer to wetlands where information from one or more criteria is accurate for the ecosystem detection, and (2) potential wetlands ecosystem that are areas vulnerable of flooding or transformed wetlands by natural or anthropic causes. These areas might have a lack of information due to the scale (1:100.000), fieldwork limitations (e.g. access difficulties, number of samples) or remote sensing constrains (e.g. clouds, topography).

This study focused on class 1 (identified wetlands) which are water bodies, or areas with hydromorphic soils and hydrophytic vegetation that present temporal flooding, and are adapted to particular wet regimes. It is estimated that Colombian inland and costal wetlands' area is around 18'000.000 ha (Figure 3), marine wetlands are not included. The Orinoco and Amazons basins have the highest percent of wetlands, 38.3% and 31.9% respectively, followed by the Caribbean (13.0%) and the Pacific (11.4%). The high and middle Magdalena area correspond to 0.6% and 3.8% overall area. The basins with less area are Cauca (0.9%) and Catatumbo (0.2%). The accuracy of this detection is the subject of an ongoing assessment.

Furthermore, the relevance of the areas detected as class 2 (potential ecosystem of wetlands) stands out for environmental planning and risk management (12'000.000 ha approximately). However, they are out of the scope of this paper.

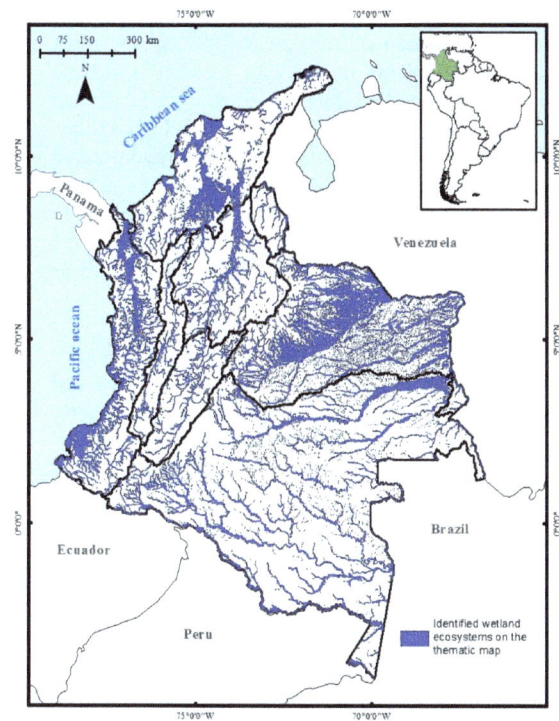

Figure 3. Wetlands identified in the thematic map at 1:100.000 in the continental territory of Colombia

3.2 Frequency flooding map

The frequency flooding map shows how many times radar detected flooding at a 50m of pixel within 5 observations. A window on the Mojana region in the north of the Magdalena basin is presented to illustrate the flood frequency map (Figure 4). Colours on the legend indicate the two main classes on the map. Open water (blue) and flood under canopy (green). Shades on the colours indicate the number of times a pixel was detected flooded. The higher the shade of the colour the higher the flood frequency.

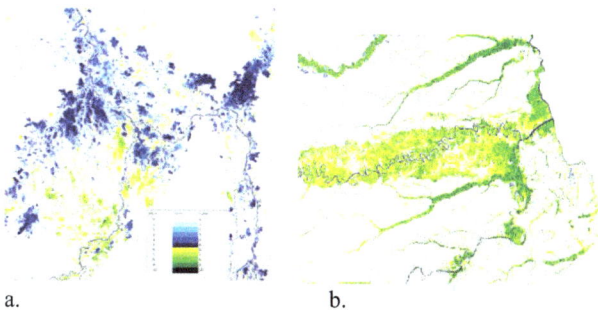

a.　　　　　　　　　　　　　b.

Figure 4. Frequency flooding map for the a. Mojana region (above) and the b. Mataven region (below). The darker the colour the higher the times the area was detected as flooded.

Flooded forest dominates in the Amazon, Pacific and Orinoco basins. Conversely, open water flooding predominates in the Caribbean and Magdalena's (higher and middle) basin. Information for Catatumbo and Cauca is not conclusive considering that their detected area is low as well as the spatial correspondence with radar (Figure 5).

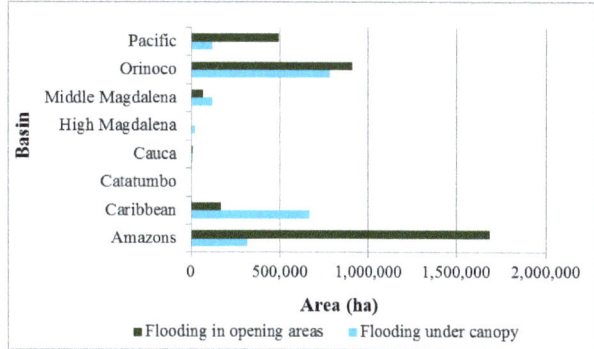

Figure 5. Area detected by radar of opening areas and under canopy in each of the Colombian basins.

For the validation process three classes were mapped from each SAR mosaic in the analysed timeline; (i) class 0: non-flooding, (ii) class 1: flooding in open areas and (iii) class 2: flooding under canopy. Hereafter a frequency-flooding map was the product that merged the information of the 6 observations. It presents information from each flooding pulse (observation) per pixel. Therefore, it is possible to read the number of times that flooding was detected in a single pixel was during the time analysis. This product was validated, by reviewing 11307 samples and the map had an overall accuracy of 85%. Higher accuracies were associated to flat terrains and mirror waters and the main errors were associated to the category of flooding under canopy, especially in areas with relieve.

The higher spatial correlation between the thematic map (class1) and the frequency flooding map is observed in lowlands. In other words, the Caribbean, Amazons and Pacific exhibit the highest correspondence with the radar detection (Figure 6) where are also most of the largest wetlands complex.

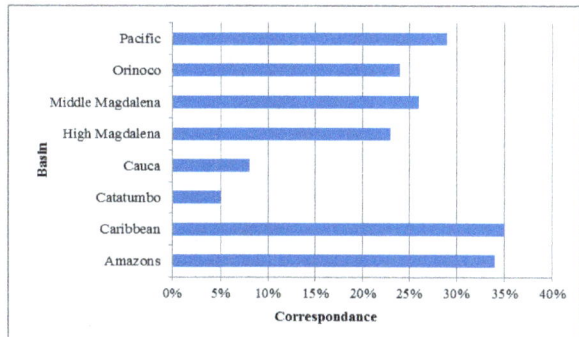

Figure 6. Spatial correspondence between the wetlands thematic map and the flooding map.

3.3 NDVI MODIS wetlands map

The NDVI map was found to be the most adequate with 45 classes based on the average separability and minimum separability criteria (Estupinan-Suárez and Florez-Ayala, 2014). The classes behave differently within basins Therefore, they represent different type of wetlands ecosystems which are mainly separated by temporal characteristics. The analysis of NDVI classes was focused on the characterization of the identified wetlands in each basin and highlight their features based on their temporal trend.

In general, class no. 1-4 detects sea water and glaciers, from class no. 5 to no. 19 the NDVI map is showing water bodies. It was observed that the classes on the 20-29 range have a good correspondence with wetlands on highland, predominates with the *Paramo* ecosystem detection. Classes from 30 to 39 identifies temporal flooding, Orinoco wetlands cover (42%) is dominates mainly by classes 37,38 and 39 which are associated to temporal shrub swamps and flooded savannahs .The evergreen vegetation is classified on the latest classes higher than no. 40 (Table 1).

Colombian basins	Percent of pixels in a no. of classes range (%)				
	1-9	10-19	20-29	30-39	40-45
Amazons	0.43	1.14	1.63	2.66	94.55
Caribbean	5.71	11.59	4.69	10.93	67.09
Catatumbo	1.95	0	0.4	8.53	89.12
Cauca	0.26	0.5	1.91	8.83	88.50
High Magdalena	2.2	6.95	8.55	22.51	59.80
Middle Magdalena	0.76	4.79	11.51	4.79	78.30
Orinoco	0.16	0.58	0.72	42.92	55.61
Pacific	2.08	1.19	1.79	6.53	88.41

Table 1. NDVI classes frequency range by basin

When the NDVI MODIS classes are analysed separately temporal differences of the wetlands are shown. For example, the dominant NDVI water mirror profiles of the Pacific and Caribbean lakes differ. In the Pacific, the opening flooding are classified as no. 13, 18, 19 while in the Caribbean as class no. 10, 12 and 16. The evergreen classes also present outcomes to separate. The hypothesis is that this separation is caused by different temporal regimes on the index. The basins with more than 80% of evergreen classes also exhibit important patterns. The Amazons is clearly dominated by class no. 44. Pacific and Cauca which are neighbour basins have similar trends, the predominant classes are 43 and 44, but Pacific also has a significant participation of class no. 40 and 45 (Figure 7). Catatumbo has an specific trend, distributed from higher to lower classes frequency as 44, 45, 43 which shows a particular composition.

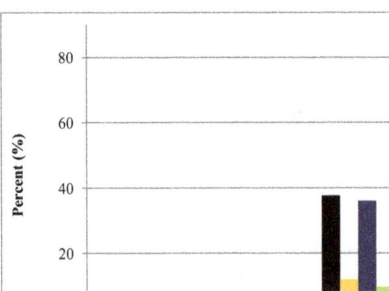

Figure 7. Percent of NDVI MODIS classes in Colombian basins. Solely from classes no. 40 to 44

The use of NDVI MODIS to detect wetlands is constrained due to their spatial resolution and optical properties. Its potential relies on its high temporal resolution that shows the temporal annual dynamic of shrub swamps and opening water bodies. In very dense vegetation, like flooded forest, its use was reduced to the main rivers and riparian vegetation.

3.4 Hydroperiods regions on Colombian's basin

Seventeen regions were identified based on hydroperiods. Figure 8 shows the geographic positions of the respective data collection points. The diagram bars show the hydroperiod for

Colombian rivers, these curves represent the discharge fluctuation (m^3/s) during the annual cycle between 1974 and 2012.

One of the largest hydrographic areas of Colombia corresponds to the Amazon, with a mean annual flow of 26800m^3/s where it is possible to find floodplains, madreviejas, marshes (small and medium size) and flooded forests. These wetlands are characterized by a monomodal hydroperiod with high values between May and June. The importance of the Amazon region is also because it has one of the RAMSAR sites the Laguna de la Cocha, located in the upper basin of the Putumayo River at an elevation of 2700 meters. Also some of the rivers of the Amazon region are born near the Colombian Massif such as Putumayo and Caqueta river.

The Orinoco basin in Colombia has a monomodal hydroperiodo, with high flows between the months of June and July. The Chingaza lacustrine system is an important site for RAMSAR and it is found in this region. In the Orinoco the wetlands can be flood plains in winter, madreviejas and permanent ponds.

Within the Magdalena River basin there are tree identified regions. The first is High Andean, which has a monomodal hydroperiod with a maximum flow value in July. Downstream the second region is Hight Magdalena, the hydrological regime is bimodal with peak flow in April and November, where there are large expanses of flooded land for rice cultivation and some hydroelectric dams. The third region is Middle Magdalena, the hydroperiod is still considered bimodal as observed in the Arrancaplumas station with peaks in May and November and a mean annual flow of 1277 m3/s over Magdalena River.

The Eastern Andean region is monomodal hydroperiod, some important wetlands in the region are Fuquene and Cucunuba Lagoon. Thanks to the physical and geographical conditions together with the contributions of the rivers of the East Andes to the Middle Magdalena region, there is a transition from monomodal to bimodal hydroperiod.

The Eastern Andean region is monomodal hydroperiod, some important wetlands in the region are Fuquene and Cucunuba Lagoon. Thanks to the physical and geographical conditions together with the contributions of the rivers of the East Andes to the Middle Magdalena region, there is a transition from monomodal to bimodal hydroperiod.

The Lower Magdalena region is characterized by a system of marshes and floodplains at the confluence of Magdalena, San Jorge and Cauca rivers (*depression momposina*). Even though the upper regions are monomodal the lower Magdalena has a monomodal trend as the Caribbean region that has a clear monomodal hydroperiod. The most important wetland in this region is the Cienaga Grande de Santa Marta RAMSAR site (RAMSAR, 2014). The Catatumbo River basin drains into the Gulf of Maracaibo and the hydroperiod is marked by two seasons of high flows in May and November.

Cauca River is divided into three regions of similar hydroperiods. The first is the Higth Cauca with monomodal regime. Then the Valle del Cauca with a bimodal regime and flow peaks in May and November. The Middle Cauca has one of the sites of RAMSAR international interest wetland complex the Otún Lagoon. According to the values recorded at the station SAN JUAN in Nechí region the hydroperiod is monomodal with highs flow in November with an average annual flow of 487 m^3/s.

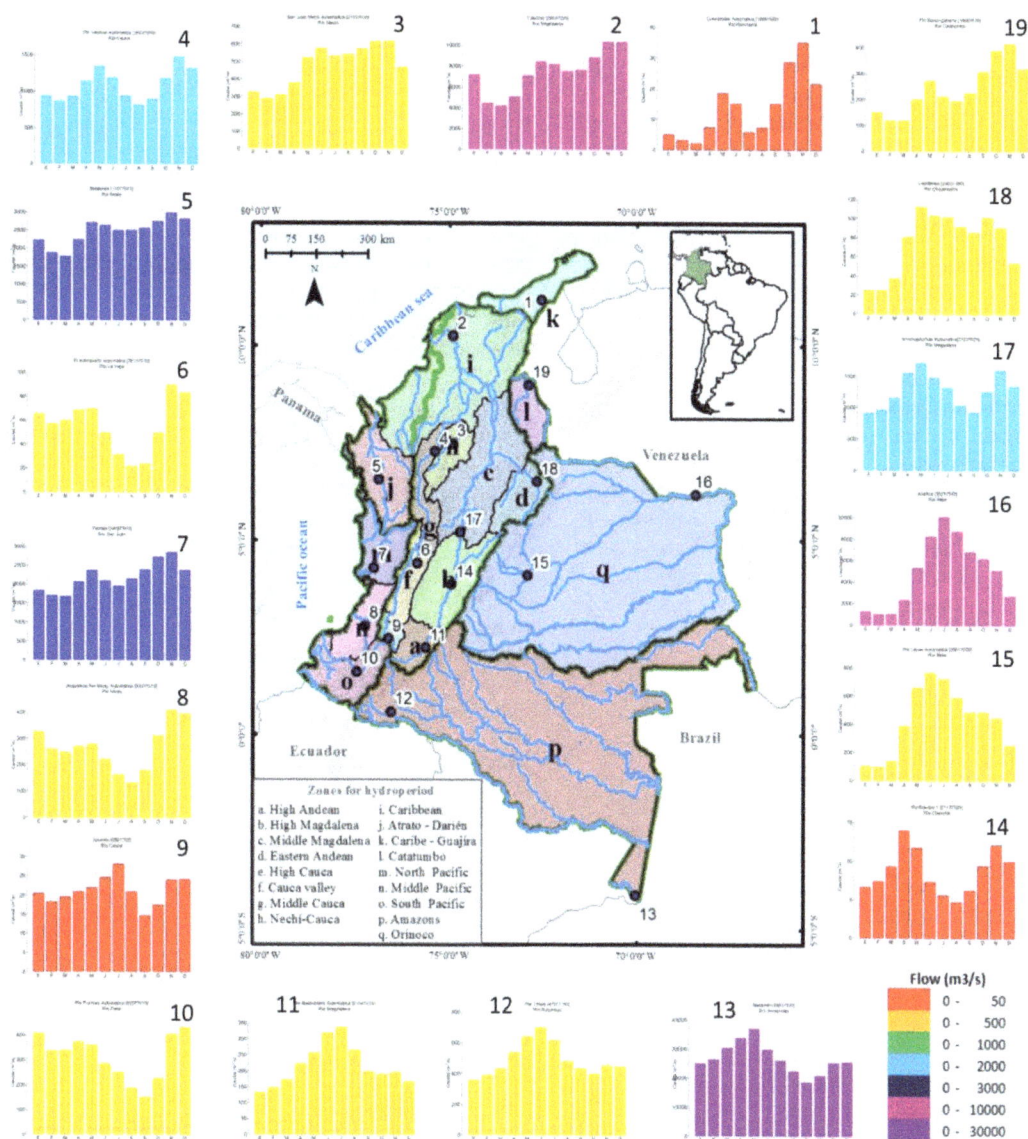

Figura 8. Hydroperiods of the hydrological zones of Colombian basins a based on flow multiannual data (1974-2004) of hydrological stations. Number indicates the geographical position of data collection points.

In Colombia the precipitation regime is heterogeneous. The region with higher precipitation is de Pacific, but the most important discharge are in the Amazon, Orinoquia and Magdalena. The Pacific region shows three hydroperiod, in the northern river basins in San Juan and Baudo (wetland of Delta Baudo River interest RAMSAR) with bimodal hydroperiod at high flow rates in the months of March and November. In the Mid and South Pacific the hydroperiod is monododal with seasonal high flow rate in November.

4. DISCUSSION

Colombia has recognized the importance of wetlands when it became a party of RAMSAR convention since 1998 and later on the legislation through a National Policy of Wetlands in 2002. By that time it was estimated an area of 20.000.000 ha for inland wetlands (MADS, 2002). But only until 2012, after the extreme flooding of *La Niña* event, a wetlands map iniciative begun. The wetlands thematic map presented in this article identified the ecosystem based on basic official national geophysical layers that were assessed and refined by thematic experts. The mapping criteria were selected on workshops with

researches and discussed environmental government agencies and are consistent with national data availability. During all the process, it was sought to incorporate the high variability of Colombian wetlands.

The largest complex of Colombia wetlands are located in lowlands which are mainly isolated areas and with lack of data as is the case of Amazon, Pacific and Orinoco regions. Radar detected that the flooding area under canopy is higher than in open areas. The flooding under canopy is associated predominantly to flooded forest in areas with high mean precipitation values as the Amazons and Pacific. This was also reaffirmed with the MODIS NDVI map results that showed a dominance of 95% of evergreen forest in the wetlands of Amazons. The radar information from Alos PalSAR is able to give a new dimension for the wetlands management in Colombia due to identifications of the flooded under the canopy which changes the management strategies that until now have been focus in the water mirrors.

The area of wetlands in Cauca, Catatumbo, High and Middle Magdalena were underestimated at the assessed scale. Firstly, the majority of wetlands on these basins are smaller than the cartographic scale (25 ha) which might be explained by the dominance of steep relief. For that reason, this information has to be seen together with the Colombian wetlands inventory to avoid misrepresentations. Complementary, Cauca and Magdalena basin hold a large percent of the country population and have been exposed to strong transformation process.

The frequency-flooding map generated with radar images constitutes an important flood analysis tool for the country, since this information provides consistent and accurate data as a monitoring baseline tool of aquatic ecosystems, as well as proper flooding information. The spatial correspondance found between the thematic map and the flooding map was lower than the reported initially by Quinones et al. (2014). This could be explained because the first analysis used the high and very high potential areas of the wetlands thematic map. Otherwise, this survey was solely focused on the identified wetlands without including potential areas. In this sense, it is necessary to explore if the areas detected by radar that correspond to potential wetlands ecosystem can be classified as identified wetlands, mainly in lowlands where radar detection is more accurate.

In addition, it is recommended that a comparison be carried out between the 50 m frequency flooding map and the 100m product (also obtained in the wetlands project) that has less spatial resolution but more observations. This could have a major number of dates that fall into the rainy season. Therein, to work with more captures of flooding could improve the correspondence between the flooding frequency maps and the thematic wetlands. Moreover, radar exposed limitations detecting mangroves, if this drawback is solved is expected that the correspondence with costal boundaries basins increase.

MODIS NDVI data has a high potential to characterize wetlands taking into account the temporal trends of vegetation and water mirrors. However, the developed methodology limits its detection to the largest water bodies. The use of more bands and the comparison between images from the dry and rainy season as suggested by Landmann et al. (2006) has still to be explored.
This study was centred on the identification and characterization of wetlands, and turns into an opportunity to accomplish Ramsar compromises adopted by the country on the convention

5. CONCLUSSION

The large extension of Colombian wetlands, their diversity and complexity requires high human and technological efforts towards the generation of proper information for decision makers. In this sense, products derived from remote sensing turns into a valuable input, guaranteeing uploaded data for wetlands management programs, conservations strategies and risk management.

The outcomes shown on this paper are unique for the country. The thematic map has a biogeophysical approach considered landforms that accumulates water, hydromorphic soils and hydrophytic vegetation which are evidence of temporal or permanent flooding. The L band-radar (Alos Palsar) of the frequency flooding map provided valuable information specially when detecting flooding under canopy, since this information it's very difficult to detect by other sensors. However the reliability of the data is higher in flat areas. The MODIS NDVI

map brought out information of the separability of wetlands including temporal information of water and vegetation. Moreover, characterization of hydroperiods at different basis zones shows the high variability of patterns and features of wetlands.

In brief, the resulting maps bring out a more complete understanding of wetlands, they included the spatial and temporal ecosystem dynamics and are a first attempt to link the hydrological information with geophysical layers integrated on the thematic map, and satellite imagery for the whole country. This frame that includes different approaches and data sources also has provided guidelines for risk management strategies and might be considered as a baseline for a wetlands monitoring program. Finally, it is recommended to develop a multi scalar approach, looking at detailed scales for small wetlands identification and at large scales to understand the spatial and temporal dynamics of wetlands complex located mostly in lowlands.

REFERENCES

Ali, A., de Bie, C. A. J. M., Skidmore, A. K., Scarrott, R. G., Hamad, A., Venus, V., & Lymberakis, P. (2013). Mapping land cover gradients through analysis of hyper-temporal NDVI imagery. *International Journal of Applied Earth Observation and Geoinformation,* 23, pp. 301-312.

Apostolakis, A. (2008). Wetland identificacion at National Level: AL. En: Fitoka E & Keramitsoglou I 2008 (Eds).

Cortes-Duque, J. and Rodríguez-Ortiz, J. 2014. Memorias simposio taller expertos. Construcción colectiva de criterios para la delimitación de humedales: retos e implicacione del país. Book compilers. Instituto de Investigaciones Biológicas Alexander von Humboldt. Bogota D.C., Colombia.

Crist, E.P and Cicone, R. C. 1984. A physically-based transformation of thematic mapper data – The TM tasseled cap. *IEEE Transactions on Geoscience and Remote Sensing,* 22, pp. 256-263.

Fitoka, E., & Keramitsoglou, I. (2008). (Eds). Inventory, assessment and monitoring of Mediterranean Wetlands: Mapping wetlands using Earth Observation techniques. EKBY & NOA. MedWet publication. (Scientific reviewer Nick J Riddiford).

de Bie, C. A. J. M., M. R. Khan, Smakhtin, V. U., Venus, V., Weir, M. J. C., & Smaling, E. M. A. (2011). Analysis of multi-temporal SPOT NDVI images for small-scale land-use mapping. *International Journal of Remote Sensing,* 32 (21), pp. 6673-6693.

Eklundh, L., & Jönsson, P. 2012. TIMESAT 3.1. Software Manual. Lund Univertisy and University of Malmo. Sweden pp82.http://www.nateko.lu.se/timesat/docs/timesat3_1_1_SoftwareManual.pdf (October 2013)

Estupinan-Suárez and Florez-Ayala, 2014. Avances en la detección de humedales en Colombia usando imágenes multitemporales (2007-2012) del Índice Normalizado y Diferenciado de Vegetación del sensor MODIS Terra. Memorias: XVI SELPER (January 2014)

Feyisa, G.L., Meilby. H., Fensholt, R., Simon, R. P. 2014. Automated Water Extraction Index: A new technique for surface water mapping using Landsat imagery. *Remote Sensing of Environment*, 140, pp. 23–35

Finlayson, C.M., D'Cruz, R. & Davidson, N.C. 2005. Ecosystems and human well-being: wetlands and water. Synthesis. Millennium Ecosystem Assessment. World Resources Institute, Washington D.C.

IDEAM and Cormagdalena. 2001. Estudio ambiental de la Cuenca Magdalena-Cauca y elementos para su ordenamiento territorial. Resumen Ejetutivo. Bogotá, D.C. http://www.pdpmagdalenacentro.org/Res.%20Ejecutivo%20Est udio%20Ambiental.pdf

IDEAM. 2005. Atlas Climatológico de Colombia. Instituto de Hidrología, Meteorología y Estudios Ambientales. Bogotá, D.C. Colombia.

IDEAM. 2010a. Estudio Nacional del Agua. Instituto de idrología, Meteorología y Estudios Ambientales. Bogotá, D. C., Colombia.

IDEAM. 2010b. Leyenda Nacional de Coberturas de la Tierra. Instituto de Hidrología, Meteorología y Estudios Ambientales. Metodología CORINE Land Cover adaptada para Colombia Escala 1:100.000. Instituto de Hidrología, Meteorología y Estudios Ambientales. Bogotá, D. C., Colombia

IDEAM and IGAC. 2001. Línea base de inundación 2001. Instituto de Hidrología, Meteorología y Estudios Ambientales e Instituto Geográfico Agustín Codazzi. Bogotá, D. C., Colombia.

IDEAM. 2013. Zonificación y codificación de unidades hidrográficas e hidrogeológicas de Colombia. Instituto de Hidrología, Meteorología y Estudios Ambientales. Bogotá, D. C., Colombia.

IGAC. 2014a. Cartografía Base, gdb escala 1:100.000. Instituto Geográfico Agustín Codazzi. Bogotá, D. C., Colombia.

IGAC. 2014b. Mapa Nacional de Geopedología 1:100.000 Instituto Geográfico Agustín Codazzi. Bogotá, D. C., Colombia

Landmann, TR, Colditz, R., & Schmidt, M. 2006. An Object-Conditional Land Cover Classification System (LCCS) Wetland Probability Detection Method for West African Savannas Using 250-Meter MODIS Observations, In Proc. 'GlobWetland: Looking at Wetlands from Space' (Ed. Lacoste H), ESA SP-634 (CD-ROM), ESA Publications Division, European Space Agency, Noordwijk, The Netherlands. En: Fitoka, E., & Keramitsoglou I., (2008).

MADS. 2002. Política Nacional para Humedales Interiores de Colombia. Estrategias para su conservación y uso sostenible. Ministerio de Medio Ambiente. República de Colombia. Bogotá D.C., Colombia.

Matthews, G.V.T. 1993. The Ramsar Convention on Wetlands: its history and development. Re-issued (2013) Ramsar Convention Bureau, Gland, Switzerland. http://archive.ramsar .org/pdf/lib/Matthews-history.pdf (January 2014)

McFeeters, S. K. 1996. The use of Normalized Difference Water Index (NDWI) in the delineation of open water features. *International Journal of Remote Sensing*, 17, pp. 1425–1432.

Mitsch, W. J. and J. G. Gosselink. 2000. Wetlands. John Wiley and Sons Inc. New York, USA.

Olofsson, P., Foody, G. M., Herold, M., Stehman, S. V., Woodcock, C. E., & Wulder, M. A. 2014. Good practices for estimating area and assessing accuracy of land change. *Remote Sensing of Environment*, 148, pp. 42-57.

Olofsson, P., Foody, G. M., Stehman, S. V., & Woodcock, C. E. 2013. Making better use of accuracy data in land change studies: Estimating accuracy and area and quantifying uncertainty using stratified estimation. *Remote Sensing of Environment*, 129, pp. 122–131.

RAMSAR. 2014. The List of Wetlands of International Importance. http://www.ramsar.org/about/wetlands-of-international-importance (March 2015).

Quiñones, MJ. 2013. Mapas de Frecuencias de inundación y tipos de vegetación para las cuencas del rio Magdalena y Cauca en Colombia. Reporte SarVision: SV-TNC-SEI Mapa Magdalena Vegetación/Inundaciones. # 80105 *15/05/2013*. pp. 26.

Quiñones, MJ., Hoekman D.H., Pedraza C.A. 2013. Flooding frequency Map/ Mapa de frecuencia de Inundaciones 2007-2010. Product of JAXA, Alos Kyoto & Carbon (K&C). SarVision-SEI-TNC collaborations. Within the JAXA-Wageningen University, agreement

Quinones, M., Vissers, M, Florez, C and Hoekman, D. 2014. Integration of Alos PalSAR data to Wetlands Mapping: An ecosystem approach. K&C product – phase 3 report. Science Team meeting # 21 Kyoto, Japan, December 3-4. 2014. http://www.eorc.jaxa.jp/ALOS/kyoto/dec2014_kc21/pdf/2-10_KC21_Quinones-Florez.pdf

van Dogen, Behn G. A., Coote M., Shanahan A., and Setiawan H. 2012. Hydroperiod classification of Cervantes Coolimba coastal wetlands using landsat time series imagery. *Int. Arch. Photogramm. Remote Sens. Spatial Inf. Sci.*, XXXIX-B8, pp. 199-202. www.int-arch-photogramm-remote-sens-spatial-inf-sci.net/XXXIX-B8/199/2012/

Vilardy, S., Jaramillo, Ú., Flórez, C., Cortés-Duque, J., Estupiñán, L., Rodríguez, J.,...Aponte, C. 2014. Principios y criterios para la delimitación de humedales continentales: una herramienta para fortalecer la resiliencia y la adaptación al cambio climático en Colombia. Instituto de Investigación de Recursos Biológicos Alexander von Humboldt. Bogotá D.C. Colombia https://www.siac.gov.co/documentos/GestCont/Carti lla_humedales_inteactivo_1.pdf

Xu, H. 2006. Modification of normalised difference water index (NDWI) to enhance open water features in remotely sensed imagery. *International Journal of Remote Sensing*, 27, pp. 3025–3033.

ACKNOWLEDGEMENTS

Our acknowledgements to *Fondo Adaptacion* and *El Ministerio de Hacienda* of Colombia that fund this project. To the *El Ministerio de Ambiente y Desarrollo Sostenible,* IDEAM and IGAC. The PALSAR product was created using radar images provided by JAXA in the frame of the JAXA K&C Initiative, in collaboration with the University of Wageningen and SarVision. TNC is acknowledged for processing funding of the K&C data in the initial stages of the project. We thank the IAvH wetlands project team and the thematic experts who contributed to generate inputs for the wetlands map.

28

LEVERAGING SPATIAL MODEL TO IMPROVE INDOOR TRACKING

L. Liu[a], W. Xu[b,*], W. Penard[c], S. Zlatanova[a]

[a] Delft University of Technology, Julianalaan 134, 2628BL, Delft, the Netherlands - (l.liu-1, s.zlatanova)
@tudelft.nl
[b] Fugro Intersite, Dillenburgsingel 69, 2263 HW, Leidschendam, the Netherlands
W.Xu@fugro.nl
[c] CGI Nederland, George Hintzenweg 89, Rotterdam, the Netherlands
w.penard@cgi.com

KEY WORDS: Spatial Model, Indoor Tracking, Semantics, Topology

ABSTRACT:

In this paper, we leverage spatial model to process indoor localization results and then improve the track consisting of measured locations. We elaborate different parts of spatial model such as geometry, topology and semantics, and then present how they contribute to the processing of indoor tracks. The initial results of our experiment reveal that spatial model can support us to overcome problems such as tracks intersecting with obstacles and unstable shifts between two location measurements. In the future, we will investigate more exceptions of indoor tracking results and then develop additional spatial methods to reduce errors of indoor tracks.

1 INTRODUCTION

Nowadays indoor navigation heavily relies on an accurate and stable positioning or localization technique. Unfortunately, existing positioning techniques are still at experimental phase (Fuchs et al., 2011; Miller, 2006). Compared with outdoor GPS tracks (recordings of positions at regular intervals), indoor tracking suffers from low accuracy, which results in a limited number of indoor tracking applications.

Although we may not get highly accurate position information in an indoor environment, tracking pedestrians still appears attractive, especially to some mission-critical scenarios (Fuchs et al., 2011). For some applications users may be more interested in moving trends than accurate coordinates.

Existing indoor positioning and navigation experiments (Miller, 2006; Spassov, 2007) show that under some conditions users (human or robot) can navigate even if individual localization accuracy may not be very high.

Meanwhile, different indoor tracking methods (Burgard et al., 1997; Girard et al., 2011; Jensen et al., 2009; Khider et al., 2012; Miller, 2006; Spassov, 2007; Thrun et al., 2005) design various strategies to avert deviation of tracks. However, they largely rely on positioning equipment, sensors and users. The results reveal varying properties in precision (Fuchs et al., 2011).

Compared with the uncertain results from combining various hardware, a more stable configuration can be achieved if the properties of the spatial model are used. The information stored in spatial models, such as coordinates, semantics of objects, topological relations between objects, etc. can provide some qualitative support for indoor localization regardless the used hardware.

In order to make a better use of indoor positioning measurements and be independent from the adopted devices, this paper is going to present a suitable spatial model and related information to qualitatively improve indoor tracking results.

2 BACKGROUND

To be able to localize a person or robot in a given indoor area, the indoor space has to be partitioned. Indoor space can be partitioned by either real building boundaries (walls) or artificial boundaries, which are the result of a subdivision/decomposition procedure. Such artificial subdivision/decomposition can be based on a regular grid (grid for short), triangulation tessellation, trapezoidal-based tessellation and Voronoi diagrams (Afyouni et al., 2013).

Grid is widely applied to indoor navigation and tracking. Li et al. (Li et al., 2010) elaborate on a grid graph model. They first overlay the building parts/ cellular units (such as a room, a wall, *etc.*) with grids and then generate a grid graph. The underlying cellular units provide semantic information to the corresponding grid cells. One grid cell of the grid graph has one and only one membership of a cellular unit, and the topological relationships among cellular units can be represented by the edges of the grid graph.

Similarly, in robot motion, a planning occupancy grid approach uses a regular matrix of equally-sized cells for autonomous navigating robots (Franz, 2005; Moravec & Elfes, 1985). In this matrix, each cell connects to its eight neighboring cells (with exception of boundary cells). A high probability value is assigned to grids in accessible/navigable spaces and a low value to grids occupied by objects.

In addition to subdivision/decomposition approach, semantic and topological information is important as well for indoor navigation. The topology of an indoor space can be modeled in either a 3D space or 2D layers (Worboys, 2011). Two types of topology are distinguished: connectivity and adjacency. Most existing spatial models utilize connectivity graph to represent indoor space topology (Domnguez et al., 2011). The semantics of a spatial model describes the basic spatial and structural concepts of indoor environments (Liu & Zlatanova, 2012; Tsetsos et al., 2006). Semantics is also referred to ontology when it is utilized for reasoning (Worboys, 2011).

With spatial information, one can conduct indoor tracking with different localization devices. Commonly indoor tracking methods includes: Dead reckoning, Grid filter, Map matching, Model-based approaches.

Dead reckoning (DR) computes a persons current location by advancing a known position with course, speed, time and distance to be travelled. DR data can be collected by inertial measurement unit (IMU) on tracking devices. The uncertainty of dead reckoning positions grows with time thus it is necessary to check the position regularly (Miller, 2006).

Grid filter is a kind of discrete Bayesian filter, which probabilistically estimates a targets location based on observations from sensors. This type of methods are widely used in the field of robotics (Burgard et al., 1997; Thrun et al., 2005). They computes the location in two phases: the prediction phase where the prior probability of location is estimated based on the previous location, a motion model and the map of tracking environment; and the update phase where the posterior probability is computed by multiplying the prior probability with a conditional probability. The conditional probability is computed according to the measurements of sensors.

Map matching assumes a user can be only located along certain routes. Some constraints on indoor environments are applied to refine estimates of the moving positions of a person inside a building. For instance, a user does not pass through walls, but only along corridors and through doorways (Miller, 2006). Basically, there are two map matching techniques: point-to-vertex matching (i.e. a measured location to a vertex in route), and point-to-edge matching (i.e. a measured location to an edge in route). An implementation of point-to-edge matching shows satisfied results in corridor environment (Spassov, 2007).

Model based methods adopt a vector model of the indoor environment to improve the estimation of user location. This method can be taken as an extension of map matching methods. They consider model features (such as walls or obstacles) (Girard et al., 2011), sensor information (e.g. speed & direction), and information from users (e.g. mean velocity and velocity variance (Khider et al., 2012)). Jensen et al. (Jensen et al., 2009) proposes a base graph model for tracking which represents the connectivity and accessibility of indoor space.

However, the existing tracking methods employ limited spatial information. To the best of our knowledge, indoor tracking research is seldom focus on integration of geometrical, topological and semantic features of indoor environments for tracking.

Based on used techniques, current types of localization systems include Angulation (angle), Lateration (distance), Fingerprinting, Inertial and motion sensors, and Neighborhood (Fuchs et al., 2011). However, we focus on location data processing rather than specific localization technique. Therefore, in this research we assume the location data are acquired and we aim at mitigating tracking errors in the data.

We take WiFi fingerprinting localization system for example: it may result in three types of tracking errors: a measured location at an incorrect space/wall/table, an incorrect moving direction between some localization results, or sudden jumps between locations (shifting back and forth) (Besada et al., 2007).

We intend to bridge spatial model and indoor tracking and show the capability of spatial model on improving indoor tracking. The grid model is chosen because it simplifies the localization. Meanwhile, geometric, topological and semantic features of the grid model are utilized to estimate the probability of a users location.

3 METHOD

This section presents how an appropriate spatial model can help on improving indoor tracking results. We concentrate on two major errors, *i.e.* 1) a track crosses indoor objects and 2) a track jumps back and forth.

3.1 Using Spatial Model

As we mentioned, we decided to adopt grid model due to its regularity and flexibility. Meanwhile we take motion direction of pedestrian into consideration as well.

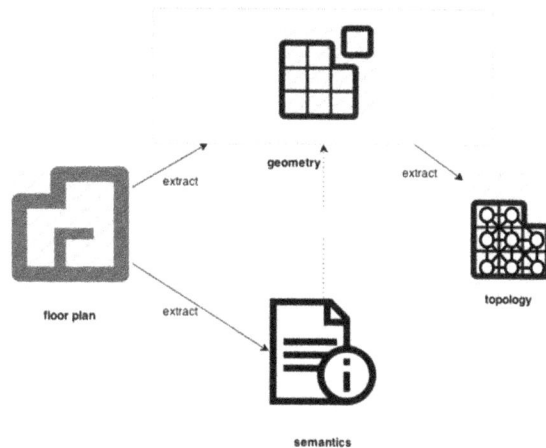

Figure 1: Spatial model construction

In the first place, we require a spatial model incorporating geometry, topology and semantics of indoor environments. In order to detect unrealistic locations, the geometry, topology and semantics of a spatial model need to be used together. Figure 1 provides the workflow of constructing the required spatial model.

Firstly, we pick digital floor plans of a building; secondly, from the original data, we extract semantics of spaces (e.g. room, door); thirdly, we keep the vector geometry of indoor spaces and discretize it to grids; fourthly, we integrate semantics and grids so that each grid cell has a clear meaning or membership; Finally, according to the computed grid model of indoor spaces, we can generate the topological model (i.e. connectivity) of the grid model.

In the next subsections, we will explain each type of information and the usages of them for tracking.

3.1.1 Geometry   At the beginning we need to link a measured location to the grid model we made. Thus we tag the center of grid with accurate coordinates. Afterwards, each measured location is mapped to its closest grid (i.e. the point-to-vertex way of map matching).

Besides coordinates, a buffer of a location is applied as well. The relevant part for a location would possibly be the deviation area of the measurement of the location (Fig. 2a). Thus a buffer of the known previous location is used to represent the search region of the current location. The size of the buffer depends on walking speed and the time interval of location determination. This implies that the human must be in the buffer area in this given time interval.

Two types of distances are introduced to estimate the correctness of the new positions: Euclidean distance and Manhattan distance. The well-known Euclidean distance is the direct length between two points in Euclidean space; while the Manhattan distance represent the sum of the absolute differences of their Cartesian coordinates (Wikipedia, 2014b) (Fig. 2b). In addition, we use an orientation vector to represent the direction from one grid cell to another (Fig. 2c).

As mentioned above, the buffer of previous computed location is used to infer the current possible location. Except the buffer, the

(a)

(b)

(c)

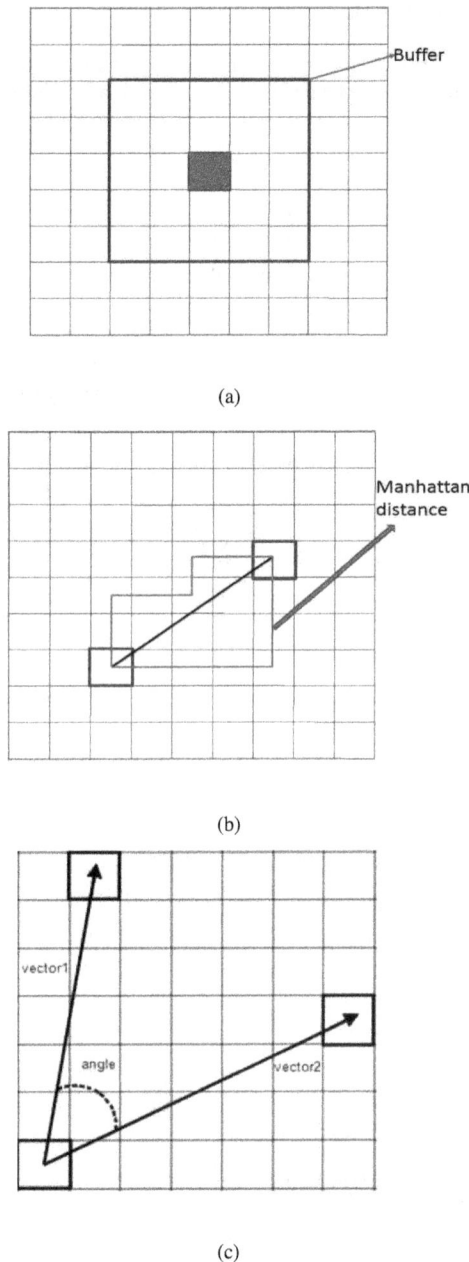

Figure 2: Buffer, Distance and Direction

previous direction and the difference of the two types of paths (Euclidean and Manhattan) are taken into consideration to compute the new location.

3.1.2 Semantics We pick out important semantics for tracking, namely, floor, space, door, and obstacle. Floor helps us to know the switch of floors; space includes room, corridor/passage and vertical passage.

We define space as a region with real boundary (e.g. walls), or a region with specific function. Space is a quite important notion since user expects to be localized in a correct space even without accurate location inside of the space.

Door is a connection between two separate spaces, namely, it is the transition from one space to another. It might happen that a user may obtain many inaccurate measurements around a door. In this case its pivotal to be aware which room the current location is.

Obstacle is defined for objects or regions, which are occupied and inaccessible. It provides constraints that a users location should not be located inside an obstacle, and the obstacle should not on a users track. In this manner, we obtain tracks outside of obstacles, *i.e.* avoiding them.

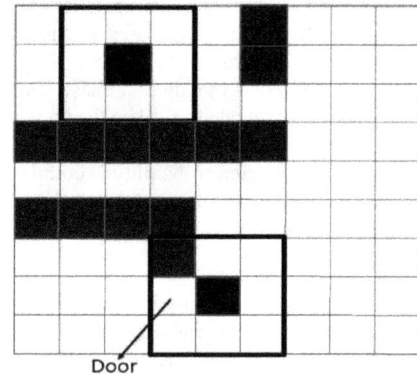

Figure 3: Neighborhood in a space and between spaces

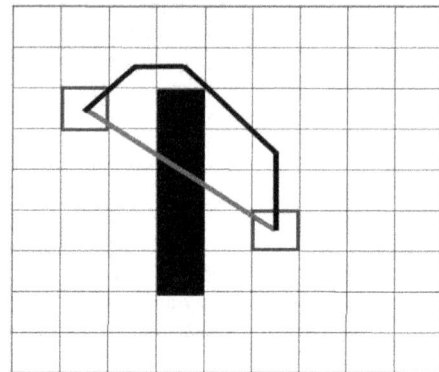

Figure 4: The difference between Manhattan shortest distance and Euclidean shortest distance

3.1.3 Topology As connectivity indicates whether a grid cell can be reached from a predefined location, its an ideal indicator for tracking. If the neighboring grid cells do not belong to any obstacle and are in the same space to a grid, then the grid connect with its neighbors. In this paper we select 8-neighborhood, which means there are at most 8 connected neighbors to one grid cell. If a grid cell and one of its neighbors are not in the same space, then they may connect by grid cells of a door between them (Fig. 3).

Based on connectivity between different grid cells, its easy to get the shortest path between two connected grid cells by employing shortest path algorithms. It is worth noting that the shortest path represents Manhattan distance.

Given two grid cells, we compute the difference between their Manhattan shortest distance and Euclidean shortest distance. The difference can indicate obstacle occurrence. If there is no obstacle between two grid cells, the difference of the two distances will be very small. Otherwise, the difference will increase due to obstacle avoidance (Fig. 4).

3.2 Improving Tracks

After we introduce the geometric, topological and semantic features of the grid model for indoor tracking, we can make full use of them. For instance, a location is taken as a possible candidate when it is accessible (semantics), connected to previous location

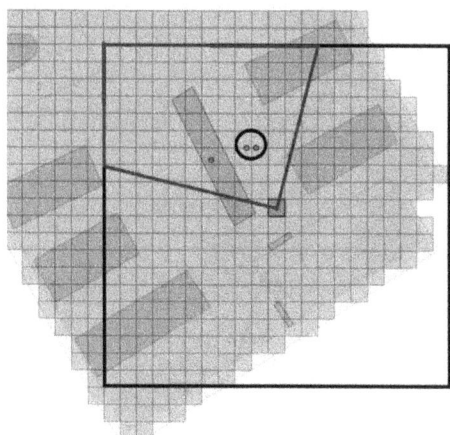

Figure 5: The Selection of measured locations

(topology), and inside a certain buffer of previous determined location (geometry).

In general, by using the previous location and previous moving direction, we could compute the probability of each grid cell to be the current location. Then we can apply the probability to a set of measured candidates of the current location. Finally the one with the highest probability is picked out.

The computation steps are as follows: Firstly, a buffer of the previous location is set up, only the grid cells inside the buffer are considered. Secondly, the probability of these grid cells are assigned. If a grid cell belongs to an obstacle, its probability is set to 0; or if the grid cell is at a distinct space from the space of the previous location, its probability is half decreased. Additionally, the difference of the two types of distances is applied. We assume that there is a negative relation between the distance difference and probability, therefore a grid cell with a larger distance difference is assigned a lower probability.

Thirdly, we filter the measured locations. If we dont have previous direction, then a moving direction is computed by averaging orientations of all location measurements at a current moment. In order to determine whether a grid cell is inside the moving direction area, we compare the angle between the orientation vector of the grid cell and the moving direction with an angle threshold. Thus if a measurement is inside the moving direction and not in obstacles, then it is selected (points in circle in Fig. 5). In other words, the candidate set is narrowed down.

In addition, we check the selected candidate locations on their moving directions. The moving directions are computed based on their connection with the previous location. In this manner, improper directions, such as the path crossing an obstacle, can be avoided.

To summarize, we employ a probabilistic method to represent the effect of spatial factors on filtering location measurements, and then pick out the most possible one. Afterwards, the computed or picked candidate locations compose a track which doesnt violate spatial constraints (e.g. no crossing over obstacle).

4 EXPERIMENT

4.1 Preparations

In this experiment a WiFi positioning system is provided by researchers of Wuhan University (Verbree et al., 2013). The system leverages WiFi fingerprinting method to localize mobile clients. In addition, we employs a magnetometer to measure orientations of a walking person.

The experiment environment is at the office of CGI Company, Rotterdam. The original data to build grid model are floor plans of CGI office. We discrete indoor spaces and objects to grid cells by intersecting the grid cells with floor plans. The grid cell size should be compatible with walking human speed. We chose 0.7m because the averaging walking speed of human is 1.4 m/s (Wikipedia, 2014a).

We hold two important assumptions during this experiment:

- Start location is known.

- Motion is in a constant pace.

In the next subsections, we will illustrate two cases of tracking. One is about walking inside a space, the other one focuses on passing between different spaces.

4.2 Case Studies

The first case is walking inside a large space. We visualized both actual motion track and measured locations of WiFi system in Figure 6a. The problems in the case include locations crossing obstacles/spaces, or even inside of obstacles (Fig. 6a). As shown in Figure 6a, the sequence of measured location are disordered, and there are some measurements located in another space, which contradicts to the real motion track.

Figure 6b presents the improvement on track after spatial model aid. The nodes represent the most possible location at each measured moment. The track is significantly improved. Errors such as inside/crossing obstacles/another space, and the wrong moving directions are now corrected.

The second case is walking between spaces. In this case, we would like to know if the correct room is selected especially around doors.

In order to distinguish the room a measurement belongs to, two conditions are proposed for determination of transferring between spaces:

- The door connecting two spaces is in the moving direction area;

- The previous computed location is at the door.

Only when the both conditions are met, the current location is confirmed in the new space (room).

We conducted a test between two spaces. Figure 7a presents the original measured locations and the actual movement route. The locations appear disordered. Figure 7b provides the improvement on the original track, and related spaces are highlighted in bold lines. Though the track includes fake zigzags due to measurement errors, the moving trend is still correct and the track doesnt lie in any obstacle. More importantly, the switching between the two spaces was detected. Around the door between the two rooms, we successfully eliminate sharp shifts (back and forth). The locations after processing are strictly successive in a time sequence.

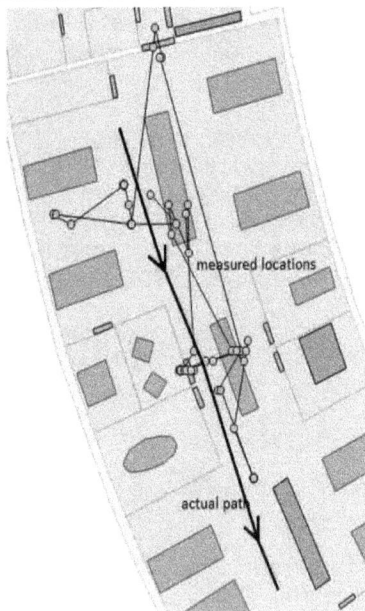

(a) before improvement

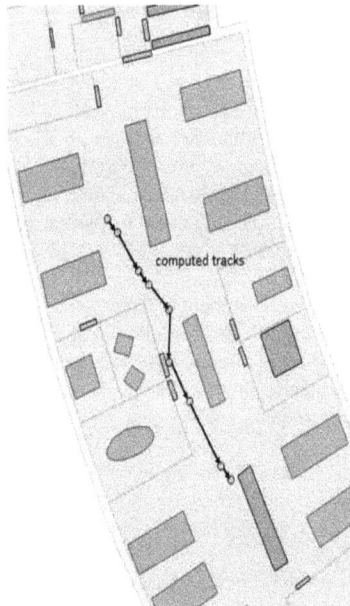

(a) before improvement

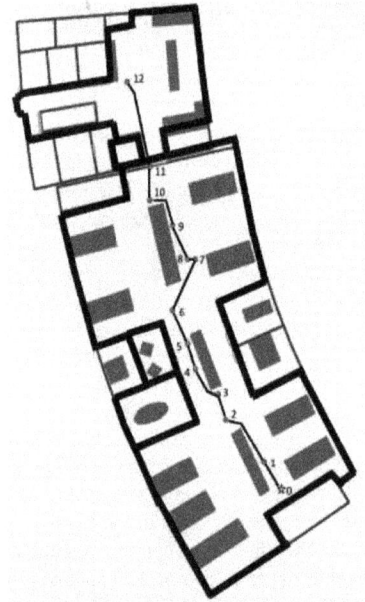

(b) after improvement

Figure 6: Walking inside a Space

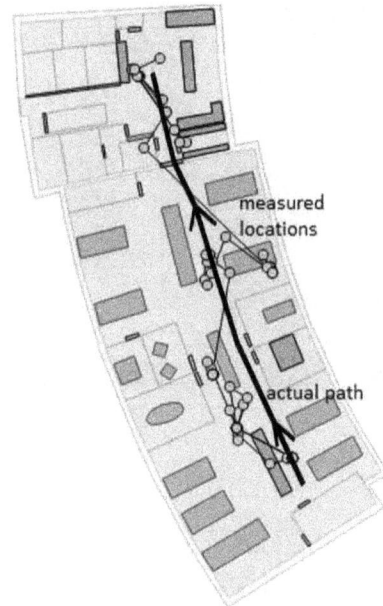

(b) after improvement

Figure 7: Walking between Two Spaces

5 CONCLUSIONS

This paper presents the usage of spatial model and related information for indoor tracking. With WiFi measurements on locations, we leverage a spatial model to qualitatively improve indoor tracking results. Our experiment demonstrates using spatial model has the potential to overcome problems such as tracks over obstacles and unstable shift between two measurements.

In the future, we will investigate more error cases during localization process. Once we collect these exceptions from tracking, we can devise more spatial constraints within grid model. Our final goal is to develop a methodology, which can mitigate tracking errors with only spatial constraints. In this manner, different measurements from distinct localization systems can be processed in a standard method, and thus we can provide improved/smooth tracks for different localization equipment.

6 ACKNOWLEDGMENTS

Our thanks to Haojun, A., Taizhou, L., Jianjian, W., Menglei , Z. of Wuhan University for allowing us to using the WiFi positioning system they had developed, and thanks to CGI company at Rotterdam for allowing us to setting up test environment in their offices.

REFERENCES

Afyouni, I., Ray, C., & Claramunt, C., 2013. Spatial models for context-aware indoor navigation systems: A survey. *Journal of Spatial Information Science* (4): 85-123.

Besada, J.A., Bernardos, A.M., Tarro, P., & Casar, J.R., 2007. Analysis of tracking methods for wireless in door localization. In: *ISWPC'07 Wireless Pervasive Computing*, 2007.

Burgard, W., Fox, D., & Hennig, D., 1997. Fast grid-based position tracking for mobile robots. In *KI-97: Advances in Artificial Intelligence*: 289300. Springer.

Domnguez, B., Garca, .L., & Feito, F.R., 2011. Semantic and topological representation of building indoors: An overview. In *Proc. of Joint ISPRS workshop on 3D city modelling & applications and the 6th 3D GeoInfo*, Wuhan.

Franz, G., Mallot, H., Wiener, J., & Neurowissenschaft, K., 2005. Graph-based models of space in architect ture and cognitive science-a comparative analysis. In *Proceedings of the 17th International Conference on Systems Research, Informatics and Cybernetics*: 3038.

Fuchs, C., Aschenbrucka, N., Martinia, P., & Wienekeb, M., 2011. Indoor tracking for mission critical scenarios: A survey. *Pervasive and Mobile Computing* 7(1): 115.

Girard, G., Ct, S., Zlatanova, S., Barette, Y., St-Pierre, J., & Van Oosterom, P., 2011. Indoor pedestrian navigation using foot-mounted imu and portable ultrasound range sensors. *Sensors* 11 (8): 76067624.

Jensen, C.S., Lu, H., & Yang, B., 2009. Graph model based indoor tracking. In *Mobile Data Management: Systems, Services and Middleware*, 2009 (MDM09). Tenth International Conference on: 122131.

Khider, M., Kaiser, S., Robertson, P., & Angermann, M., 2012. The effect of maps-enhanced novel movement models on pedestrian navigation performance. In *The 12th annual European Navigation Conference (ENC 2008)*.

Li, X., Claramunt, C., & Ray, C., 2010. A grid graph-based model for the analysis of 2d indoor spaces. *Computers, Environment and Urban Systems* 34(6): 532540.

Liu, L., & Zlatanova, S., 2012. Towards a 3D network model for indoor navigation, In *Urban and Regional Data Management, UDMS Annual 2011*: 79-92. Taylor and Francis Group, Boca Raton, London.

Miller, L.E., 2006. Indoor navigation for first responders: a feasibility study. *tech. report*, National Institute of Standards and Technology, Wireless Communication Technologies Group.

Moravec, H., & Elfes, A., 1985. High resolution maps from wide angle sonar. In *Proceeding of the IEEE International Conference on Robotics and Automation*: 116-121. St. Louis, MO.

Spassov, I., 2007. Algorithms for map-aided autonomous indoor pedestrian positioning and navigation. *EPFL Thesis*, Lausanne.

Thrun, S., Burgard, W., & Fox, D., 2005. *Probabilistic robotics*. MIT press.

Tsetsos, V., Anagnostopoulos, C., Kikiras, P., & Hadjiefthymiades, S., 2006. Semantically enriched navigation for indoor environments. *International Journal of Web and Grid Services* 2(4): 453478.

Verbree, E., Zlatanova, S., van Winden K.B.A., van der Laan, E.B., Makri, A., Taizhou, L., & Haojun, A., 2013. To localise or to be localised with WiFi in the Hubei museum?, In *Int. Arch. Photogramm. Re-mote Sens. Spatial Inf. Sci.*, XL-4/W4: 31-35.

Wikipedia, 2014a. Preferred walking speed. `http://en.wikipedia.org/wiki/Preferred_walking_speed` (29 August 2014).

Wikipedia, 2014b. Taxicab geometry. `http://en.wikipedia.org/wiki/Manhattan_distance` (26 August 2014).

Worboys, M., 2011. Modeling indoor space. In *Proceedings of the 3rd ACM SIGSPATIAL International Workshop on Indoor Spatial Awareness*: 1-6. ACM.

Permissions

List of Contributors

S. J. Tang
State Key Laboratory of Information Engineering in Surveying Mapping and Remote Sensing, Wuhan University, Wuhan, P.R. China

Q. Zhu
State Key Laboratory of Information Engineering in Surveying Mapping and Remote Sensing, Wuhan University, Wuhan, P.R. China
Faculty of Geosciences and Environmental Engineering of Southwest Jiaotong University, Chengdu, P.R. China

W. W. WANG
Shenzhen research center of digital city engineering, Shenzhen, P.R. China

Y. T. ZHANG
State Key Laboratory of Information Engineering in Surveying Mapping and Remote Sensing, Wuhan University, Wuhan, P.R. China

Can Li
State Key Laboratory of Information Engineering in Surveying, Mapping and Remote Sensing, 129 Luoyu Road, Wuhan, China

Xinyan Zhu
State Key Laboratory of Information Engineering in Surveying, Mapping and Remote Sensing, 129 Luoyu Road, Wuhan, China

Wei Guo
State Key Laboratory of Information Engineering in Surveying, Mapping and Remote Sensing, 129 Luoyu Road, Wuhan, China

Yi Liu
School of Geodesy and Geomatics, Wuhan University, 129 Luoyu Road, Wuhan, China

Liang Huang
State Key Laboratory of Information Engineering in Surveying, Mapping and Remote Sensing, 129 Luoyu Road, Wuhan, China

S. Pazhanivelan
Tamil Nadu Agricultural University, Coimbatore, Tamilnadu, India

P. Kannan
Tamil Nadu Agricultural University, Coimbatore, Tamilnadu, India

P. Christy Nirmala Mary
Tamil Nadu Agricultural University, Coimbatore, Tamilnadu, India

E. Subramanian
Tamil Nadu Agricultural University, Coimbatore, Tamilnadu, India

S. Jeyaraman
Tamil Nadu Agricultural University, Coimbatore, Tamilnadu, India

Andrew Nelson
International Rice Research Institute (IRRI), Los Banos 4031, Philippines

Tri setiyono
International Rice Research Institute (IRRI), Los Banos 4031, Philippines

Francesco Holecz
Sarmap, Purasca 6989, Switzerland

Massimo barbieri
Sarmap, Purasca 6989, Switzerland

Manoj Yadav
Deutsche Gesellschaft für Internationale Zusammenarbeit (GIZ) GmbH, New Delhi 110029, India

Ruochen SI
Center for Spatial Information Science, The University of Tokyo, Kashiwa, Japan

Masatoshi ARIKAWA
Center for Spatial Information Science, The University of Tokyo, Kashiwa, Japan

C. H. Hardy
Department of Mechanical Engineering Science, University of Johannesburg, Johannesburg, South Africa

A. L. Nel
Department of Mechanical Engineering Science, University of Johannesburg, Johannesburg, South Africa

D. M. Ermakov
Space Research Institute of RAS (IKI RAS), 84/32 Profsoyuznaya str, Moscow, 117997, Russian Federation

E. A. Sharkov
Space Research Institute of RAS (IKI RAS), 84/32 Profsoyuznaya str, Moscow, 117997, Russian Federation

A. P. Chernushich
Institute of Radioengineering and Electronics of RAS, Fryazino department (FIRE RAS), Vvedenskogo sq., 1, Fryazino, Moscow region, 141120, Russian Federation

Kazuo Oda
Asia Air Survey Co., Ltd

Satoko Hattori
Asia Air Survey Co., Ltd

Hiroyuki Saeki
Asia Air Survey Co., Ltd

Toko Takayama
Asia Air Survey Co., Ltd

Ryohei Honma
Asia Air Survey Co., Ltd

J. Sánchez
EOLAB, Parc Científic Universitat de València, Catedrático José Beltrán, 2. 46980 Paterna (Valencia), Spain

F. Camacho
EOLAB, Parc Científic Universitat de València, Catedrático José Beltrán, 2. 46980 Paterna (Valencia), Spain

R. Lacaze
HYGEOS, Toulouse, France

B. Smets
VITO, Belgium

J. Leckey
National Aeronautics and Space Agency (NASA) Langley Research Center (LARC), Bldg. 1202 MS 468, Hampton, Virginia 23681 USA

A. P. Fernandes
CESAM & Department of Environment and Planning, University of Aveiro, 3810-193 Aveiro, Portugal

M. Riffler
Oeschger Centre for Climate Change Research, University of Bern, Switzerland
Remote Sensing Research Group, Department of Geography, University of Bern, Switzerland

J. Ferreira
CESAM & Department of Environment and Planning, University of Aveiro, 3810-193 Aveiro, Portugal

S. Wunderle
Remote Sensing Research Group, Department of Geography, University of Bern, Switzerland

C. Borregoa
CESAM & Department of Environment and Planning, University of Aveiro, 3810-193 Aveiro, Portugal

O. Tchepel
CITTA, Department of Civil Engineering, University of Coimbra, 3030 - 788 Coimbra, Portugal

N. Kussul
Space Research Institute NAS Ukraine and SSA Ukraine, Department of Space Information Technologies and Systems, Kyiv, Ukraine

S. Skakun
Space Research Institute NAS Ukraine and SSA Ukraine, Department of Space Information Technologies and Systems, Kyiv, Ukraine

A. Shelestov
Space Research Institute NAS Ukraine and SSA Ukraine, Department of Space Information Technologies and Systems, Kyiv, Ukraine

M. Lavreniuk
Taras Shevchenko National University of Kyiv, Kyiv, Ukraine

B. Yailymov
Space Research Institute NAS Ukraine and SSA Ukraine, Department of Space Information Technologies and Systems, Kyiv, Ukraine

O. Kussul
National Technical University of Ukraine "Kyiv Polytechnic Institute", Kyiv, Ukraine

S. Bontemps
Université catholique de Louvain, Earth and Life Institute, Belgium

M. Boettcher
Brockmann Consult GmbH, Hamburg, Germany

C. Brockmann
Brockmann Consult GmbH, Hamburg, Germany

G. Kirches
Brockmann Consult GmbH, Hamburg, Germany

C. Lamarche
Université catholique de Louvain, Earth and Life Institute, Belgium

J. Radoux
Université catholique de Louvain, Earth and Life Institute, Belgium

M. Santoro
Gamma Remote Sensing, Switzerland

E. Van Bogaert
Université catholique de Louvain, Earth and Life Institute, Belgium

U.Wegmüller
Gamma Remote Sensing, Switzerland

M. Herold
Wageningen University, the Netherlands

F. Achard
Joint Research Centre, Italy

F. Ramoino
European Space Agency, European Space Research Institute, Italy

O. Arino
European Space Agency, European Space Research Institute, Italy

P. Defourny
Université catholique de Louvain, Earth and Life Institute, Belgium

J. Baade
Department of Geography, Physical Geography, Friedrich-Schiller-University Jena, 07737 Jena, Germany

C. Schmullius
Department of Geography, Earth Observation, Friedrich-Schiller-University Jena, 07737 Jena, Germany

M. Barbarella
Civil, Chemical, Environmental and Materials Engineering Department (DICAM), University of Bologna, Viale Risorgimento 2, 40136 Bologna, Italy

M. De Giglio
Civil, Chemical, Environmental and Materials Engineering Department (DICAM), University of Bologna, Viale Risorgimento 2, 40136 Bologna, Italy

N. Greggio
Interdepartmental Centre for Environmental Science Research (CIRSA), Lab. IGRG, University of Bologna, Via S. Alberto 163, 48100 Ravenna, Italy

L. Panciroli
Civil, Chemical, Environmental and Materials Engineering Department (DICAM), University of Bologna, Viale Risorgimento 2, 40136 Bologna, Italy

H. Bach
Vista Geowissenschaftliche Fernerkundung GmbH, Gabelsbergerstr. 51, 80333 München, Germany

P. Klug
Vista Geowissenschaftliche Fernerkundung GmbH, Gabelsbergerstr. 51, 80333 München, Germany

T. Ruf
Vista Geowissenschaftliche Fernerkundung GmbH, Gabelsbergerstr. 51, 80333 München, Germany

S. Migdall
Vista Geowissenschaftliche Fernerkundung GmbH, Gabelsbergerstr. 51, 80333 München, Germany

F. Schlenz
Ludwig-Maximilians-Universität München, Luisenstraße 37, 80333 München, Germany

T. Hank
Ludwig-Maximilians-Universität München, Luisenstraße 37, 80333 München, Germany

W. Mauser
Ludwig-Maximilians-Universität München, Luisenstraße 37, 80333 München, Germany

A. Qayyum
Centre for Intelligent Signal and Imaging Research (CISIR),Department of Electrical and Electronic Engineering, Universiti Teknologi PETRONAS 31750 Tronoh, Perak, Malaysia

A. S. Malik
Centre for Intelligent Signal and Imaging Research (CISIR),Department of Electrical and Electronic Engineering, Universiti Teknologi PETRONAS 31750 Tronoh, Perak, Malaysia

M. N. M. Saad
Centre for Intelligent Signal and Imaging Research (CISIR),Department of Electrical and Electronic Engineering, Universiti Teknologi PETRONAS 31750 Tronoh, Perak, Malaysia

M. Iqbal
Centre for Intelligent Signal and Imaging Research (CISIR),Department of Electrical and Electronic Engineering, Universiti Teknologi PETRONAS 31750 Tronoh, Perak, Malaysia

F. Abdullah
Centre for Intelligent Signal and Imaging Research (CISIR),Department of Electrical and Electronic Engineering, Universiti Teknologi PETRONAS 31750 Tronoh, Perak, Malaysia

W. Rahseed
Centre for Intelligent Signal and Imaging Research (CISIR),Department of Electrical and Electronic Engineering, Universiti Teknologi PETRONAS 31750 Tronoh, Perak, Malaysia

T. A. R. B. T. Abdullah
Universiti Tenaga Nasional ,43000 Kajang ,Selangor, Malaysia

A. Q. Ramli
Universiti Tenaga Nasional ,43000 Kajang ,Selangor, Malaysia

H. H. Jaafar
Faculty of Agriculture and Food Sciences, American University of Beirut

F. A. Ahmad
Faculty of Agriculture and Food Sciences, American University of Beirut

R. Kumazaki
ITU, Department of Landscape Architecture Science, Tokyo University of Aguriculture, 1-1-1 Sakuragaoka, Setagaya, Tokyo, 156-8502, Japan

Y. Kuniia
ITU, Department of Landscape Architecture Science, Tokyo University of Aguriculture, 1-1-1 Sakuragaoka, Setagaya, Tokyo, 156-8502, Japan

K. Grant
Bavarian State Research Center for Agriculture (LfL), Institute for Crop Science and Plant Breeding, 85354 Freising, Germany

R. Siegmund
GAF AG, 80634 Munich, Germany

M. Wagnerb
GAF AG, 80634 Munich, Germany

S. Hartmanna
Bavarian State Research Center for Agriculture (LfL), Institute for Crop Science and Plant Breeding, 85354 Freising, Germany

Q Zhana
School of Remote Sensing and Information Engineering, Wuhan University, Wuhan 430079, China
Research Center for Digital City, Wuhan University, Wuhan 430072, China

F. Meng
School of Remote Sensing and Information Engineering, Wuhan University, Wuhan 430079, China
Research Center for Digital City, Wuhan University, Wuhan 430072, China

Y. Xiaoa
School of Urban Design, Wuhan University, Wuhan 430072, China
Research Center for Digital City, Wuhan University, Wuhan 430072, China

Bicen Li
Beijing Institute of Space Mechanics and Electricity (BISME), Beijing, China

Lizhou Hou
Beijing Institute of Space Mechanics and Electricity (BISME), Beijing, China

Hong-wei Zhang
CMA Henan Key Laboratory of Agro-meteorological Safeguard and Applied Technique, Zhengzhou 450003, China
Henan Meteorological Administration, Zhengzhou 450003, China

Huai-liang Chen
CMA Henan Key Laboratory of Agro-meteorological Safeguard and Applied Technique, Zhengzhou 450003, China
Henan Meteorological Administration, Zhengzhou 450003, China

M.V. Akinina
Ryazan state radioengineering university, Department of Space Technology, Ryazan, Russia, 390005, Gagarina Str. 59/1

N.V. Akininaa
Ryazan state radioengineering university, Department of Space Technology, Ryazan, Russia, 390005, Gagarina Str. 59/1

A.Y. Klochkovb
Ryazan state radioengineering university, Ryazan, Russia, 390005, Gagarina Str. 59/1

M.B. Nikiforov
Ryazan state radioengineering university, Department of Electronic Computers, Ryazan, Russia, 390005, Gagarina Str. 59/1

A.V. Sokolova
Ryazan state radioengineering university, Department of Electronic Computers, Ryazan, Russia, 390005, Gagarina Str. 59/1

S. Asam
Institute for Applied Remote Sensing, EURAC Research, Viale Druso, 1, 39100 Bolzano, Italy
Department of Remote Sensing, Institute of Geography, University of Wuerzburg, Oswald-Külpe-Weg 86, 97074 Wuerzburg, Germany

D. Klein
German Aerospace Center (DLR), German Remote Sensing Data Center (DFD), Oberpfaffenhofen, 82234 Weßling, Germany

S. Dech
German Aerospace Center (DLR), German Remote Sensing Data Center (DFD), Oberpfaffenhofen, 82234 Weßling, Germany

S. Park
Geography Education, Pusan National University, Busan, Korea

T. Yamamoto
Dept. of Civil Engineering, Shibaura Institute of Technology, 3-7-5 Toyosu, Koto-ku, Tokyo, Japan

M. Nakagawa
Dept. of Civil Engineering, Shibaura Institute of Technology, 3-7-5 Toyosu, Koto-ku, Tokyo, Japan

A. Klisch
University of Natural Resources and Applied Life Sciences (BOKU), Institute of Surveying, Remote Sensing and Land Information (IVFL), Peter-Jordan-Straße 82, 1190 Wien, Austria

C. Atzberger
University of Natural Resources and Applied Life Sciences (BOKU), Institute of Surveying, Remote Sensing and Land Information (IVFL), Peter-Jordan-Straße 82, 1190 Wien, Austria

L. Luminari
National Drought Management Authority (NDMA), Lonrho House -Standard Street , Nairobi, Kenya

R. R. Colditz
National Commission for the Knowledge and Use of Biodiversity (CONABIO), Av. Liga Periférico-Insurgentes Sur 4903, Parques del Pedregal, Tlalpan, CP 14010, Mexico City, DF, Mexico

R. M. Llamas
National Commission for the Knowledge and Use of Biodiversity (CONABIO), Av. Liga Periférico-Insurgentes Sur 4903, Parques del Pedregal, Tlalpan, CP 14010, Mexico City, DF, Mexico

R. A. Ressl
National Commission for the Knowledge and Use of Biodiversity (CONABIO), Av. Liga Periférico-Insurgentes Sur 4903, Parques del Pedregal, Tlalpan, CP 14010, Mexico City, DF, Mexico

Benjamin Brede
Laboratory of Geo-Information Science and Remote Sensing, Wageningen University, Droevendaalsesteeg 3, 6708PB, Wageningen, The Netherlands

Jan Verbesselt
Laboratory of Geo-Information Science and Remote Sensing, Wageningen University, Droevendaalsesteeg 3, 6708PB, Wageningen, The Netherlands

Loïc P. Dutrieux
Laboratory of Geo-Information Science and Remote Sensing, Wageningen University, Droevendaalsesteeg 3, 6708PB, Wageningen, The Netherlands

Martin Herold
Laboratory of Geo-Information Science and Remote Sensing, Wageningen University, Droevendaalsesteeg 3, 6708PB, Wageningen, The Netherlands

L. M. Estupinan-Suarez
Alexander von Humboldt Institute for Research on Biological Resources (IAvH), Scientific and Applied Projects Assistance Office, The Wetlands Project, Cll. 72 No 12-65 piso 7 Edf. Skandia, Bogota, Colombia

C. Florez-Ayala
Alexander von Humboldt Institute for Research on Biological Resources (IAvH), Scientific and Applied Projects Assistance Office, The Wetlands Project, Cll. 72 No 12-65 piso 7 Edf. Skandia, Bogota, Colombia

M. J. Quinones
SarVision Application in Remote Sensing, Agro business Park 10 6708 PW Wageningen, The Netherlands

A. M. Pacheco
Alexander von Humboldt Institute for Research on Biological Resources (IAvH), Scientific and Applied Projects Assistance Office, The Wetlands Project, Cll. 72 No 12-65 piso 7 Edf. Skandia, Bogota, Colombia
IAvH and Institute of Hydrology, Meteorology and Environmental Surveys (IDEAM) partnership

A. C. Santos
National University of Colombia, Faculty of Engineering, Cr. 45 No 26–85 Edf. 453, Bogota, Colombia

L. Liua
Delft University of Technology, Julianalaan 134, 2628BL, Delft, the Netherlands

W. Xub
Fugro Intersite, Dillenburgsingel 69, 2263 HW, Leidschendam, the Netherlands

W. Penard
CGI Nederland, George Hintzenweg 89, Rotterdam, the Netherlands

S. Zlatanova
Delft University of Technology, Julianalaan 134, 2628BL, Delft, the Netherlands